Metaphysics and Cognitive Science

METAPHYSICS AND COGNITIVE SCIENCE

Edited by
Alvin I. Goldman
and
Brian P. McLaughlin

OXFORD
UNIVERSITY PRESS

OXFORD
UNIVERSITY PRESS

Oxford University Press is a department of the University of Oxford. It furthers the University's objective of excellence in research, scholarship, and education by publishing worldwide. Oxford is a registered trade mark of Oxford University Press in the UK and certain other countries.

Published in the United States of America by Oxford University Press
198 Madison Avenue, New York, NY 10016, United States of America.

CIP data is on file at the Library of Congress
ISBN 978–0–19–063967–9

9 8 7 6 5 4 3 2 1

Printed by Sheridan Books, Inc., United States of America

Contents

Part VI Sortals and Natural Kinds

Part VII Debunking and Cognitive Science

Contributors

Mark Baker is Distinguished Professor of Linguistics and Cognitive Science at Rutgers University. He received his PhD in linguistics in 1985 from MIT and taught at McGill University in Montreal for twelve years before moving to Rutgers in 1998. He specializes in the syntax and morphology of less-studied languages, particularly those of the Americas, Africa, and Asia, with the goal of understanding the similarities and differences among languages on the largest scale, and their implications for an understanding of human cognition. He has written five research monographs, as well as numerous journal articles, and one book for a popular audience (*The Atoms of Language*, Basic Books, 2001). He is also coeditor of *The Soul Hypothesis: Investigations into the Existence of the Soul* (Continuum, 2011).

Craig Callender is Professor of Philosophy and Codirector of the Institute for Practical Ethics at the University of California, San Diego. Specializing in philosophy of science, he is especially interested in foundational issues concerning physics and time. He is the author of *What Makes Time Special?* (Oxford University Press, 2017).

Christopher Frugé is a PhD student in philosophy at Rutgers University. His research primarily focuses on metaphysics and ethics, and his interests extend to relevant epistemological issues.

Susan A. Gelman is Heinz Werner Distinguished University Professor of Psychology at the University of Michigan. She studies concepts and language in young children. She is especially interested in how children organize their experience into categories, how categories guide children's reasoning, how children discover and reason about nonobvious aspects of the world, and the role of language in these processes. She is a fellow of the National Academy of Sciences, the American Academy of Arts and Sciences, the Association for Psychological Science, the American Psychological Association (Division 7), and the Cognitive Science Society. Her honors include the Developmental Psychology Mentor Award of the American Psychological Association and the G. Stanley Hall Award for Distinguished Contribution to Developmental Psychology (APA, Division 7).

Alvin I. Goldman is Board of Governors Professor of Philosophy Emeritus at Rutgers University. His 160-plus papers, four monographs, several collections and festschrifts on his work span individual and social epistemology, action theory, philosophy of mind, cognitive science, and political philosophy. A past fellow of the Center for Advanced Study in the Behavioral Sciences, he is a member of the American Academy of Arts and Sciences. He has served as president of the American Philosophical Association and the Society for Philosophy and Psychology. Mark Johnston is Walter Cerf Professor of Philosophy at Princeton University, where he wrote his PhD, "Particulars and Persistence," under the supervision of Saul Kripke. He is the author of many influential and widely reprinted articles in ontology, philosophy of mind, philosophical logic, and ethics, along with *Saving God* (Princeton University Press, 2009) and *Surviving Death* (Princeton University Press, 2010), two works that present a novel form of religious naturalism. Two volumes of his collected papers, *Human Beings* and *The Obscure Object of Hallucination*, will soon be forthcoming with Princeton University Press. He is currently working on a book entitled *The Manifest*, which explains how the world of lived experience can be very much as it appears to be, despite the discoveries of the physical and biological sciences.

Daniel Z. Korman is Professor of Philosophy at the University of California, Santa Barbara. He is the author of numerous articles in metaphysics. Much of his research has been devoted to the defense of a conservative metaphysics of material objects, according to which our natural, intuitive judgments about what objects there are and aren't are more or less correct.

Nick Leonard is a graduate student in philosophy at Northwestern University. His research focuses on reasoning and rationality from both philosophical and psychological perspectives. His dissertation develops a view of epistemic rationality according to which there can be genuine rational indeterminacy; that is, it can be indeterminate which principles of rationality one should satisfy and thus indeterminate which doxastic attitudes one is permitted or required to have. He hopes that this view provides a systematic way of resolving epistemic dilemmas, points to a general decision-making procedure for agents who must act in the face of indeterminacy, and reveals an unexplored phenomenological dimension of being subject to conflicting rational requirements.

Sarah-Jane Leslie is 1943 Professor of Philosophy at Princeton University and the dean of the Graduate School. She is the author of numerous articles in philosophy and psychology and has published in journals such as *Science, PNAS, Philosophical Review*, and *Noûs*. Much of her work focuses on how we categorize and generalize information about the world around us. As part of this project, she examines how the language young children hear shapes their perception of social groups, and how this relates to stereotyping and prejudice. Most recently, she has been examining gender gaps in educational and career choices, and her work on the topic was named as one of 2015's most interesting scientific findings by *Edge*. She has delivered the Gareth Evans Memorial Lecture at Oxford University, the Daniel Greenberg Lecture at Reed College, and was the 2015 recipient of the Stanton Award from the Society for Philosophy and Psychology. Her work has been covered extensively in the media, including by the *New York Times*, the *Washington Post*, and the *Wall Street Journal*. She has appeared on NPR, WHYY, and CBC Radio.

John McCoy is an Assistant Professor of Marketing at the Wharton School of the University of Pennsylvania. His research seeks to undersand and model human judgement and decision making, and to improve crowd wisdom.

Brian P. McLaughlin is Distinguished Professor of Philosophy and Cognitive Science at Rutgers University and director of the Rutgers Cognitive Science Center. He has numerous publications in the philosophy of mind, metaphysics, and philosophical logic. His work in metaphysics has focused mainly on supervenience and emergence. His work in the philosophy of mind has focused mainly on consciousness, mental causation, self-knowledge and content-externalism, self-deception, delusions, the nature of cognitive architecture, the computational theory of mind, color perception, and visual perception of geometrical properties.

Friederike Moltmann is Research Director at the French Centre Nationale de la Recherche Scientifique (CNRS) and in recent years has been visiting researcher at New York University and visiting professor at the University of Padua. Her research focuses on the interface between natural language semantics and philosophy. She received a PhD in 1992 from the Massachusetts Institute of Technology and taught both linguistics and philosophy at various universities in the United States, the United Kingdom, France, and Italy. She is author of *Parts and Wholes in Semantics* (Oxford University Press, 1997) and *Abstract Objects and the Semantics of Natural Language* (Oxford University Press, 2013).

Shaun Nichols is Sherwin Scott Professor of Philosophy at the University of Arizona. His research is at the intersection of philosophy and psychology, and it has been funded by the National Institute of Health, the Templeton Foundation, and the Office of Naval Research. He has published three monographs and over one hundred articles.

L. A. Paul is Professor of Philosophy and Cognitive Science at Yale. Recent books and articles include *Causation: A User's Guide* (coauthored with Ned Hall, Oxford University Press, 2013), *Transformative Experience* (Oxford University Press,

2014), "What you can't expect when you're expecting," *Res Philosophica* (2015), and "Transformative treatments" (coauthored with Kieran Healy), *Noûs* (2017).

Peter Railton is Kavka Distinguished University Professor and Perrin Professor of Philosophy at the University of Michigan, Ann Arbor. His main research has been in ethics, metaethics, moral psychology, and the philosophy of science.

Lance J. Rips is Professor of Psychology at Northwestern University. His research centers on human reasoning and concepts. He is the author of *Lines of Thought* (Oxford University Press, 2011), *The Psychology of Proof* (MIT Press, 1994), and *The Psychology of Survey Response* (with Roger Tourangeau and Kenneth Rasinski, Cambridge University Press, 2000). He has edited *Reasoning: Studies of Human Inference and Its Foundations* (with Jonathan Adler, Cambridge University Press, 2008) and *Similarity and Symbols in Human Thinking* (with Steven Sloman, MIT Press, 1998).

David Rose is a McDonnel Postdoctoral Fellow in the Philosophy, Neuroscience and Psychology Program at Washington University, St. Louis. He mainly works on issues at the intersection of metaphysics and cognitive science and epistemology and cognitive science and has published on these issues in venues such as the *Journal of Philosophy, Noûs, Philosophical Studies*, and *Synthese*.

Jonathan Schaffer is Distinguished Professor of Philosophy at Rutgers University, New Brunswick. He holds a PhD from Rutgers in 1999, and has since worked at the University of Houston, the University of Massachusetts–Amherst, and the Australian National University, before returning to Rutgers in 2011. He works mainly in metaphysics, with interests in epistemology, language, mind, and science.

Tomer Ullman is Assistant Professor of Psychology at Harvard University. He received his PhD in cognitive science from MIT in 2015, and his BS in physics and cognitive science from the Hebrew University of Jerusalem in 2008. His research interests include intuitive theories, with a focus on intuitive physics, intuitive psychology, and computational models of cognitive development.

Frédérique de Vignemont is a CNRS research director at the Jean Nicod Institute in Paris. Her research is at the intersection of philosophy of mind and cognitive science. Her major current works focus on the first person, body schema, agency, empathy, and more recently on pain. She has just published *Mind the Body* (Oxford University Press, 2018), a philosophical exploration of bodily self-awareness, as well as *The Subject's Matter* (MIT Press, 2017), an interdisciplinary collected volume on the body and the self. She is one of the executive editors of the *Review of Philosophy and Psychology*.

Dean Zimmerman earned a bachelor's degree from Minnesota State University–Mankato, and a PhD in philosophy from Brown University. He has taught at the University of Notre Dame, Syracuse University, and Rutgers University, where he started the Rutgers Center for the Philosophy of Religion and serves as chair of the philosophy department. Zimmerman is founding editor of *Oxford Studies in Metaphysics* (now coedited with Karen Bennett), and coeditor of *Oxford Studies in the Philosophy of Religion*. He has coedited several other books, and his publications include over fifty articles in scholarly journals and books.

Metaphysics and Cognitive Science

Introduction

Alvin I. Goldman and Brian P. McLaughlin

I.1 METAMETAPHYSICS

This volume is addressed, in the first instance, to the methodology of metaphysics. In particular, it illustrates how the methodology of metaphysics can be enriched with the help of cognitive science. Many parts of the volume show how specific metaphysical issues—or the enterprise of metaphysics in general—can profit from the findings of cognitive science. We view the contents of the volume as a considerable departure from standard analytical metaphysics.

How should metaphysicians go about pursuing the main goal of metaphysics, presumably the goal of limning the nature of reality (to borrow Quine's formulation), or determining which kinds of objects and properties are fundamental and how they are related to less fundamental ones?

Certain metaphysicians, including very eminent ones, highlight the role of common sense. David Lewis does this in the course of his defense of modal realism. In *On the Plurality of Worlds*, he writes:

> In trying to improve the unity and economy of our total theory . . . I am trying to accomplish two things. . . . I am trying to *improve* that theory, that is to change it. But I am trying to improve *that* theory, that is to leave it recognizably the same theory we had before. . . . [A] theory cannot earn credence just by its unity and economy. What credence it cannot earn, it must inherit. It is far beyond our power to weave a brand new fabric of adequate theory *ex nihilo*, so we must perforce conserve the one we've got. A worthwhile theory must be credible, and a credible theory must be conservative. It cannot gain credence . . . if it disagrees with too much of what we thought before. And much of what we thought before was just common sense. Common sense is a settled body of theory . . . and I presume that we are reasonable to believe it. (1986, 134)

Lewis's stance here suggests the frequently endorsed methodology of "reflective equilibrium." Appealing to "intuitions" rather than to common-sense "beliefs," this method says that a philosophical position is justified just in case it is part of the most coherent systematization of one's (own) total body of intuitions. Whatever exactly intuitions are, they are definitely not products of empirical science.

Many contemporary analytical metaphysicians endorse the methodology of "reflective equilibrium," but at the other end of the spectrum are philosophers who assert that science has a crucial, and indeed essential, role to play in metaphysics. According to Quine, for example, determining what exists is a matter of figuring out what beliefs are required to explain the truth of our best-confirmed theories. Since the theories to which Quine refers are scientific theories, it becomes clear that science has a central role to play in metaphysics. At the very far end of this spectrum is the view that science is metaphysics enough.

Granting the weaker assumption that science has a role to play in metaphysics, the question arises: Which branches of science are crucial for metaphysics? Almost everyone would agree that, at least in selected metaphysical domains, physical science is what is called for. One cannot seriously study space or time without looking to physics. But can cognitive science contribute even to the study of space and/or time? We do not endorse the bold claim that the study of *every* metaphysical issue would profit by some sort of appeal to cognitive science. But a number of contemporary metaphysicians maintain that some aspects of time, for example, can be helpfully explored with the assistance of cognitive science (cf. Callender, this volume, and see Goldman 2015).

A traditionally central metaphysical question that has again become a focus of inquiry in recent years is whether God exists. Many metaphysicians agree with LaPlace that physics has no need of that hypothesis, and they take that as adequate grounds for rejecting the God hypothesis. Some physicists, however, appeal to facts revealed by physics to offer support to the God hypothesis. Stephen Hawking, for instance, advances a "fine tuning" argument to support the existence of God based on cosmological considerations. In brief (and without endorsement), the argument runs as follows:

> If the rate of expansion one second after the Big Bang had been smaller by even one part in one hundred thousand million million, the universe would have recollapsed before it ever reached its present state.

Thus, human beings would not exist unless someone (or something) as powerful and accurate as God existed. Therefore (probably) he does.

However, can cognitive science play a role in addressing the question whether God exists? Obviously, cognitive science can shed light on psychological questions, including ones that adjoin metaphysical questions. For example, cognitive science might help explain why people *believe* in God. But the metaphysical question is whether God exists, not why people believe in God. Might evidence from cognitive science bear on that metaphysical issue? There is now a rapidly growing subfield of cognitive science called the cognitive science of religion. Many of the researchers in this field argue that evidence from cognitive science provides us with reason to doubt that there is a God. The chapters in this volume by Baker and Zimmerman (chapter 5) and by Goldman (chapter 6) fall within the cognitive science of religion and take opposing views on that matter.

I.2 METAPHYSICAL TARGETS AND METHODS

Metaphysics is a difficult subject not only because its questions are often subtle and even arcane, but because there are so many types of arguments and types of evidence for addressing those questions. Consider some of the questions that are its targets. First, metaphysics may ask about the existence or nonexistence of some putative entities or properties. With respect to the putative entity, God, for example, or other supernatural things, it may ask whether such entities exist or not. With respect to other entities, properties, relations, etc., there may be a wide variety of possible characters or statuses for these properties or relations to have. With respect to the number ten, for example, or the color red, a pair of metaphysicians might agree that the number or color exists but disagree about their characters. Red, for example, might be considered a categorical property by some metaphysicians, but a dispositional property by others. In the latter case, red would be exemplified by an entity's disposition to give rise to a certain type of visual experience in a certain class of subjects when the subjects are present under certain lighting conditions. Dispositionality, or response-dependence, is a kind of metaphysical status postulated by many metaphysicians. It is obviously a different issue than existence versus nonexistence; and this complicates the easily expandable scope of metaphysics. A wide variety of terms have been introduced by metaphysicians to accommodate the characters of assorted properties or relations. These include dispositionalism, expressivism, fictionalism, projectivism, constructivism, etc.

A significant portion of the work of metaphysics is to identify suitable descriptors of the foregoing kinds, which can generate an assortment of ways to categorize or describe the world. Such metaphysical work, however, doesn't in itself provide any *methods* or types of *evidence* for deciding which of these classifications are best for assorted preanalytic categories. This is an *epistemic task* for the metaphysician (or community of metaphysicians) to execute. Different possible methodological choices might be made. On the one hand, metaphysicians may opt for commonsense experience as their methodological guide. Alternatively, they may appeal to physical science or cognitive science as a guide. Or some combination of the foregoing (family of) sciences might be utilized. As we have underscored, the present volume focuses on cognitive science and the reasons it offers a promising methodology (or family of methodologies) for trying to tackle a variety of metaphysical questions.

We do not hold that cognitive science should be the central or primary methodology for all of metaphysics. Cognitive science receives special attention here not because it should be ranked as the *chief* method (for all, or even parts) of metaphysics, but merely because its contributions have hitherto been largely neglected by mainstream metaphysicians, especially as an explicitly identified methodology. Aspects of its promise and its potential challenges are considered in the chapters that comprise this volume.

I.3 OVERVIEWS OF THE CHAPTERS

Chapter 1: Craig Callender, "Time Lost, Time Regained"

Callender starts with a dramatic tale of how cognitive science was actually "born" with an episode in the metaphysics of time. He writes:

> Immanuel Kant's claim that time is imposed on experience by our cognitive architecture ("nothing but the form of inner sense") loomed over research into time in the 19th century. In this context figures such as Johann Friedrich Herbart, . . . Hermann von Helmholtz, Wilhelm Wundt, Ernst Mach, and William James performed and discussed experiments on time perception, much of which would now be classified as cognitive science or psychophysics. Most understood their psychological work as bearing on the truth or falsity of Kant's bold thesis—and hence saw cognitive science as relevant to temporal metaphysics.

Callender goes on to tell us:

> After a brief setup, I describe the interplay between cognitive science and the three modes of time identified by Kant, namely, duration, succession and simultaneity. I then sketch the beginnings of a solution to one of the main puzzles in the metaphysics of time, the so-called flow of time.

Callender notes that in our experience, it seems that time flows, that is, that there is, as he puts it, a tripartite structure {past, present, future} that updates itself. If, however, general relativity is correct, then there is no simultaneity, and the flow of time is, arguably, an illusion. General relativity thus seems to be in conflict with our experience of time, with what Callender calls "manifest time." This is a salient point of tension between what Wilfrid Sellars (1962) called "the manifest image" of the world and "the scientific image" of the world. In the chapter, Callender assumes that general relativity theory is correct and complete as concerns the metaphysics of time. He maintains that "to fully explain manifest time, we need to look at the physical objects in our environment and our cognitive engine and see what actually produces our simultaneity experiences."

Callender appeals to a variety of results in cognitive science research to undermine the claim that there is "a psychological now" and to make a case that the flow of time is indeed an illusion. Moreover, he defends a novel account of the source of this illusion. The illusion of the flow of time, he argues, is due to another illusion spawned by our cognitive architecture, namely, the illusion of an enduring self.

Chapter 2: Jonathan Schaffer, "Cognitive Science and Metaphysics: Partners in Debunking"

Schaffer begins by noting that it seems, at first blush, that cognitive science could not be relevant to *realist metaphysics*, i.e., a subject concerning the genuine nature of the world. It could only be relevant to what the "folk" (nonexperts on the street) *think* about metaphysics, not to metaphysical truths. Cognitive science, after all, is a family of disciplines that study how the mind works. There is no guarantee—certainly not from the start—that the way the mind works will give us any insight into the way the extramental world or universe works, or into the fundamental nature of reality. An understanding of how the mind works, it might be argued, is as irrelevant to realist metaphysics as folk physics is to physics. Schaffer goes on, however, to endorse a large part of Goldman's position on this subject (presented in several papers, especially Goldman 2015). In particular, Schaffer agrees that cognitive science *is* relevant, *even* to "realist" metaphysics.

Schaffer underscores that some arguments in metaphysics are premised on intuitions. Cognitive science is relevant in assessing what we find intuitive, and so in assessing such arguments. He agrees with Goldman that cognitive science is relevant to metaphysics via the debunking project, the project of appealing to work in cognitive science to show that the relevant intuitions (or beliefs) behind a metaphysical view result from an unreliable process.

Schaffer offers two revisions—or elaborations—of Goldman's picture. First, the fact that there is an evidential role for cognitive science isn't specific to metaphysics; it is generic to any inquiry that invokes intuitions. Second, metaphysics itself plays a crucial role alongside cognitive science within the debunking project. Work in cognitive science can debunk a metaphysical view by showing that the relevant intuitions (or beliefs) result from an unreliable process. But, he says, real metaphysics plays a role, too, in such debunking, specifically in assessing whether intuitions fit reality, and whether a module tracks or fails to track reality. Thus, he notes: "To assess what is and what is not reliable as an indicator of reality, one needs some conception of what is out there in reality. By reliabilist lights, debunking requires comparing the cognitive output to the metaphysical facts, in order to measure their degree of correlation." Thus, in place of Goldman's "one way" slogan, that cognitive science is *relevant* to metaphysics, Schaffer prefers to say that cognitive science and metaphysics are *partners* in the debunking project.[1]

[1] Although he doesn't address the issue, presumably he would also consider them partners in the "vindicating" or "unbunking" projects that Nichols and Frugé discuss in chapters 4 and 15, respectively.

Focusing on the epistemological issue, Schaffer stresses that there is no "external vantage point" from which to critique metaphysics. Assessment of the intuitions that are relevant to metaphysics is partly a metaphysical assessment. He agrees with what he calls "*Goldman's liaison*" that intuitions, experiences, and judgments on which metaphysics relies should be probed or tested by a scientific study of the "cognitive engine." Indeed he sketches a trio of interconnected factors—(*a*) the psychological aspect, (*b*) the metaphysical aspect, and (*c*) the epistemological aspects—that can all jointly contribute to a suitably revisionary metaphysics that enables debunking projects.

Chapter 3: Peter Railton, "Moral Metaphysics, Moral Psychology, and the Cognitive Sciences"

As Railton notes: "Recent years have seen a succession of troubling questions launched from cognitive science about the origin, nature, or rationality of moral thought and practice, ranging from questions about the possible evolution of moral motivation, to the relation between skilled motor control and virtue, to the epistemic status of moral intuitions." The bulk of this work has been aimed at undermining moral realism. In response, Railton appeals to work in cognitive science to contribute to defending a kind of naturalistic moral realism that he calls "tracking realism," according to which we have capacities to detect objective value or rightness, against some of the prominent challenges of the sort in question.

Railton responds specifically to two different lines of objection to tracking moral realism that rely on work in cognitive science. One appeals to dual-process theory and our different intuitive reactions to different kinds of Trolley cases. The other appeals to work in evolutionary psychology. Railton argues not only that these lines of objection fail to show that belief in tracking moral realism is unjustified, but also that a proper understanding of the relevant cognitive factors in fact provides some reason to believe tracking moral realism. He thus invokes work in cognitive science to contribute to making a case that moral intuitions may constitute a source of bona fide evaluative information. This in turn provides at least some reason to resist debunking theories of morality, and indeed makes a contribution toward vindicating tracking moral realism.

Chapter 4: Shaun Nichols, "Debunking and Vindicating in Moral Psychology"

Shaun Nichols examines the role cognitive science can play in debunking and in vindicating our intuitive or common-sense metaphysical commitments. He examines, in particular, process debunking and process vindicating. He notes: "A process-debunking argument aims to show that an agent's belief that p is generated by a process that does not justify the belief that p, and as a result, the agent is not justified in believing that p." The idea is that "people are unjustified in believing that p to the extent that their belief is based on a process that is epistemically defective for believing that p." But evidence concerning the process that generates an agent's belief that p can also play a vindicating role. He says: "The basic idea of process vindication is just that beliefs are vindicated to the extent that they are based on rational processes." The fact that the belief is based on rational processes provides a pro tanto justification for the belief.

Nichols goes on to offer a process vindication argument for moral beliefs that appeals to a relation between consensus and objectivity judgments, making a case that it is rational to take consensus as evidence for objectivity. This and certain facts about moral consensus, he maintains, partially vindicate moral judgment rather than debunking it.

Chapter 5: Mark Baker and Dean Zimmerman, "On Perceiving God: Prospects for a Cognitive Science of Religious Experience"

Baker and Zimmerman's chapter falls squarely within the "cognitive science of religion" (CSR). The main proposal in their chapter, however, is a very significant departure from typical CSR claims.

Traditional CSR writers have argued that a propensity to believe in God is the byproduct of other propensities that serve evolutionary purposes unrelated to God. They propose, for example, that people have a hyperactive agency detection device (HADD) that overresponds to various stimuli so as to attribute purposes and other mental states to nonanimate objects. Such oversensitivity may be beneficial because avoiding an imaginary predator is better than failing to recognize a real one. The HADD device may also explain why people so readily impute mental states to gods, even if they don't observe things behaviorally manifesting purposes or other god-linked mental states. Baker and Zimmerman think that there is a (preferable) "road less traveled" (toward belief in God or gods) than the one traveled in standard CSR. For one thing, they think that the concept of God is a *simple* one, not a compound one built out of other concepts. Rejecting the idea that the concept of God is a mental tool primarily useful for other purposes, they propose that people have specialized cognitive mechanisms for mapping distinctive experiential states onto beliefs about God. These domain-specific computations could provide an empirical basis for arguing that we have a god-faculty—a faculty "built" for the perception of God—that is part of our cognitive system. They write:

> These beliefs may well be the outputs of a cognitive subsystem aimed at truth, operating in the sort of environment for which it was selected, not accidental byproducts of mental tools useful mainly for other purposes.

Chapter 6: Alvin I. Goldman, "God and Cognitive Science: A Bayesian Approach"

Goldman begins by examining and critiquing a number of attempts to use cognitive science to support the existence of God, including Baker and Zimmerman's attempt to establish the plausibility of a module for God perception. A God-perception module, they say, would feature a reliable counterfactual-sustaining pattern in which a subject is caused to believe that God exists when the subject undergoes certain experiences. Ultimately, however, their only claim is that there is no obstacle, in principle to God's standing in a perceptual relation to typical religious experiences. What this means, however, is merely that *if God exists* there is no obstacle to someone standing in a perceptual relation to God. But this does nothing to show (with the help of cognitive science) that there actually *is* a God that stands in such a relation.

In the positive part of Goldman's chapter, he advances a Bayesian model of how evidence from cognitive science could lower the probability of God's existence from some initial starting point. Suppose someone has evidence for the widespread *belief* in God's existence and uses this "consensus gentium" to assign a nontrivial probability to God's existence. Suppose he then learns from cognitive science that this consensus gentium could easily have arisen not from God's actual existence but because of psychological propensities (that humans actually possess) that would engender theistic belief despite the absence of a divine being. Bayesian reasoning would then lead him to a substantial reduction in the posterior probability of God's existence.

Chapter 7: Mark Johnston and Sarah-Jane Leslie, "Cognitive Psychology and the Metaphysics of Meaning"

Leslie and Johnston appeal to work on human categorization to illuminate the metaphysics of meaning. They distinguish two kinds of concepts. What they call "phi-concepts" are what are expressed by terms, and they take these to be the meanings of terms. A phi-concept can be a constituent of a more complex phi-concept, and phi-concepts have a compositional semantics, in that the meaning of a complex phi-concept is a function of the meanings of its constituent phi-concepts and how they are combined. The study of these concepts has been a central part of the business of formal semantics. Formal semantics invokes such concepts in providing individuation conditions for languages, in specifying truth conditions for sentences in a language, and in characterizing analytical validity in a language. What they call "psi-concepts" are the psychologically real heuristics we in fact deploy in applying terms. These are the kinds of concepts with which theories of human categorization are concerned. It is, they maintain, a misleading metaphor to think of word usage as guided by a grasp of word meaning, that is, by a grasp of phi-concepts. Rather, usage is guided by psi-concepts. Phi-concepts do not generate usage, they merely register usage in an idealized way.

Leslie and Johnston argue that only confusion arises from conflating these two kinds of concepts. To take one case they use to illustrate this, Jerry Fodor and Ernest LePore (1996) have argued that prototype representations are not concepts, because concepts are compositional and prototype representations are not. If, for instance, something is both a pet and a fish, then it is a pet fish. But although a goldfish is a prototypical pet fish, it is neither a prototypical fish nor a prototypical pet. This line of objection to prototype theory, Leslie and Johnston maintain, conflates phi-concepts and psi-concepts. Prototype representations are indeed not compositional, but the fact that phi-concepts are compositional is no objection at all to the prototype theory of psi-concepts. (Leslie and Johnston in fact favor a hybrid view of psi-concepts, one that combines aspects of prototype theory with aspects of a theory-theory view.)

Relying on work in cognitive science, Leslie and Johnston make a case that the generalizations deployed in psi-concepts are typically generics, such as "A cheetah can outrun a man" or "Birds fly," rather than universal generalizations. They thus claim: "The heuristics which typically guide our use of terms . . . are properly formulated in generic terms." They also point out the bearing of the distinction between psi-concepts and phi-concepts on a plethora of issues in metaphysics and semantics.

Chapter 8: Friederike Moltmann, "Natural Language and Its Ontology"

Moltmann tells us:

Natural language, it appears, reflects in part our conception of the world. Natural language displays a great range of types of referential noun phrases that seem to stand for objects of various ontological categories and types, and it also involves constructions, categories and expressions that appear to convey ontological or metaphysical notions, for example ontological categories of various sorts, plurality, quantity, identity, causation, parthood, truth, and existence.

She goes on to present the project of discerning the ontology implicit in natural language. That ontology, she maintains, is a rich one, involving artifactual, derivative, abstract, and even nonexistence entities. She points out that the ontology can differ significantly not only from the reflective ontology proposed by a metaphysician, but also even from the ontology that a nonphilosophical speaker of the language would take as credible upon careful reflection. Natural language ontology needn't be reflective of what there really is, or even of what a reflective speaker believes there to be. It is, however, she claims, revealing of the ontological commitments of our cognitive system, to the extent that they are reflected in language. The project of discerning the ontology implicit in natural language is part of linguistics, more specifically of natural language semantics, and so is part of cognitive science. It is thus a branch of metaphysics that is part of cognitive science.

In the course of engaging in the project, Moltmann points out affinities between both Peter Strawson's (1959) project of descriptive metaphysics and Kit Fine's (2017) project of naive metaphysics. One thing that is distinctive about her project is the systematic use of tools of theoretical linguistics in uncovering natural language ontological commitments. In doing natural language ontology, she tells us, linguistically reflective intuitions take priority over common-sense intuitions.

Chapter 9: John McCoy, L. A. Paul, and Tomer Ullman, "Modal Prospection"

McCoy, Paul, and Ullman explore how we interpret, represent, and understand self-involving possibilities, especially possibilities for oneself. This involves what they call "prospection," and they distinguish two kinds of prospection, saying:

> As we'll define it, *temporal prospection* is the act of representing or assessing one's own future (or present, or past) experiences. *Modal prospection* is the act of representing or assessing one's own possible experiences. Such prospection, more generally, is an act involving the assessment of various sorts of self-involving possibilities.

They maintain that we have a mechanism, a "self-simulator," that functions to enable us to make intuitive judgments about self-involving possibilities.

Moreover, they explore the ways that individuals reflect on and decide about whether they would undergo various kinds of fantastical transformative experiences, experiences that would be dramatically life-changing, such as becoming a vampire, or consulting an all-knowing oracle. Such explorations, they maintain, help us to understand, among other things, how individuals assess self-involving counterfactual scenarios.

Chapter 10: Frédérique de Vignemont, "Against Phenomenal Parsimony: A Plea for Bodily Feelings"

As Vignemont points out, a number of authors have recently denied that our sense of agency (of being the agent of our actions) and bodily ownership (our sense of our body being our own body) have a distinctive phenomenology. They maintain there is no need to posit feelings of agency or feelings of bodily ownership in order to explain how we know we are at the origin of our actions, and that the body in which we experience sensations is our own body. Vignemont argues against this position. She appeals to pathological disorders such as delusions of control (delusions that one's movements are entirely controlled by an external force) and somatoparaphrenia (a kind of delusion in which a patient denies ownership of a limb or an entire side of their body) to make a case that there are distinctive feelings of agency and of bodily ownership. The idea is that these delusions occur when those feelings are absent.

She argues, moreover, that the feeling of agency and the feeling of bodily ownership will figure in the best explanation of our sense of agency and of bodily ownership.

Chapter 11: Lance J. Rips and Nick Leonard, "Does the Identity of an Object Depend on Its Category? The Role of Sortals in Thought"

Most of the essays in this volume focus on how work in cognitive science can influence work in metaphysics. But the influence can go in the other direction as well, and in fact, in one instance, has. As Rips and Leonard point out,

> Some prominent cognitive theories have adopted an intriguing idea from metaphysics: The conditions of identity and individuation of objects come from the meaning of sortal nouns—count nouns, such as "dog" or "cup." According to this sortalist theory, Rover's identity over time and his distinctness from Fido depend on the meaning of "dog."

They characterize the sortalist view in metaphysics, and then say how some cognitive scientists have adapted and modified the theory in order to explain people's judgments about the identity and individuation of objects. They call the resulting view "psychosortalism." They make an empirical case that we should reject psychosortalism. More specifically, they make a case that there is no reason to believe that people have a distinguished class of basic-level concepts that supply criteria for resolving questions of identity and individuation.

Chapter 12: Susan A. Gelman, "What the Study of Psychological Essentialism May Reveal about the Natural World"

Gelman tells us:

> Psychological essentialism is an intuitive belief that certain categories, such as *tiger*, *gold*, or *woman*, share not just observable features, but also a deeper, nonobvious reality, that there exists some inherent, internal, immutable substance or quality (the "essence") that causes the characteristics that category members share, and that words in ordinary language reveal these natural kinds. . . . Thus, tigers share more than a certain size, gait, striped fur, and ferocity, but also internal parts, brain structure, temperament, instincts, as well as an innate, unchanging tiger "essence." That hidden essence might be presumed to be genes, or blood, or something as-yet-unknown.

Psychological essentialism, she argues, distorts how we think about natural kinds. She maintains that there is a compelling body of evidence from cognitive science research that people exhibit systematic errors in how they think about natural kinds. These errors result from people holding an essentialist conception of natural kinds. The essentialist conception, she maintains, leads people to underestimate variability within a kind, to view category boundaries as objectively correct, and to mistakenly assume that all members of a kind share a causal essence.

In the course of concluding, she states:

> Natural kinds sit at the intersection of world and mind. They represent the human attempt to organize and make sense of reality, but must be viewed through the distorting lenses of our own cognitive limitations and heuristics. Our concepts of natural kinds have our human fingerprints all over them.

Chapter 13: Daniel Z. Korman, "Debunking Arguments in Metaethics and Metaphysics"

Suppose that a range of representations can be explained without reference to what they purport to represent. Does the availability of such explanations undermine the justificatory status of those representations, and thus debunk them? In addressing this question, Korman focuses on material-object metaphysics and our beliefs about what midsized objects there are, and attempts to draw lessons from the moral debunking literature for the metaphysical object debunking literature, lessons that he maintains generalize.

Korman rejects the idea that evolutionary debunking arguments won't arise for our beliefs about midsized objects since those beliefs can be given an evolutionary vindication. He says: "Our object reactions cannot be vindicated on the grounds that their evolutionary explanations make indispensable reference to their accuracy. They don't." Thus, the relevant representations can be explained without reference to what they represent. He then defends an alternative strategy for vindicating the beliefs. The strategy involves what he calls "bootstrapping" from the very beliefs targeted by the debunking argument. The idea behind bootstrapping is that "even when some phenomenon can be explained without reference to a certain range of entities or facts, we may still sometimes have good reason to prefer a less parsimonious explanation that does make reference to them." In some cases, he argues, the less parsimonious explanation is a better explanation. He next explains why debunking arguments framed in terms of the "insensitivity" of targeted beliefs—the fact that we would have them even if they are false—are ineffective against a bootstrapping defense of the beliefs. He then goes on to defend an explanatory constraint on bootstrapping, and shows how a certain kind of response to debunking arguments, one widely offered in the metaethics literature and recently advanced in the literature on objects, violates the constraint.

In the concluding section of the chapter, Korman tells us:

> Here is what I hope to have shown. Even what would seem to be some of our most secure beliefs, our perceptual beliefs about midsized objects, are not safe from debunking by cognitive science and evolutionary psychology. But even when cognitive scientists or evolutionary psychologists are able to explain some range of beliefs without reference to the associated range of facts, we needn't accept that these explanations are the full story. For we are often entitled to bootstrap to an expanded explanation that does cite the associated facts. That said, one's bootstrapping had better culminate in embracing some such explanatory connection, for conceding that there is no connection would undermine one's entitlement to bootstrap.

Chapter 14: David Rose, "Cognitive Science for the Revisionary Metaphysician"

Revisionary metaphysicians espouse metaphysical views that conflict with folk intuitions; that's why the views count as revisionary. Rose claims that revisionary metaphysicians thereby incur a dialectical obligation. The obligation is to explain where the folk go wrong. He maintains that one way revisionary metaphysicians can discharge that obligation is by appeal to evidence from cognitive science about the sources of the folk intuitions in question. The sources can be such that those intuitions can be discounted by the revisionary metaphysician.

Rose provides an illustration of this in the case of folk intuitions about composition (when two objects compose a third object) and persistence (what changes an object can and cannot survive). He appeals to evidence from cognitive science that those folk intuitions arise from a benighted, promiscuous teleology, and that beliefs based on them are thereby unjustified.

Chapter 15: Christopher Frugé, "Unbunking Arguments: A Case Study in Metaphysics and Cognitive Science"

Several chapters in this volume either endorse or respond to debunking arguments that make use of results from cognitive science. Frugé maintains that results from cognitive science can also sometimes play a vindicating or "unbunking" role as concerns a metaphysical view.

Subtleties aside, according to process reliabilism, a belief is justified if it is produced by a reliable process. Cognitive science could of course reveal that certain of our metaphysical beliefs are produced by a certain kind of process, and that process could prove to be reliable. Given process reliabilism, it would follow that the beliefs are justified when so produced. To determine, however, that the belief-producing process in question is reliable, we would have to determine that it tends to produce true beliefs. Thus, it may seem that we have to vindicate the metaphysical beliefs in question by appeal to our independent (of the cognitive process) determination that the metaphysical beliefs are true. Frugé, however, explicates a strategy, one he calls a "cross-domain strategy," for providing a cognitive-science-based undebunking argument. The leading idea is that in some cases, we can make a case that, given that a cognitive process that we have independent reason to believe is reliable is indeed reliable, we have good reason to believe that a certain other process that produces the metaphysical beliefs in question is also reliable, and so reason to believe that those metaphysical beliefs are justified when so produced. After rejecting a defense by Timothy Williamson (2007) of our modal beliefs, Frugé goes on to use the cross-domain strategy in a cognitive-science-based unbunking argument for the reliability of certain modal beliefs.

I.4 CONCLUSION

The chapters in this volume by no means cover all of the metaphysical issues on which cognitive science can be brought to bear. There is, for instance, a large literature bringing cognitive science to bear on the metaphysics of color (which is a reason we didn't include a chapter on that much-covered topic). The volume illustrates, however, how cognitive science can be fruitfully brought to bear on a wide range of metaphysical issues. And there are, we believe, many ways in which cognitive science is relevant to metaphysics that remain largely unexplored.

We don't claim that science is metaphysics enough. Science itself leaves many metaphysical questions unanswered. But metaphysicians should consult science when doing metaphysics. That's a lesson most metaphysicians have taken to heart where physics is concerned. The overarching point of this volume is that, in many instances, cognitive science should also be consulted.

ACKNOWLEDGMENT

We wish to thank Sam Carter for preparing the index.

REFERENCES

Fine, K. (2017). Naïve metaphysics. *Philosophical Issues* 27 (1): 98–113.

Fodor, J. A., and LePore, E. (1996). The pet fish and the red herring: Why concepts aren't prototypes. *Cognition* 58 (2): 243–76.

Goldman, A. I. (2015). Naturalizing metaphysics with the help of cognitive science. In K. Bennett and D. Zimmerman, eds., *Oxford Studies in Metaphysics*, vol. 9. New York: Oxford University Press, 171–213.

Lewis, D. (1986). *On the Plurality of Worlds*. Oxford: Basil Blackwell.

Sellars, W. (1962). Philosophy and the scientific image of man. In R. Colodny, ed., *Frontiers of Science and Philosophy*. Pittsburgh, PA: University of Pittsburgh Press, 35–78.

Strawson, P. (1959). *Individuals: An Essay in Descriptive Metaphysics*. London: Methuen.

Williamson, T. (2007). *The Philosophy of Philosophy*. Oxford: Blackwell.

PART I
WHAT MIGHT BE THE ROLE OF COGNITIVE SCIENCE IN METAPHYSICS?

1 Time Lost, Time Regained

Craig Callender

Cognitive science was born with a relevance to the metaphysics of time. Immanuel Kant's claim that time is imposed on experience by our cognitive architecture ("nothing but the form of inner sense") loomed over research into time in the nineteenth century. In this context figures such as Johann Friedrich Herbart, Gustav Theodor Fechner, Karl von Vierordt, Rudolf Hermann Lotze, Hermann von Helmholtz, Wilhelm Wundt, Ernst Mach, and William James performed and discussed experiments on time perception, much of which would now be classified as cognitive science or psychophysics. Most understood their psychological work as bearing on the truth or falsity of Kant's bold thesis—and hence saw cognitive science as relevant to temporal metaphysics.

Today research on time in cognitive science, neuroscience, and psychology is enjoying a renaissance. The last twenty years have witnessed an explosion of interest in the field. Due to increasing specialization and philosophy of time's long "linguistic turn," this work's connections to metaphysical issues are not as transparent as they once were. That doesn't mean, however, that such connections don't exist. This essay will describe some ways in which it has relevance. If I am right, cognitive science and related fields are at least as important as physics to temporal metaphysics. The mechanisms revealed by this research help us regain the time "lost" by physics, and in so doing, indirectly confirm some hypotheses in the metaphysics of time. After a brief setup, I describe the interplay between cognitive science and the three modes of time identified by Kant, namely, duration, succession, and simultaneity. I then sketch the beginnings of a solution to one of the main puzzles in the metaphysics of time, the so-called flow of time.

1.1 COGNITIVE SCIENCE AND TEMPORAL METAPHYSICS

It will be useful to borrow the "Kantian" framework Goldman (2015) uses when thinking about cognitive science's relationship to metaphysics. Goldman lumps together the experiences, representations, and intuitive judgments that we have about the world into a category called "common-sense experiences." Examples include seeing colors but also judgments about object persistence and number of objects.

What explains our common-sense experience? Goldman assumes that our common-sense experiences are a function of extramental reality, our cognitive engines, and cultural influences. In schema form:

$$CSE = f(R, COGEN, CUL),$$

where the abbreviations are the natural ones. The point Goldman wants to stress is that our common-sense experience is often not best explained by reality alone ($CSE = f(R)$) but instead reality plus our cognitive engine.

As a simple example consider the well-known Rotating Snakes illusion by Akiyoshi Kitaoka.[1] It presents a very powerful impression of wheels moving. The wheels are not actually moving, or so our "theory of reality" tells us. Here our cognitive engine comes to the rescue of our preferred theory of reality, for the illusion prompted scientists to discover mechanisms linking asymmetric luminance and the resulting differences in neuronal firing and adaptation rates to illusory motion. These mechanisms enable us to explain the gap between experience and what would otherwise be the best metaphysics of the image (i.e., no motion).

Let me make a few observations. First, Goldman's goal with his framework is to show that cognitive science *can* be relevant to our understanding of reality. This claim is true and uncontroversial. The principle of total evidence dictates that we take all of our knowledge into account—including cognitive science—when forming beliefs about the world. Second, ideally we would amend the schema in a few ways. We ought to add the other sciences, such as physics and biology, we ought to clarify the "reality" category because we are part of it, and we may also wish to tease the category of common-sense experience into subcategories, separating the cognitive and the perceptual (to the extent possible). Third, our judgments are holistic. What we hold fast depends on context and our relative confidence in each hypothesis.

Restricted to time, common-sense experience is essentially identical with what I call *manifest time* in Callender (2017). See figure 1.1. Manifest time is not simply our temporal experience but a kind of regimented common-sense model of time, a model that psychology suggests we come to in late childhood. Tied to notions of identity, agency, freedom, and selfhood, it is tremendously important to us. Mellor (2001) appropriately calls it the *time of our lives*, as it is the model employed as we navigate the world.

[1] The illusion is available here: http://www.ritsumei.ac.jp/%7Eakitaoka/rotsnake.gif.

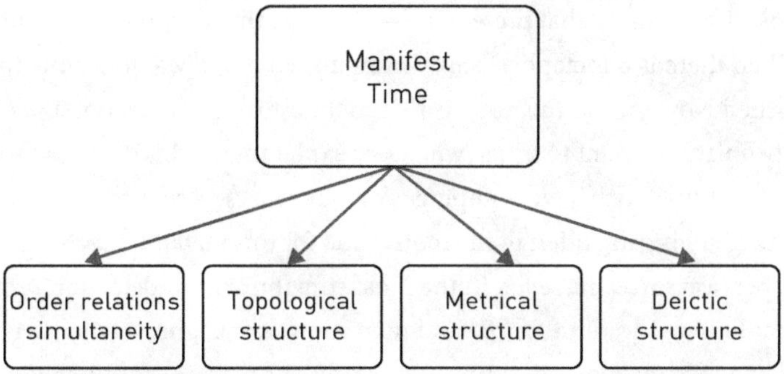

FIGURE 1.1 Manifest time. Manifest time is a somewhat rough-around-the-edges model of time that we naturally come to in development. The model treats time as having—at least—topological, metrical, and deictic structure.

Applied to manifest time, Goldman's schema is

$$Manifest\,time = f\left(R, COGEN, CUL, etc.\right),$$

where I've added "etc." to include the additional sources of knowledge one might draw upon, e.g., biology. In the Rotating Snakes illusion, we're confident in the background theory informing us that nothing really moves in the picture. That mismatch between reality and experience is one reason why it's an illusion. In the metaphysics of time, by contrast, the million-dollar question is whether there is mismatch. What do we substitute for *R*? Do we assume that physics is essentially correct about time? If so, then manifest time and temporal reality disagree over some features. Or do we claim that physics misrepresents or incompletely represents time and replace it with a metaphysical model including features found in manifest time? If so, then allegedly manifest time is *detecting* a property of temporal reality.

Either way, I want to claim, we cannot ignore our cognitive engines and the other sciences. If there is mismatch, as in illusions, then cognitive science and other sciences obviously become relevant in explaining the mismatch. If no mismatch, philosophers are sometimes tempted to think our cognitive engines are irrelevant. This conclusion is wrong. There is no path straight from temporal reality to manifest time, nor from manifest time straight to temporal reality.

This point is an important one for the metaphysics of time. In my opinion the field regularly makes both mistakes and effectively assumes that *Manifest time* = *f* (*R*). Going from left to right in the schema, the field often "discovers" temporal reality by reflection upon manifest time. The example of the Rotating Snakes illusion displays the danger here: by ignoring the contributions of our cognitive faculties, one may attribute to reality features that aren't there, e.g., motion in a still picture. Going from right to left, some philosophers assume that if a property is in reality, then its appearance in manifest experience doesn't require explanation.

Suppose, for example, that presentism—the claim that only present events exist—is true. Then there is a metaphysically distinguished Now. Manifest time includes a distinguished Now too, so the hope is that manifest time is detecting the Now. The job is only finished, I want to insist, when one explains how this detection works. No one has shown how such a Now impinges on our senses or shapes our judgments, a problem threatening to undercut the motivation for this metaphysics.[2] Our cognitive engines are always relevant, even in the most straightforward detection properties.

In what follows I want to fix "*R*" and then show how cognitive science is relevant if we're going to explain manifest time. Indirectly this partly tests our choice of *R*, for if cognitive science (and etc.) can't step up and account for manifest time, then that is a mark against our choice (*if* other theories of *R* can account for manifest time). I'll assume that physics is correct and complete when it comes to temporal metaphysics. This is a big and controversial assumption—of course I may be wrong. Yet it's worth seeing if we *can* explain manifest time without contradicting physics. Plenty of impressive evidence for physical time exists, and sticking with it is the most parsimonious option, as we need the temporal structure it posits to get back the rich successes of physics. By contrast, the evidence for the model described by manifest time is a bit of a mixed bag when it comes to features on which the manifest and physical models most famously conflict. What gets explained are primarily intuitions and it's not even clear that the explanation is very powerful (see section 1.2).

With this default position, we ask: why do creatures model time as manifest if it is fundamentally physical time? Answering this question leads to cognitive science, neuro science, and psychology, but also evolution, our typical environments, higher-level physics, development, and much more. I'll illustrate this with three modes of time, highlighting some of the surprising cognitive science involved. Then I'll tackle the notorious flow of time, commonly said to be an illusion by physicists. I'll sketch the beginning of a theory of flow that relies on our cognitive engines and much more.

1.2 SUCCESSION, DURATION, SIMULTANEITY

Neither manifest nor physical time is a commitment to one feature. Our concepts of time are multifaceted, committing us to order relations (e.g., one event being earlier than another), topological properties (e.g., time being open, continuous, connected,

[2] If the only property the Now has different from all other moments is bare existence, then you will have a hard time explaining how our senses detect it and our judgments notice it. This is a point I make in Callender 2011, 2017. Miller (2013) puts the point succinctly: "In making the privileged present empirically undetectable, it becomes very difficult to see how the presence of such a present could be the explanation for our temporal phenomenology, the very thing that motivates both views to posit a privileged present in the first place." Stepping back, the complaint is similar to Benacerraf's (1973) "no epistemic access" objection to Platonism about mathematical objects. Presumably an answer to Benacerraf, or to Miller and me, will rely in part on our cognitive engines.

oriented), and metrical features (e.g., the duration of events). Our common-sense picture crucially adds a flowing deictic structure of past, present, and future all relative to a distinguished moment (discussed below). Perhaps manifest time also includes immediate judgments about time, such as that the future is in front of us—in which case culture is relevant (Nùñez and Cooperrider 2013).

Kant famously claimed that the mind imposed upon time three modes, succession, duration, and simultaneity. To warm up, let's see how cognitive science relates to these three core features of time. We'll find that temporal features are just like everything else we represent, namely, that there are plenty of examples of mismatch even when we have reason to think that a feature is part of reality and also represented, e.g., shape.

Start with succession, or temporal order. Unlike Kant, we're assuming that physics gets the essence of time right. By "physics" I'll mean general relativity, as that is our best science tackling space-time itself. Does relativity posit objective time order? Not for all events. See Figure 1.2. It does, however, posit an invariant and genuine time order between *timelike*-related events—events connectable by subluminal signals. Not all the events I perceive are timelike related to each other. However, I am a worldline of timelike-related events, so the temporal order of these perceptual events is fully objective. In addition, given the speeds involved, the time it takes to perceive, and that I'm often tracking enduring timelike objects, typically the objects of my perceptions are timelike related to one another too. I perceive the baseball pitch prior to the hit. Relativity agrees: not only is my perception of the pitch really before my perception of the hit, but indeed the pitch really is before the hit.

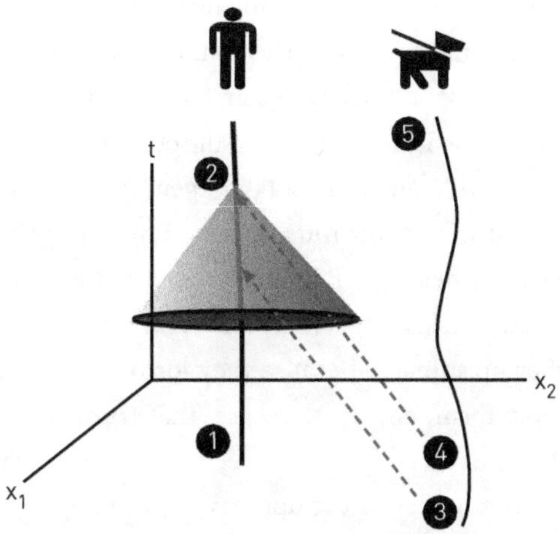

FIGURE 1.2 Person and dog in relativistic space-time. The past light cone at event 2 is partially drawn. Order: invariant time order exists on timelike worldlines such as the person and dog. Hence event 1 < event 2 and event 3 < event 4 < event 5. Event 2 is not before or after event 5. Event 3 happens prior to event 4 for the dog, and the person sees it so. Duration: invariant duration holds between events 1 and 2 and also among events 3, 4, and 5. How long the signals took to get from the dog to person is also invariant. Simultaneity: no pairs of events are objectively simultaneous.

Confident in relativity and mindful of the fact that creatures who get local temporal order badly wrong probably don't live long, I'm expecting that manifest time order is often a (fallible) detection of objective temporal order. Then the schema is

Manifest time order = f (relativistic order, COGEN, etc.),

where we suspect that manifest order is the result of detecting relativistic order.

That is not always the case, of course. Sometimes our experiences of temporal order don't mirror objective order. To take an unrealistic though vivid example, suppose I'm stargazing and see a supernova in the east and then a moment later witness a supernova in the west. The two supernova events are not timelike related and therefore temporally unordered, despite my ordered experiences. To explain cases like this we would have to resort to the invariant temporal order along my worldline. In addition, note with Kant that thunder is heard after lightning is seen, despite the two experiences being produced by the same event. Here again experienced order finds no counterpart in external events themselves.

Locally manifest temporal order tends to track objective temporal order. Regarding non-timelike-related events (like the supernova example), this is so simply because light is so fast. Imagine snapping your left and right fingers, trying to make the snaps non-timelike related. It's possible. But you will never know without the very best detectors that money can buy. Regarding signal discrepancies (like thunder and lightning) our local success is due to the discrepancies in signal speeds having less time to grow, but also from the brain's hardware and software using clever features to bind together signals from the same event source (see Callender 2017; Harris et al. 2010). For instance, fast auditory processing helps compensate for light's greater speed than sound, and mechanisms such as temporal recalibration help shift signal streams toward one another to aid binding. Our cognitive and perceptual architecture often gets it "wrong" (shifts an information stream) to get it "right" (recover the objective spatiotemporal map).

Experienced time order can be mistaken even without signal discrepancies. Suppose we present you with a low tone, followed by a noise burst, followed by a high tone. What will you hear? As Benussi pointed out over a century ago, typically one will hear instead a low tone, high tone, and then a noise burst. One hears the low-high together presumably because they form a natural gestalt, or psychologically salient whole form, and in this case, that trumps order (Benussi 1913; Holcombe 2013). A century later, scientists seem to have vindicated Kant's thought that judgments of causation can affect impressions of temporal priority. In an experiment that would cause Hume to roll in his grave (Bechlivanidis and Lagnado 2016), participants were presented with three objects, one of which starts moving before its cause. Participants claimed to have seen a temporally reordered sequence that matched the causally consistent sequence instead of the actual one.

Kant held that temporal order is imposed by the mind, not the world. Perhaps a modern Kantian might use these experiments to support Kant's order idealism. For me, impressed by the suite of evidence from the sciences and overall experience for objective time order, I instead hold fast to objective temporal order (among timelike-related events) and try to explain why our cognitive system may sometimes fail to register it. I learn that the brain seeking causation and gestalt can sometimes shape our impression of the space-time manifold, just as Rotating Snakes taught us about the odd effects of asymmetric luminance, patterns of neuronal decay, and motion. Helmholtz in 1881 (Helmholtz 1962) wrote:

> It is just those cases that are not in accordance with reality which are particularly instructive for discovering the laws of the processes by which normal perception originates. (251)

This is exactly right. Mismatches between our experienced order and actual order help us understand how we track order as well as we do. Cognitive science helps us explain both detection of and departure from objective order, together aiding our theory that temporal order (along timelike paths) is objective.

Much the same can be said about duration. Duration is given by the proper time in relativity. Evaluated along timelike worldlines, the proper time is identical to the relativistic metric and therefore invariant. Physics—and not merely convention or psychology—assigns a temporal distance between my birth and my typing this sentence. Duration is not just in our heads. For the same reasons as above, we expect that

$$\textit{Manifest duration} = f\left(\textit{physical duration}, \textit{COGEN}, \textit{etc.}\right)$$

will obtain. Manifest duration, we suspect, is a fallible detector of physical duration.

Again we find mismatch between the two. Durations measured by our heads are not in agreement with objective duration or even other heads. A film might seem long to you but short to a companion. Retrospective judgments about duration seem to vary with almost every imaginable variable (e.g., caffeine, alcohol, attention, excitement). Even short-term immediate impressions of duration can result in mistakes, as in the Oddball Effect. Present subjects with a stream of repeated stimuli, such as a picture of a shoe, all of the same duration. Randomly introduce a deviant object, such as an image of an alarm clock. Although the presentation of the alarm clock lasts as long as any one of the shoes, subjects tend to overestimate its duration (Pariyadath and Eagleman 2007). And as with order, judgments of causation can interfere with experienced duration (Eagleman and Holcombe 2002; Schutz and Kubovy 2009).

Despite all this variability, it seems clear that we're tracking objective duration. We use all sorts of temporal cues when estimating duration and with their use often can perform amazingly well. For long durations, even minus temporal cues, Campbell (1990) estimates that we judge each objective hour to be 1.12 subjective hours. For short durations Mauk and Buonomano (2004) claim we're usually within 10% of objective duration. Clearly a signal is fallibly being detected—in the absence of cues, at least by monitoring one of our many biological clocks. How these biological clocks work is still something of a mystery, but there are many models developed. And we can test whether, say, prediction or attention makes an oddball an oddball, thereby helping us discover the mechanisms at work. Once again, cognitive science explains both detection of and departure from reality in common-sense experience.

Simultaneity is different. If relativity is right, there is no temporal structure corresponding to simultaneity. It's just not there. Manifest time, by contrast, does appear to include simultaneity. I snap my left and right fingers together. I can tell whether they happened at the same time or one after another. The extent to which this simultaneity is phenomenological is not entirely clear. Does simultaneity "pop out" (for some evidence that it does in limited situations, see Van der Burg 2008)? Or is it a second-order judgment based on comparing two events? If we teased apart judgments and phenomenology within "manifest time," it might be debatable where simultaneity is best classified. Nonetheless, however understood, subjective simultaneity belongs in the wide category of manifest time.

How do we explain manifest simultaneity if reality doesn't include objective simultaneity? Ask physicists and they will tell you that the physics of objects traveling at low relative velocities compared to the speed of light (like us and everything we encounter) can be approximated with classical physics. Classical physics possesses invariant simultaneity structure. That observation is true and important, yet all it really says is that physical systems can sometimes be modeled *as if* simultaneity existed.[3] To fully explain manifest time, we need to look at the physical objects in our environment and our cognitive engine and see what actually produces our simultaneity experiences. This task initially seems an uphill one because cognitive science reveals a truly unexpected amount of intersubjective variability.

Consider the case of patient PH (Freeman et al. 2013). PH is a retired pilot whose vision and hearing became desynchronized due to a small stroke. PH will hear you say "Hi!" and only after a consciously noticeable lag see your lips move in the appropriate manner. Experiments confirm his subjective experience of hearing people speak before seeing their lips move. He was tested with a battery of time order judgment tasks. These are tasks where a subject is asked to report whether two stimuli occur before or after each other, e.g., what came first, the visual signal or the sound? Throughout the test the stimuli are presented across a range of discrepancies, from

[3] For much more detail on the physics, see Callender 2017; Hartle 2005.

sound first to light first. For PH, lip movements lagged voices. To generate subjective simultaneity for him, voices need to arrive more than 200 ms after lip movements.

Interestingly, the lag had to go in the opposite direction roughly 200 ms to maximize the McGurk illusion for PH. The McGurk illusion occurs when mismatching lip movements cause one to hear the wrong phoneme. For example, mismatched lip movements can change an audio input of /ba/ to a heard /da/. Here one is reporting on some phenomenology—hearing a /ba/ or /da/—and not simply judging before or after as in a time order judgment task. Given the time order tasks, one would expect that the illusion was maximized for PH when the sounds were delayed after the lip movements. Not so! To maximize the illusion for PH, the mismatched lip movements need to arrive more than 200 ms *after* the sounds. PH's "now" isn't as unified as we might think.

If you run into PH you might not agree with him on what happens at any one moment. For you the noise "Hi!" and the lip movements appropriate to that might be in sync, whereas for him they may not be. Though not tested, perhaps you will also disagree on whether a finger-snap happened at the same time as the snap sound. And you also may disagree on what you heard, a /ba/ or a /da/, given some asynchronous gaps in signals. PH is something of an outlier. Perhaps the most surprising finding about PH, however, is that he is not much of one. Healthy subjects were given the same tests by Freeman et al., and the intersubjective variability was very high. Some subjects in the time order tasks needed lag or lead times as high as 200 ms too. Most also showed this strange pattern of a negative correlation between the time order judgment task on subjective simultaneity and what maximizes the McGurk illusion (and a similar Stream-Bounce Illusion). Many subjects even consciously noticed the lag times they reported. All of this fits what we've seen elsewhere in many other studies. For example, using simultaneity judgment tasks—which ask subjects whether two stimuli are synchronous or not—Stone et al. (2001) found that the point that maximized judgments of synchrony varied considerably. For maximum synchrony, one observer needs sounds to precede flashes by 21 ms, whereas another needs flashes to precede sounds by 150 ms.

In general, we're not grabbing together as subjectively simultaneous precisely the same events. Pick a moment and take a perfectly accurate inventory of what is experienced at that moment. You and the person next to you may have very different events on your lists. And given the way asynchronous inputs affect phenomenology—as the McGurk illusion shows—the experience may be different even if you agree on inventory: for instance, your friend may hear /ba/ when you hear /da/.

Where should we find manifest simultaneity in all of this variability? Fortunately, in ecologically valid situations (i.e., ordinary life), we don't much notice this intersubjective variance. Most of the events we discuss and care about are more coarse-grained than the blips and flashes presented to subjects in a lab. Think of a finger-snap. The middle finger slides down the face of the thumb and then makes the noise when slapping the fatty part of the thumb near the palm. In an ecologically valid setting, you're not really going to notice whether the sound came at the same time as you saw the finger slap the fat part of the thumb. Suppose a person in a room snaps her fingers and you hear a snap sound, 100 ms before or after the visual impression of the snap. First, thanks to the processing differences between audio and visual perception, plus a large and malleable window of temporal integration, you may well experience the visual impression and the sound together. Second, as mentioned, even if you fail to bind them together, if the lag is slight and you're not looking for it, there is very little chance that you won't think the finger-snapping caused the snap sound. Third, as events get longer, our points of subjective simultaneity tend to converge. One study increased stimuli length from 9 ms to 40 ms, and then from 40 ms to 500 ms (see Boenke et al. 2009). As duration increased, agreement on simultaneity increased.

PH and intersubjective differences generate interest. Nevertheless, the bigger story here is that despite all this variability, when time scales get longer, events are salient, distances are small, relative velocities tiny compared to the speed of light, and much more (see Callender 2017; Hartle 2005), agreement tends to happen. Physical reality does not contain simultaneity. Cognitive science plus supplementary features of nonfundamental physics, local environments, and so on help explain manifest simultaneity.[4]

In sum, in all three cases—simultaneity, duration, order—our cognitive engines are relevant to our physical or metaphysical hypotheses about reality. The role they play depends on background theory. In the first two cases where we believe common-sense experience detects a property in the world, our cognitive engine helps explain how the detection works and also departures from it. In the third case where we believe there is no such property in the world, our cognitive engines help answer the question of why we nonetheless feel that there is such a property, namely simultaneity. In all three cases our cognitive engines play an important role in the holistic judgments underlying our theory of reality.

1.3 TEMPORAL DEICTIC STRUCTURE

One of manifest time's most important properties is temporal deictic structure. See Figure 1.3. In our conceptualization of time, we can characterize temporal

[4] We're essentially following the method Christoph von Sigwart employed in his monumental *Logik* (1873; von Sigwart 1895). There he made a case for an objective time system in addition to Kant's subjective time. How would we find a time system common to all? His answer: simultaneity. Objective simultaneity is obtained by "reducing the Now of one man to comparison and coincidence with the Now of others." You and I are both simultaneously conscious of the same fact, e.g., a bird singing. This correspondence, Sigwart says, must be due to "external perceptions which are shared by all, and which occur simultaneously for all." Unfortunately for Sigwart, subjectively we're not always aware of the same facts, and objectively, relativity has no simultaneity structure. Still, his methodology is sound: we're *almost* simultaneously aware of the same content, and this is good enough to ground manifest if not physical simultaneity.

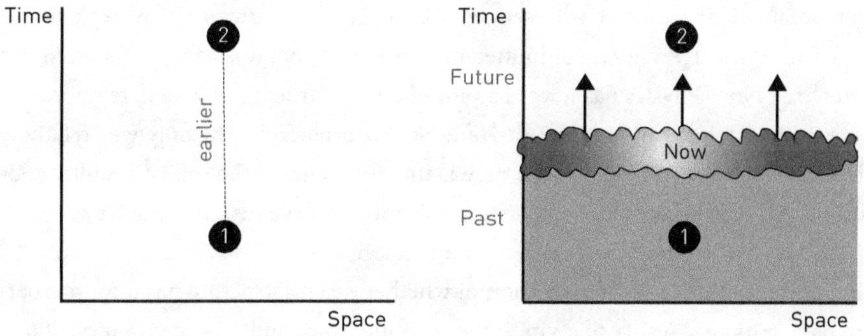

FIGURE 1.3 Sequential vs deictic time. Sequential or B-series time is depicted on the left. Event 1 < event 2, and a certain duration exists between them. A deictic or A-series time is depicted on the right. Superimposed onto sequential time is a distinguished Now that imparts a tripartite structure onto time of past, present, and future. The Now moves toward later events. Relative to the Now depicted, event 2 is future, while event 1 is past.

relationships either by reference to the present moment, or Now, or simply to another moment in the time series. The former conceptualization leads to a classification of events in terms of past, present, and future, whereas the latter leads to one in terms of the earlier than relation. Over a century ago McTaggart (1908) dubbed the first conceptualization an *A-series* and the second a *B-series*. Meanwhile in semantics and cognitive linguistics the first is sometimes called *deictic time*, referring to its need for a deictic center, the Now, which is typically the time of utterance, and the second is often called *sequence time*. The telltale difference between the two conceptualizations is that deictic expressions change truth value depending on when they are said, whereas sequential expressions do not. For instance, "Socrates will drink hemlock" was true only before Socrates drank hemlock but false afterward, whereas "Nixon is before Carter" is always true.

It's hard to overestimate how significant deictic temporal structure is to our lives. Our language, thought, and behavior are all tuned to it. Finding out that the meeting is five minutes from Now motivates action in a way that finding out noon is five minutes later than 11:55 a.m. will not. Agency is understood in terms of this time series. The past is fixed and the future open. We don't think earlier events are fixed unless those moments are prior to the Now. Our preferences are deeply tied to this structure too: all else being equal, no one cares about the past headache as much as the impending future one; and when discounting the value of distant future goods, it's distant *future* not distant *later*, as our discount functions march in sync with the Now. Deeper than all that, we think that *we*—our selves—are wholly present in the Now. Unlike spatial parts like hands, we don't consider our temporal parts to be parts. No, we think we're entirely squashed into each Now.

Physics does not require temporal deictic structure in order to succeed. There is no distinguished Now in physics. Without a Now, there is no temporal deictic center, no past, future, or flow. This rejection of temporal deictic structure is commonly held to be a consequence of relativity, but that thought rests on confusing temporal deictic structure with sequence time, i.e., conflating the Now with simultaneity. Classical physics is committed to simultaneity but was never committed to a preferred Now. Physics has never employed deictic structure—spatial, temporal, or spatiotemporal. No Now, Here, or Here-Now is privileged. The only way relativity has made life "harder" for deictic time is that it eliminates the sets of simultaneous events that one might choose to distinguish and "animate" as a flowing Now.

Given deictic time's importance, the metaphysics of time literature since 1908 has been dominated by the question of whether deictic structure has a counterpart in basic temporal reality that vindicates its importance and use (see Dainton 2001). So-called *tense* theorists, impressed with the significance of temporal deictic structure, propose scores of temporal models incorporating a flowing Now, including presentism, becoming, branching. Differing in detail, they are unified in proposing what is supposed to be an objective counterpart for the temporal deictic center found in manifest time. In essence, these models assume physics is wrong about time, either through a sin of misrepresentation or omission. Research is dominated by arguments for and against tensed models, most of it focusing on the coherency of these models.

In this essay we're trying to give physics a run for its money when confronted by manifest time. Is it possible to explain manifest temporal deictic structure while assuming that it isn't fundamentally in the world? In terms of our schema, we're assuming the usual on the right-hand side and hoping that temporal deictic structure emerges on the left,

$$temporal\,deictic\,structure = f\left(relativity, COGEN, etc.\right),$$

much as it did for manifest simultaneity.

Explaining manifest simultaneity was less daunting. Facts about signal speeds, low relative velocities, temporal integration windows, etc., more or less straightforwardly gave us approximate physical and psychological simultaneity—and eventually manifest simultaneity. With deictic structure, it's not so obvious how to "build" the deictic from the nondeictic. If we think about it linguistically, mindful of what we know about the essential indexical (Perry 1979), it seems to be a nonstarter.

1.3.1 Explaining the Flow

In approaching this difficult question, it's important to sharpen our understanding of the goal. The issue isn't so much getting the deictic out of the sequential but instead why the temporal deictic frame is viewed as an objective window on reality, whereas the spatial one isn't. As we've seen, everyone tends to agree, more or less, on the inventory of (big, salient) events scattered across sequence time. The same is true of what we might call sequence space. We agree on adjacency, size, shape, distance, and more. First base is 90 ft in a straight line from home plate, for instance. We can insert or occupy a deictic center in either sequence time or space, or both. The Now picks out a special moment of sequence time. The spatial here, or better, a particular orthonormal triad of vectors, picks out the origin of a spatial frame. With respect to this spatial center, we speak of forks being on the right, knives on the left, the table in front, the painting behind, and so on. The spatial here plays an important role in language, thought, and behavior too. Yet we're not tempted to view this window as objective, whereas its almost irresistible to objectify the window associated with the Now. So, yes, we want to explain why we use temporal deictic structure, but the really crucial target is why we believe it has a worldly counterpart when we don't think spatial deictic structure has any.

Once we have an explanation of that question, two other targets emerge: accounting for the flow of time and for the fixity of the past and openness of the future. Many people mean many things by "flow of time." Here I simply mean that the tripartite structure {past, present, future} updates itself. Psychologically I don't think this is detachable from the tripartite structure itself. If that structure didn't update, there wouldn't be reason for an organism to employ it. For purposes of analysis, however, we can tease the two apart and explain one and then the other. I won't have time to tackle it in detail, but we also model the past as fixed and unchanging, whereas the future is understood to be open, ripe with possibility.

Summing up, we want to answer the following three questions:

1. Why is the temporal deictic frame viewed as an objective window on reality whereas the spatial one isn't?
2. Why does it monotonically update in one direction unlike the spatial one?
3. Why do we model the past as fixed and the future as open?

We'll begin with the first.

The question why we think the Now mirrors a feature of reality and the Here does not is easy to answer—at least at first pass. We have massive systemic agreement about the Now but massive systemic disagreement about the spatial here. The fork is on *my* right, but if you're sitting across from me, it's on *your* left. The rug is *below* me; but to someone standing on her head, *above* her. The model of space that objectifies a spatial perspective couldn't make it through dinner, never mind a lifetime of navigation. But the Now? We all agree on it. No doubt this wide intersubjective agreement tempts us to objectify the Now and the wide intersubjective disagreement about spatially perspectival discriminations does not similarly tempt us to objectify it.

That observation is important and correct, but it merely moves the bump in the rug to a new location. Now we want to know: *why* do we have so much intersubjective agreement about the Now?

A big part of the answer appeals to what we might call the *hard facts of life*, namely, very basic features of our physical environments and biological and psychological structure. When space-time is divided into space and time, space is three-dimensional and time one-dimensional. Barring objects in our way, we can travel back and forth in each of the three spatial dimensions but due to the lack of time travel we cannot travel back and forth in time. These two deep features of the physical world have ramifications for our movements in time and our mutual orientations. In particular, we can't rotate in time, whereas we can in space. By rotating 180 degrees, we can swap both left-and-right and back-and-front. Without a temporal counterpart of these actions, we can't bring ourselves through sheer rotation to disagree with reference frames in time. There is also a massive asymmetry between what we can know and cause along the temporal directions that is not mirrored in the spatial directions. At any event, I can only cause events in the future light cone of that event, not past light cone; similarly, what I can know about the past is very different from what I can know about the future. Neither of these deep features has spatial counterparts. Combine this observation with the fact that our brains are hardwired with significant differences between time and space. Some are so basic that we typically don't notice them. For instance, at any given place, I can have different mental states at different temporal parts, yet at any given time I don't have different mental states in different spatial parts. My left foot and right arm don't each have different beliefs at a time, for instance.

Together these hard facts help constrain the deictic centers for time in a way they don't for space. Deictic structure imposes an egocentric reference frame upon sequence time and space. The question is why we disagree on deictic centers spatially and not temporally. What I've called the hard facts of life explain a lot of that. There is widespread agreement, spatially and temporally, on the nondeictic structure of space-time. The difference is that with time, due to the hard facts of life mentioned previously, we're locked into a common temporal perspective. Using memory or anticipation, we can escape backward or forward in time. But in terms of immediate experience we're all stuck in the same perspective. We don't have spatial parts with different mental states that could disagree at any one time, and we can't move into positions of differing temporal perspective. It is as if we were all chained to one side of an indefinitely long table. Then we could talk of the right and left side of the table and be tempted to think they mark objective categories in nature. Without disagreement, it would be the natural hypothesis.[5] In the case of time, with all of us locked into the same perspective, it is. *Due to this "locking," the temporal deictic frame inherits the agreement found among observers on events in sequence time.* The hard

[5] Compare with Shoemaker (1996)'s point about the property heavy-to-lift. If I associate with bodybuilders and the infirm, then I'll notice the disagreement over what is heavy to lift and regard the property relationally. But if I associate with people who are similarly strong, I may be tempted to regard it nonrelationally. The same thing is happening with the Now, where here I'm associating with people like me perceptually (and otherwise).

facts of life represent significant constraints that break what is otherwise a symmetrical situation between space and time.

If the preceding is correct, then we have an explanation for why we tend to objectify the temporal deictic frame and not the spatial one. See Figure 1.4. The deictic center provides us with a tripartite division of the world into past, present, and future. The hard facts of life then force agreement on sequential time onto deictic time. Absent such spatial locking, we disagree over deictic space and aren't tempted to paint it onto the world.

Turn to flow. Manifest time is also committed to the temporal deictic center updating itself monotonically with time. We don't merely agree upon temporal perspectives at each moment of time; we also believe that *our* temporal perspective is *moving* toward the future. Some enduring thing is crawling up the worldline.

The best earlier attempts to explain the flow are memory-based accounts. These theories, defended by Mellor and others, point out that memories are constantly accumulating up one's worldline. Because we recall having memories, they can also have a nice nested structure that suggests temporal flow. But these theories seem to be missing something. Accumulation is important, yet there is nothing surviving change moving up the worldline.

If we stick with physics, of course, nothing is literally moving up the worldline. That doesn't mean we don't model things as if they do. To represent change, as Kant noted, we must "represent something as retaining its identity through the change" (Prosser 2016, 173). What is it that retains identity through time? It could be all objects, as Prosser believes, but I prefer to concentrate on what psychologists call the *sense of self*. With this self we can help ourselves to the Buddhist-inspired idea,

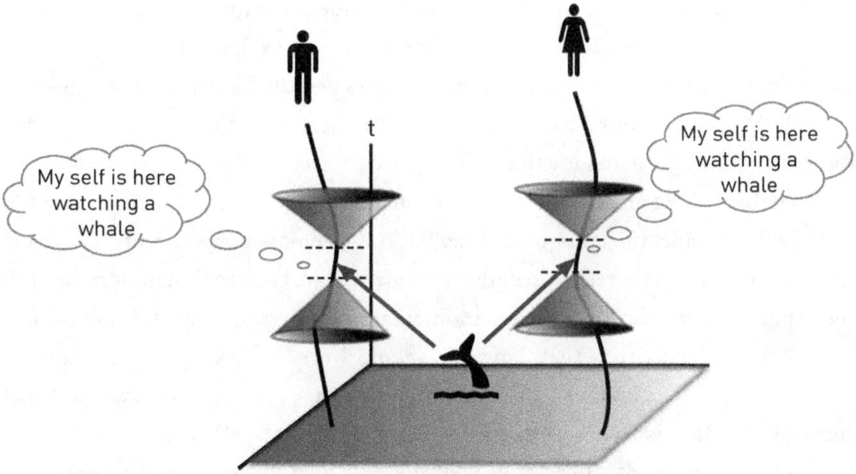

FIGURE 1.4 Building the Now. A man and woman watch a whale jump. Given signal speeds, processing times, and so on, they each have the jump event in their immediate experience, not memory. That is also where they consider themselves to be, as the sense of self is created from memories (and other psychological states)—it's the leading edge of memory. This situation holds for each psychological moment. In this way the man and woman share a common Now.

nicely captured by Velleman (2006b), that the "illusion" of the flow of time is due to the "illusion" of an enduring self. I find this theory appealing, and it can easily be connected to the preceding thoughts. My self is created (epistemologically and psychologically, at least) from my immediate experiences and memories. The self I form is one I regard as enduring, one that is selfsame through time and wholly present at each moment. At each moment, my immediate experiences are the leading edge of this self, the point where my memories run out. But this self changes continually, as moment by moment the threads of identity are weaved, the self retaining identity through time. To borrow a term we'll encounter in a moment, we model ourselves as *moving egos* that endure.

Of course the nature of this self is highly controversial, both metaphysically (what facts, if any, determine who is genuinely who across time) and practically (what facts determine practical determinations of personhood across time). Since we're focusing on why creatures model a certain way, not whether that modeling is correct, we need only focus on the practical problem. Here narrative theories of self are tempting. These are theories that attribute identity to the "story" we tell about ourselves. These theories are often advertised as being superior to traditional accounts based on psychological continuity and connectedness, for they add the self-creation that we employ but that traditional accounts lack.

What type of self do we need? The literature on selves, and in particular, narrative selves, is messy and runs across many academic fields. Very little agreement exists on what narration involves. The selves discussed range from very "thin" ones, momentary flickers of nonconceptual first-person content (Zahavi 2005), to highly conceptualized and even socially dependent "thick" conceptions, such as that one in high school is telling the story of a Goth and not an Emo (Schechtman 2007). In between are theories like Dennett's (1992), which identifies the self as a fiction, like a center of gravity, a story determined by events that best explains what happens, and Velleman's (2006a), which in contrast to Dennett's identifies a self real enough to act causally. Due to the complexity of the issue, the question of the type of self needed for flow requires greater study than we can give it here.

Nonetheless, the answer is bounded from below very clearly: the selves need to be capable of reidentification over time. Lacking the possibility of reidentification, we lose identity and therefore an enduring moving ego. So very thin notions of a self incapable of reidentification, such as momentary first-person content, wouldn't help us. Anything thicker than that, however, should work. Strawson (2004) complains about narrative theories that if I narrate whenever I just get a cup of coffee, then that sense of narration is trivial. Perhaps that complaint is aptly directed at some of his targets. When the point is to get time flow, however, that trivial sense may be enough; after all, time flows as we get coffee, not just when we construct elaborate narratives about ourselves as surfing philosophers. A thin notion of narration is probably all we need (perhaps so thin as to be questioned whether it qualifies as narration and not traditional psychological continuity and connectedness). A question that may be interesting to consider is whether thicker notions of narration provide one with a richer notion of flow. Velleman thinks not, but the question deserves investigation.

Finally, manifest time paints the future as open, ripe with possibility, and the past as fixed and dead. Why? There are plenty of important temporal asymmetries in sequence time. Two prominent ones are the causal asymmetry, the fact that causes typically precede their effects, and the knowledge asymmetry, the fact that we know "more" about the past than future. Now graft an evolving self onto a world containing these asymmetries. This self provides us with a temporal perspective of past, present, and future sliding along sequence time. The self has a *sense of agency* over some of its actions. When it comes to it, the self chooses the blue shirt, not the red shirt, at the shop. Due to the causal asymmetry, its choices will be effective in one direction, again and again, moment by moment. With the temporal perspective intimately associated with these choices—the self makes decisions in the now—the self will learn not only that decisions *later* have effects and *earlier* do not but also that events in what it considers *the future* are alterable and in what it considers *the past* aren't (Campbell 1994). For more along these lines, see Ismael 2011, and for how this connects with the temporal asymmetries of our emotions and preferences, see Suhler and Callender 2012.

Those, in brief, are the core ideas. Very coarsely put, we add an "indexical" element to the world, a temporal deictic center, and the physical, biological, and psychological constraints do the rest. The hard facts of life, plus facts about signal speeds, typical distances to one another, our cognitive engines, and more, together "lock" us in a shared temporal perspective, a view of the world as having a past, present, and future. Because it is shared and agreed upon, we paint this perspective on the world, unlike our spatial perspectives. Much the same causes us to regard the past as "dead." Physical temporal asymmetries plus our cognitive engines constrain their temporal deictic center to see the entire future (and not just later events) as alterable, unlike past events. Evolutionary and cognitive pressures demand a sense of self. The hard facts of life then constrain this self to change very differently in time than in space, leading to the notion that the temporal perspective updates itself as the self does and makes the past direction unalterable.

1.3.2 Empirical Connections?

What would be nice to know is the extent that empirical science can be brought to bear on the preceding claims. Regarding the now and the dead past, I think empirical connections already have been made. What is less clear is what confirmation there might be for the new idea that the self is involved with the flow of time. Right now I see three potential points of contact, although readers may recognize others. I'll briefly discuss two and mention one.

First, many claims are made about distorted experiences or conceptions of time and self among patients with various psychiatric or brain disorders, e.g., memory impairment among schizophrenics. For instance, Martin et al. (2014) explore the possibility that disturbances in the conception of self are associated with problems in time processing among schizophrenics. Given all the confounds associated with the disorder, however, drawing definitive conclusions from these and other clinical populations is extremely challenging. In addition, many of these studies concern duration perception and other features unrelated to deictic time. See Craver et al. (2014) for a cautionary tale. For these reasons I haven't delved into this research yet.

Second, and directly connected to our work, is research in cognitive linguistics. Deictic structure has lately been a very active area of research there. Work looking for evidence of cultural variation distinguishes between two cognitive frames we employ when speaking and thinking about deictic time, the Moving Time frame and the Moving Ego frame (Clark 1973; Nùñez and Cooperrider 2013). The difference lies in whether time is flowing toward or away from speakers. Is New Year's Day approaching? Or are we approaching New Year's Day? Experiments show great variation in what frames we adopt. These experiments are very interesting, but for our purposes, what is important is that we can switch back and forth between these two frames. This may require some effort, but it is not hard to do. How is this fact relevant to our theory of flow? Our theory states that time flow arises from the reidentification of the self through time. The theory gives us a Moving Ego. One might then wonder why we attribute to time itself dynamic qualities. The answer is that we can and do flip back and forth between Moving Ego and Moving Time frames. Cognitive linguists backs up the intuition that once we get the ego moving, time's flow comes for free.[6]

Third, one may look at children's development of temporal concepts. Look at the roots of our mature A-series conception of time. We master an A-series conception of space or time when we can decenter. Decentering occurs when one adopts a different deictic center than one's own. With spatial decentering, I can say that the salt on *my* left is on *your* right. In doing so I adopt your spatial reference frame. No doubt I do this because I must be able to reconcile multiple spatial perspectives (given all the disagreement). Being able to do so in time means adopting different nows. The salt shaker was empty, now is full, and in the future will be empty again. The ability to shift this Now along sequential time is the crucial component of the temporal A-series. How do we accomplish this feat? In the spatial case we have

[6] McTaggart seems to have noticed the difference between the Moving Time and Moving Ego frames: "It is very usual to present Time under the metaphor of a spatial movement. But is it to be a movement from past to future, or from future to past? . . . If the events are taken as moving by a fixed point of presentness, the movement is from future to past, since the future events are those which have not yet passed the point, and the past are those which have. If presentness is taken as a moving point successively related to each of a series of events, the movement is from past to future. Thus we say that events come out of the future, but we say that we ourselves move towards the future" (1908, 470).

all the disagreement, disagreement we lack (in the moment) temporally. The answer (roughly, and cutting a longer story short) must lie in our autobiographical memories, our memories "from the inside." We can't rotate around as we can in space and experience different spatial perspectives, but we can remember previous times that used to be Now. We can also anticipate times that will be Now. Thanks to these memories and anticipations, we entertain multiple conflicting temporal perspectives. Reconciling this conflict leads to understanding a Now shifting along sequential time.

If the picture sketched is on the right track, we might expect correlations among our (a) abilities to form a sense of self, (b) ability to temporally decenter, and (c) abilities to form autobiographical memories. The reasons to expect links are that self-creation draws on autobiographical memories for its story, hence a connection between (a) and (c), and that seeing oneself as temporally extended is plausibly crucial to temporal decentering, hence a connection between (a) and (b). As children develop, their abilities to form memories improves, leading eventually to a more mature conception of the self; with this self they can temporally decenter, leading to the conception of time as deictic or atheoretic. Whether this sketches the correct causal arrows remains to be seen, and other hypotheses may well predict correlations among (a), (b), and (c). Disentangling these abilities and coming to accept some diagnostic task as representative of each one is theoretically tricky (but there are plenty of proposals and tests for each already, e.g., the delayed self-recognition task (Povinelli 2001). One could then search for correlations in performance on each task. Right now we have the tantalizing fact that interesting developments occur in all three abilities at roughly age four years old.

1.4 CONCLUSION

Illusory motion sends scientists on a search for its causes. The result is new science, the discovery of mechanisms and a better understanding of our system of motion detection. If physics is right about time, then the same process should occur for the explanation of manifest time. I examined how this works with succession, duration, and simultaneity. I then turned to the knotty problem of the so-called illusion of the flow of time. If illusory—or if not strictly an illusion but at least the failure to detect some fundamental feature of time—then it too should send scientists on a search for its causes. By dubbing it an illusion physicists take the flow and put its explanation on the desks of cognitive scientists, yet cognitive scientists don't know it's been placed on their desks. The job falls in the gap between the two desks. If anything like the theory of flow developed here is correct, this result is unacceptable: the explanation of temporal flow demands an all-out interdisciplinary attack. Philosophers knowledgeable of the psychological and physical sciences—and who have large desks—can help.

ACKNOWLEDGMENTS

Many thanks to Jonathan Cohen, Charles Sebens, and an audience at NYU Philosophy for comments and to Erik Curiel for the title suggestion.

REFERENCES

Bechlivanidis, C., and Lagnado, D. (2016). Time reordered: Causal perception guides the interpretation of temporal order. *Cognition* 146: 58–66.

Benacerraf, P. (1973). Mathematical truth. *Journal of Philosophy* 70 (19): 661–79.

Benussi, V. (1913). *Psychologie der Zeitauffassung*. Winter: Heidelberg.

Boenke, L., Deliano, M., and Ohl, F. (2009). Stimulus duration influences perceived simultaneity in audiovisual temporal-order judgment. *Experimental Brain Research* 198 (2–3): 233–44.

Callender, C. (2011). Time's ontic voltage. In A. Bardon, ed., *The Future of the Philosophy of Time*. London: Routledge, 73–98.

Callender, C. (2017). *What Makes Time Special?* New York: Oxford University Press.

Campbell, J. (1994). *Past, Space and Self*. Cambridge, MA: MIT Press.

Campbell, S. (1990). Circadian rhythms and human temporal experience. In R. A. Block, ed., *Cognitive Models of Psychological Time*. Hillsdale, NJ: Erlbaum, 101–18.

Clark, H. (1973). Space, time, semantics and the child. In T. Moore, ed., *Cognitive Development and the Acquisition of Language*. New York: Academic Press, 27–63.

Craver, C. F., Kwan, D., Steindam, C., and Rosenbaum, R. S. (2014). Individuals with episodic amnesia are not stuck in time. *Neuropsychologia* 57: 191–95.

Dainton, B. (2001). *Time and Space*. Montreal: McGill Press.

Dennett, D. C. (1992). The self as a center of narrative gravity. In *Self and Consciousness: Multiple Perspectives*. Hillsdale, NJ: Lawrence Erlbaum Associates, 103–15.

Eagleman, D., and Holcombe, A. (2002). Causality and the perception of time. *Trends in Cognitive Science* 6: 323–26.

Freeman, E. D., Isper, A., Palmbaha, A., Paunoiu, D., Brown, P., Lambert, C., and Driver, J. (2013). Sight and sound out of synch: Fragmentation and renormalization of audiovisual integration and subjective timing. *Cortex* 49: 2875–87.

Goldman, A. I. (2015). Naturalizing metaphysics with the help of cognitive science. In K. Bennett and D. W. Zimmerman, eds., *Oxford Studies in Metaphysics*, vol. 9. New York: Oxford University Press, 171–213.

Harris, L. R., Harrar, V., Jaekl, P., and Kopinska, A. (2010). Mechanisms of simultaneity constancy. In R. Nijhawan, and B. Khurana, eds., *Space and Time in Perception and Action*. New York: Cambridge University Press, 232–53.

Hartle, J. B. (2005). The physics of now. *American Journal of Physics* 73 (2): 101–9.

Helmholtz, H. (1962). *Popular Lectures on Scientific Subjects*. New York: Dover.

Holcombe, A. (2013). The temporal organization of perception. In J. Wagemans, ed., *The Oxford Handbook of Perceptual Organization*. New York: Oxford University Press, 820–42.

Ismael, J. (2011). Decision and the open future. In A. Bardon, ed., *The Future of the Philosophy of Time*. London: Routledge, 149–68.

Martin, B., Wittmann, M., Franck, N., Cermolacce, M., Berna, F., and Giersch, A. (2014). Temporal structure of consciousness and minimal self in schizophrenia. *Frontiers in Psychology* 5 (1175): 1–12.

Mauk, M., and Buonomano, D. (2004). The neural basis of temporal processing. *Annual Review of Neuroscience* 27: 307–40.
McTaggart, J. M. E. (1908). The unreality of time. *Mind* 17 (68): 457–74.
Mellor, D. H. (2001). The time of our lives. *Royal Institute of Philosophy Supplement* 48: 45–59.
Miller, K. (2013). Presentism, eternalism, and relativity physics. In H. Dyke and A. Bardon, eds., *A Companion to the Philosophy of Time*. John Wiley and Sons, 345–64.
Nùñez, R., and Cooperrider, K. (2013). The tangle of space and time in human cognition. *Trends in Cognitive Science* 17 (5): 220–29.
Pariyadath, V., and Eagleman, D. (2007). The effect of predictability on subjective duration. *PLOS ONE* 2: e1264–.
Perry, J. (1979). The problem of the essential indexical. *Noûs* 13 (1): 3–21.
Povinelli, D. (2001). The self: Elevated in consciousness and extended in time. In C. Moore, and K. Lemmon, eds., *The Self in Time: Developmental Perspectives*. Hillsdale, NJ: Lawrence Erlbaum Associates, 75–95.
Prosser, S. (2016). *Experiencing Time*. Oxford: Oxford University Press.
Schechtman, M. (2007). Stories, lives, and basic survival: A refinement and defence of the narrative view. In D. Hutto, ed., *Narrative and Understanding Persons*. New York: Cambridge University Press, 155–78.
Schutz, M., and Kubovy, M. (2009). Causality and cross-modal integration. *Journal of Experimental Psychology: Human Perception and Performance* 35 (6): 1791–810.
Shoemaker, S. (1996). *The First-Person Perspective and Other Essays*. New York: Cambridge University Press.
Stone, J. V., Hunkin, N. M., Porrill, J., Wood, R., Keeler, V., Beanland, M., Port, M., and Porter, N. R. (2001). When is now? Perception of simultaneity. *Proceedings of the Royal Society B: Biological Sciences* 268: 31–38.
Strawson, G. (2004). Against narrativity. *Ratio* 17 (4): 428–52.
Suhler, C. L., and Callender, C. (2012). Thank goodness that argument is over: Explaining the temporal value asymmetry. *Philosopher's Imprint* 12 (15): 1–16.
Van der Burg, E. (2008). Pip and pop: Non-spatial auditory signals improve spatial visual search. *Journal of Experimental Psychology: Human Perception and Performance* 34 (5): 1653–65.
Velleman, J. D. (2006a). The self as narrator. In *Self to Self: Selected Essays*. New York: Cambridge University Press, 203–23.
Velleman, J. D. (2006b). So it goes. *Studies in Social Justice*: 1–23.
von Sigwart, C. (1895). *Logic*. Google (Stanford).
Zahavi, D. (2005). *Subjectivity and Selfhood: Investigating the First-Person Perspective*. Cambridge, MA: MIT Press.

2 Cognitive Science and Metaphysics

Partners in Debunking

Jonathan Schaffer

George Bealer does it. Roderick Chisholm does it a lot. Most philosophers do it openly and unapologetically, and the rest arguably do it too, although some of them would deny it. What they all do is appeal to intuitions in constructing, shaping, and refining their philosophical views (Komblith 1998, 129).

Is cognitive science relevant to metaphysics? From the perspective of a realist metaphysician aiming to describe the objective structure of reality itself, it may be natural to think that cognitive science is largely irrelevant. It may seem that cognitive science could at most shed light on *folk metaphysics*, which concerns the contours of our conception of the world and not the structure of reality itself. Folk metaphysics may seem as irrelevant to real metaphysics as folk physics is to real physics.

Yet Goldman (1992, 1993, 2007, 2015; also Paul 2010a) argues that cognitive science is relevant to metaphysics. I take his core point to be that some arguments in metaphysics are premised on intuitions, and that cognitive science is relevant to assessing what we find intuitive, and whether a given intuition should be respected or debunked. In short: cognitive science is relevant to metaphysics via *the debunking project*. For instance, suppose one believed—following Kripke (1980)—in the modal essentiality of origins on the basis of intuitions about certain thought experiments, but one then discovered that those very intuitions were produced by a cognitive module one had independent reason to believe was broken. Then one's basis for believing in the essentiality of origins would be debunked. To put this point in other terms: folk metaphysics *is* evidentially relevant to real metaphysics (in a way that folk physics is not relevant to real physics). And so cognitive science, by illuminating folk metaphysics, casts indirect light on real metaphysics.

I am convinced. I agree with Goldman that cognitive science is relevant to metaphysics via the debunking project. But I offer two adjustments (or perhaps just elaborations) to Goldman's picture. First, I take the relevance of cognitive science to be not specific to metaphysics, but rather generic to any intellectual inquiry that invokes intuitions.

Second—and perhaps more important—I take metaphysics itself to play a crucial role alongside cognitive science within the debunking project. For a crucial part of the debunking project is selecting *when to debunk*, and a crucial part of selecting when to debunk is assessing *whether the intuition fits reality* (or when the module is *failing to track reality*). So in place of Goldman's one-way slogan that cognitive science is relevant to metaphysics, I prefer to say that cognitive science and metaphysics serve as *partners* within the wider debunking project. Visually the difference between Goldman and me may be depicted as in figures 2.1 and 2.2 (with arrows representing relevance).

I do not know whether Goldman would disagree with my picture, or regard it as just an elaboration of his own. I am after all agreeing with Goldman that there is a pathway of relevance from cognitive science to metaphysics via the assessment of intuitions (the "debunking" box). I am then adding, first, that the debunking project does not specifically target metaphysics but rather has a generic impact on any intellectual inquiry that invokes intuitions. But that addition is clearly in line with Goldman's (1992, 1–6) overarching idea that cognitive science proves relevant throughout philosophy.

I am also adding a second root node for metaphysics, representing my claim that cognitive science and metaphysics are partners in the debunking project. *If* I have any major disagreement with Goldman (note the "if"), it is here. On my picture there is no "external vantage-point" from which to critique metaphysics, nor is there any prospect that cognitive science can take on the debunking project alone. The assessment of intuitions relevant to metaphysics is in part a metaphysical assessment. This proves relevant to the extent to which one can appeal to cognitive science to defeat naive realism, and proves relevant to the places where I criticize Goldman for overreaching against naive realism. It also proves relevant to general issues of philosophical methodology, and the not-always-acknowledged place of metaphysics therein.

FIGURE 2.1 Goldman's picture (the relevance of cognitive science to metaphysics)

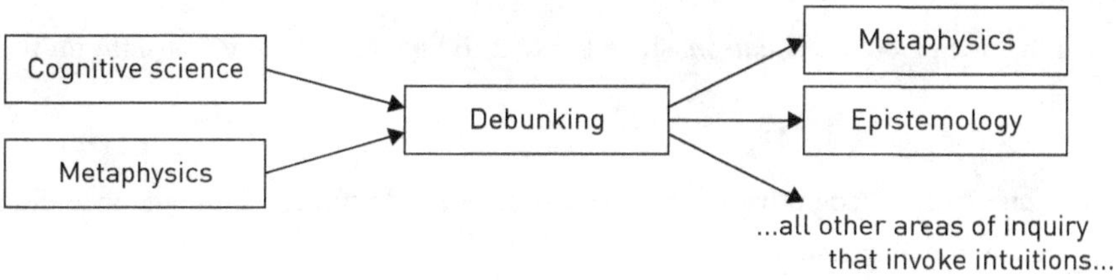

FIGURE 2.2 Schaffer's picture (metaphysics as a partner in the debunking project)

If I have any criticism to make of Goldman's work on cognitive science and metaphysics, it is not that he overplays the role of cognitive science, but that he underplays the role of metaphysics.

Before proceeding, I pause to avert two potential misunderstandings. First, no one is denying that there may be other connections between cognitive science and metaphysics. For instance, if—as Wittgenstein (1969, 36e) imagines—our skulls were filled with sawdust, this presumably would make trouble for some metaphysical theories of mind such as mind-brain identity theories. I only mean to follow Goldman in tracing out one particular connection (via the debunking project). Second, no one is assuming that there is any true unity to cognitive science or metaphysics. Perhaps these disciplines are unified merely by family resemblances and academic politics. Again I only mean to follow Goldman in detailing a particular task of debunking intuitions, and then to further elaborate the connection(s) with one's background picture of reality.

Overview: In section 2.1 I make the general case for metaphysics as a required partner to cognitive science in the debunking project, for providing an external standard to assess intuitions. In sections 2.2 and 2.3 I consider the specific case studies of color, temporal passage, and spatial unity. These illustrate the general role of metaphysics in debunking, while also shedding more light on the interplay between cognitive science and metaphysics.

2.1 METAPHYSICS FOR DEBUNKING

2.1.1 Goldman's Liaison

Is cognitive science relevant to metaphysics? I think that cognitive science is generically relevant to any field of inquiry featuring arguments premised on intuitions, because cognitive science has a role to play in assessing what we find intuitive, and whether a given intuition should be respected or debunked (which I am labeling "the debunking project").

On this point I follow Goldman (2015, 177), who writes:

> What lies "behind" commonsense intuitions, experiences, and judgments (which metaphysics use as *prima facie* guides to the nature of reality) should be probed, including what can be gleaned from the scientific study of the cognitive engine.

I call this idea *Goldman's liaison,* since I take it to be Goldman's core insight on the matter:

Goldman's liaison: Cognitive science is relevant to debunking intuitions, including those used in metaphysics.

I do not claim that *Goldman's liaison* is original with Goldman. Indeed the background idea of looking to the mind to critique our naive view of reality perhaps runs through the history of philosophy. But I take Goldman to be the leading contemporary exponent and developer of the view.[1]

My use of the terms "intuitions" and "debunking" may be misleading, for I mean both very broadly. Under "intuitions" I include common-sense beliefs and sensory seemings, since—as Goldman (2015, 181) is explicit—these are all cognitive outputs whose evidential bearing on metaphysics needs to be assessed:

> The metaphysician's initial evidence bearing on the theories would be a set of ordinary experiences, intuitions, or beliefs about the domain that each theory tries to accommodate in its own way. A basic assumption here is that such experiences, intuitions, and beliefs are examples of evidence that metaphysicians (legitimately) use when weighing competing theories.

Likewise under "debunking" I include all assessments of these cognitive outputs, where the assessment could in the end be one of respecting the output, or even in principle (Goldman, personal communication) one of *boosting* the evidential force of the output. I use "the debunking project" and speak of "debunking intuitions" as metonyms for the more general project of assessing cognitive outputs for evidentiary force, in part because this aspect of the project tends to attract the most interest.

It might help to distinguish *Goldman's liaison* from other theses in the vicinity. To begin with, given the holism of confirmation, everything is relevant to everything. Following Duhem and Quine, Fodor (1983, 105) notes: "[T]he facts relevant to the confirmation of a scientific hypothesis may be drawn from anywhere. . . . In principle, our botany constrains our astronomy, if only we could think of ways to make them connect." *Goldman's liaison* goes beyond an utterly generic claim of the form "Everything is relevant to everything; a fortiori, cognitive science is relevant to metaphysics." It identifies a definite way in which cognitive science and metaphysics connect, namely via the assessment of intuitions.[2]

[1] Paul (2010a, 470) has more recently defended a similar view, on which metaphysical judgments are informed by "ordinary judgments" about what causes what or what is a part of what, such that "one role for cognitive science in ontology is to identify places where our ordinary judgments might not be appropriately generated, as with illusions." And also: "Given the role of ordinary judgments in ontological modeling, we need to know if facts about our cognitive apparatus result in certain sorts of perceptual or judgmental bias."

[2] Of course if one thinks that intuitions are used in every intellectual inquiry, then *Goldman's liaison* will apply everywhere, but it will still apply in a specific way (through intuitions). That said, insofar as cognitive faculties are used in every intellectual inquiry (think of the role of perception and reasoning in science and mathematics), there is also a generalization: *Cognitive science is relevant to assessing cognition, and hence relevant to all fields of inquiry.* For present purposes I leave open whether intuition should count as a distinctive cognitive kind.

There is also a sense in which cognitive science might be thought to have something very specific to say just about metaphysics. For instance, in a broadly "Kantian" vein one might think that there are certain specific questions—"metaphysical" as said with a sneer—which lie beyond the ken of human cognition. If there is such a very specific sort of relevance, *Goldman's liaison* does not attempt to describe it. *Goldman's liaison* identifies a definite way in which cognitive science and metaphysics connect, but only by in effect grouping metaphysics under the more general heading of *inquiries featuring arguments premised on intuitions*. When the metaphysician argues from intuitions about when mereological composition occurs, she is—from the perspective of *Goldman's liaison*—doing the same thing as when the epistemologist argues from intuitions about when knowledge is present, when the philosopher of language argues from intuitions about what a word means, or when the ethicist argues from intuitions about what actions are required. In all these cases one can ask whether the intuition should be respected or debunked.

Goldman's liaison is compatible with any combination of views on these other theses. I myself would accept the general holism of confirmation, but would not accept the claim that there is something specifically defective about metaphysics (see Bennett 2016).[3] But for present purposes these should by regarded as largely independent matters. Though since I argue that metaphysics plays a crucial role alongside cognitive science within the debunking project, I am in effect arguing that those who would reject metaphysics thereby bar themselves from participating in the debunking project in the first place. (The cost of rejecting metaphysics just got higher.)

Goldman's liaison does embed a particular conception of at least some lines of metaphysical inquiry, namely as being inquiries that draw on intuitions. But I think it does so in a highly uncontroversial way. First of all, no assumptions are made about what intuitions are. I am using the term "intuition" in the broadest sense, including naive beliefs, or dispositions to believe, or sui generis states of seeming true.[4] My own view is that we enjoy beliefs, dispositions to believe, and states of seeming true, and that each of these is invoked in various places in metaphysics, and that each of these is open to assessment and potential debunking. But *Goldman's liaison* requires no stand on the matter.

Second, no assumptions are made as to whether intuitions feature in the content of a given premise or merely as the rationale for the premise. Perhaps the relevant premise in a case of modality might be "intuitively, it is not possible to have water

[3] See also Paul 2012 for the case that metaphysics has some distinctive subject matters but no distinctive methodology. As she (2012, 9) puts the claim: "Both fields [metaphysics and science] can be understood as relying on modeling to develop and defend theories, and both use a priori reasoning to infer to the best explanation and to choose between empirical equivalents." She (2012, 28) concludes: "Metaphysics stands side by side with natural science as an important and legitimate developer of our conceptual schemes."

[4] See Pust 2012: section 1 for a useful overview of different conceptions of intuitions.

without H_2O" or perhaps the relevant premise is merely "it is not possible to have water without H_2O" where the justification for this premise would be that it is intuitive.[5] My own view is that both are eligible to serve as premises, and that each of them is open to assessment and potential debunking. But again Goldman's liaison requires no stand on the matter.

It is only assumed that intuitions, whatever they may be, are used in some way or another. Even this may seem controversial. For instance Cappelen (2012) argues that intuitions play no real role in philosophy, and that there is merely a bad stylistic reflex in the discipline now to insert "intuitively" in place of stating a reason. My own view is that philosophers have tended to use "intuitively" to mean intuitively, and that this has an evidential meaning requiring a kind of immediate judgment.[6] But in any case Cappelen would say that one should just replace "intuitively" with a direct evaluation of the (worldly, nonpsychologized) claim at issue. *Goldman's liaison* can still come in here, since there is still room to assess and perhaps debunk our capacities for a direct evaluation of a given worldly claim. So in that sense I take *Goldman's liaison* to use "intuition" in an inessential way (cf. Cappelen 2012, 224–27), merely as a pointer to whatever exactly is going on, in metaphysics and elsewhere, when philosophers assert premises which they call "intuitive."

Think what you will of intuitions. I take it that everyone needs to make sense of the prospect of *debunking*, in which certain beliefs (typically folk beliefs) are shown to be evidentially defeated. The friend of *Goldman's liaison* is really committed to two essential claims:

- Cognitive science is relevant to debunking, and
- Some claims in metaphysics are potentially open to debunking.

Talk of intuitions is only coming in as a mediator between these two claims: the claims in metaphysics that are potentially open to debunking are *intuitions,* and cognitive science is relevant to the debunking of these very intuitions. So anyone who would make sense of debunking can make sense of *Goldman's liaison*, whether or not they like to posit intuitions as playing this mediating role.

2.1.2 The Debunking Project

Evolution suggests that human cognition is a powerful but flawed tool. On the one hand it is plausible that many of our cognitive faculties evolved to help us with

[5] See Williamson 2007, esp. 210–14, for a discussion of the perils of "psychologizing the evidence," in which we "conceive the evidence in philosophy as consisting of psychological facts, such as the fact that we believe that there are mountains in Switzerland, not the fact that there are mountains in Switzerland" (234).

[6] I credit this point to Jennifer Nagel, from her comments on Cappelen's book at a 2014 Central American Philosophical Association symposium.

the four Fs (feeding, fighting, fleeing, and reproduction), and plausible that this pressured our ancestors towards reliably tracking the environment. On the other hand it is equally plausible that many of our cognitive faculties evolved to give us quick and dirty heuristics reliable only for limited purposes in evolutionarily salient contexts.

In particular, some of our intuitions seem insightful. People naturally intuit, when shown two hands, that—in the words of Moore ([1939] 1993, 166)—"here is one hand and here is another." People naturally intuit that 2 + 3 = 5. Barring radical skepticism, it seems that our native endowments actually guide us in the right direction in these cases. But some of our intuitions seem to be crude superstitions. People intuit that the earth is flat, and that animate spirits dwell in trees, rivers, and rocks. If science has taught us anything at all, it seems that some of our naive prejudices deserve to be debunked.

Thus consider the intuition that animate spirits dwell in trees, rivers, and rocks, as a paradigm case of an intuition that needs debunking. I take it that the debunking project, as applied to this case, has at least three connected components. First, there is a psychological story to tell—in this case involving what Boyer (2001; cf. Guthrie 1993; Bloom 2007) calls a "hypertrophy of social cognition"—in which we tend to overattribute psychological agency (our "theory-of-mind module" delivers many false positives). In this vein Bloom (2007, 149) notes:

> The classical demonstration here is that of Heider and Simmel (1944), who made a simple movie in which geometric figures—circles, squares, and triangles—moved in certain systematic ways, designed, based on the psychologists' intuitions, to tell a tale. When shown this movie, people instinctively described the figures as if they were specific people (bullies, victims, heroes) who have goals and desires.[7]

But secondly—and most crucially for my purposes—there is a background metaphysical picture in play, informed in part by science, against which one judges the intuitions. I presume that it is false that animate spirits dwell in trees, rivers, and rocks. I presume that it is false that the geometric shapes in Heider and Simmel's movie are agents with real goals and desire. There is nothing internally incoherent about accepting these animist intuitions as respectable, and holding that rocks and movie images of triangles are really agents. *It is only because the intuition conflicts with my background metaphysical picture that I seek to debunk it.* Likewise I do not seek to debunk the intuition that here is a hand, or that 2 + 3 = 5, because in fact I think that these intuitions fit the world.

Third, given the psychological story and the background metaphysical picture, there is an epistemic account to be given of how the animistic intuition loses evidential

[7] See Arico et al. 2011 (esp. sec. 2) for a useful overview of the psychological basis for agency attribution.

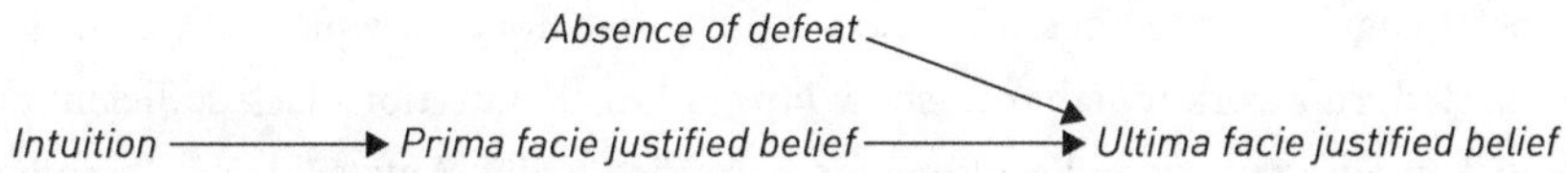

FIGURE 2.3 Epistemological backstory for defeat

standing. For the sake of definiteness I will take on an epistemological backstory on which intuitions provide prima facie justification for belief, which converts to ultima facie justification in the absence of ultimately undefeated defeaters.[8] The loss of evidential standing involved in debunking can then be identified with *defeat.*[9] So the picture looks as given in figure 2.3.

This is not the only way the epistemological story may be told. For instance, for those—no doubt including Goldman (1979) himself—who favor a more reliabilist-oriented backstory, there may be ways to think of defeat in reliabilist terms (see Nichols 2014, esp. section 2 on "process debunking"; see generally Grundmann 2009).[10] Or for those—including myself—who favor a relevant alternative theory, one may think of defeat in terms of the relevance of alternatives in which the mind is misreading the world (see Lewis 1996, 558). The epistemological backstory may even be told in more minimal terms, by replacing "loss of evidential standing" with the weaker idea of some lowering of rational posterior credence. Goldman (2015)—at least for one cluster of cases (the projections: section 2.3.1)—takes up a Bayesian model with some limited constraints on rational priors, to associate debunking with at least some dampening of rational posterior credence.

The epistemological backstory is more of a fixed background. It provides the machinery of debunking, into which the psychological and metaphysical inputs are fed, and an epistemic verdict on a given intuition is then issued as output. (For this reason I do not include an "epistemology" box in the input to "debunking" in my picture of the terrain.) Nothing I say in the main text should depend on the details of the epistemological backstory. I am only supposing that the epistemological backstory has the resources to say why the psychological story about the hypertrophy

[8] See Pollock 1987 for a seminal treatment, and Grundmann 2011 for a useful overview of these issues. I am remaining neutral on whether there can be purely "factive" external defeaters beyond the ken of the subject, or whether the subject needs to have the belief ("doxastic" defeater), or whether it merely needs to be the case that the subject *should* have the belief ("normative" defeater). In the cases under discussion in the main text, I am concerned with the epistemic status of certain intuitions about the metaphysical nature of reality, for those who have all the relevant beliefs.

[9] Kahane (2011, 106; cf. Mason 2010) associates debunking explanations with undermining defeaters: "Debunking arguments are arguments that show the causal origins of a belief to be an undermining defeater," though as will emerge in section 2.2.2 there are also cases with the structure of rebutting defeaters. And as Nichols (personal communication) has suggested, there may even be cases of debunking which undermine any claim to prima facie justification, before issues of defeat even come into consideration. I am not presupposing that there is any deep epistemic unity to debunking.

[10] As Cohen (1984, 282–83) points out, while internalists and externalists tend to disagree on *why* a given process is epistemically defective, they still tend to agree as to *which* processes are defective.

of social cognition, together with the metaphysical story on which animism is to be rejected, can work together to show how animistic intuitions lack sufficient evidential standing to engender ultima facie justified belief. Any viable epistemology should find a way to make sense of debunking.[11]

What emerges from this paradigm case of a belief that needs debunking are three crucial and interconnected aspects of the debunking project:

- *The psychological aspect*: Debunking requires identifying the psychological mechanisms involved in the production of the intuition and showing them to work in a certain way.
- *The metaphysical aspect*: Debunking requires painting in the background metaphysical picture so as to establish that the psychological mechanism is *going wrong* by working in this way.
- *The epistemological aspect*: Debunking requires showing how an intuition produced by psychological mechanism gone wrong in this way thereby loses evidential standing.

For present purposes the crucial point is the second bulleted point. Even given a fixed epistemic backstory, cognitive science cannot go it alone when it comes to debunking, but requires metaphysics as a partner. What follows is primarily an elaboration of this point.

2.1.3 Metaphysics for Selectivity

Given that human cognition is a powerful but flawed tool, one should expect *selective debunking*. Human cognition is not a complete disaster. Barring radical skepticism, some of our intuitions deserve respecting. But human cognition is equally not a perfect oracle. If science has taught us anything at all, some of our intuitions deserve debunking. I have already suggested some candidates for intuitions to be respected (Moore's "here is one hand and here is another," 2 + 3 = 5), as well as intuitions to be debunked (the earth is flat, animate spirits dwell in nature). But leaving these specific candidates aside, I take it as nonnegotiable that some intuitions should be respected but others should be debunked. No other attitude fits a naturalistic conception of human cognition as a powerful but flawed tool.

(Indeed the very prospect of an unselective story raises self-undermining concerns. An unselective story in which all intuitions are to be respected is self-undermining, given—as I take to be the case—that among our intuitions is the very intuition that some of our intuitions deserve debunking. But worse, an unselective

[11] Also I am not supposing that there is anything *distinctive* to debunking. For instance, White (2010) argues that the epistemic action in debunking derives from the more general phenomenon of disagreement. For present purposes I take no stand on the matter.

story in which all intuitions are to be debunked is self-undermining, given that the debunking project is itself premised on certain intuitions, including an intuitive epistemological backstory about defeat.)[12]

The need for selective debunking invites the question: *When should an intuition be respected and when should it be debunked?* One *bad* answer—bad because unselective—is that an intuition can be debunked when one can tell a cognitive story about how it arises. This is unselective because of course there is always some cognitive story to be told about every cognitive output, intuitions included. Cognitive outputs are not miracles. They all have causal etiologies through our cognitive engines.

A second *bad* answer—also bad because unselective—is that an intuition should be debunked when it can be triggered in error. This is unselective because every human cognitive process is fallible and can be triggered in error. We can make mistakes even about simple visual beliefs and simple arithmetical beliefs. If perfection is needed for evidential standing then nothing stands.

A better version of this second answer is that an intuition should be debunked when it is *often* triggered in error, or more generally when it is *unreliable* as an indicator of reality. (This fits the reliabilist backstory of Goldman 1979, and also Goldman's [2015] analogy between debunking intuitions and discovering when a gas gauge is stuck.) This offers a more selective answer. But it is one that requires a metaphysical component, for to assess what is and what is not reliable as an indicator of reality, one needs some conception of what is out there in reality. By reliabilist lights, debunking requires *comparing* the cognitive output to the metaphysical facts, in order to measure their degree of correlation.

But leaving reliabilism aside, on any reasonable theory of debunking, the information that a given intuition does not fit reality should suffice for debunking. This is selective (not every intuition fails to fit reality), and it defeats any prima facie justification the intuition provides.

I am claiming that metaphysics is needed to answer the question of when to debunk. An intuition should be debunked when it fails to fit the world, by failing to find a place in the background metaphysical picture, as presumably informed by science. Thus I say:

Metaphysics for selectivity: Determining whether a given intuition deserves debunking requires determining whether it fits reality.

When I spoke of *the metaphysical aspect* of the debunking project (section 2.1.2), it was by way of helping render debunking selective.

[12] See Korman 2009 for a similar style of self-undermining concern against global challenges to intuition.

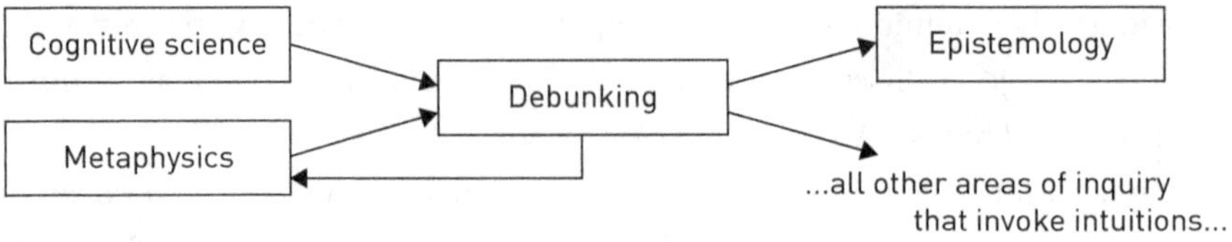

FIGURE 2.4 Schaffer's picture redrawn with a loop

But the need for a background metaphysical picture comes out not only in explaining why we debunk in particular cases (for example, naive animism: section 2.1.2), but also in the very idea of *assessing* a folk theory on anything more than internal grounds of coherence. A crucial part of that further assessment is in terms of an external comparison with reality. The background metaphysical picture is providing the external standard against which folk conceptions may be judged tor fit. In short: our intuitions may be assessed for internal coherence and for external fit with reality. The latter is a metaphysical assessment.

What is this background metaphysical picture, and where does it come from? The picture is our conception of the objective and intrinsic nature of reality. I leave open how this should be understood, so long as it is understood as whatever is making us opt to debunk intuitions of naive animism.[13] The picture needs to be informed by science, but science itself just gives us models and equations, which themselves stand in need of metaphysical *interpretation*. Or perhaps better: science itself is continuous with and commingled with metaphysical inquiry. So the background metaphysical picture itself must draw on intuitions (including those used in preferring a given interpretation of science). In that sense one is seeing a kind of feedback loop in which an initial intuitive metaphysical picture, augmented with information from physics and cognitive science (inter alia), is being used to assess itself and reach a wider reflective equilibrium. Such a loop is implicit in my picture diagrammed in figure 2.2, in which I label two boxes as "metaphysics," but equally could have drawn things with a single "metaphysics" box and a loop, as in figure 2.4.

In any case, invoking the metaphysical aspect of the debunking project as part of the answer to the question of when to debunk has two immediate consequences. First it resituates metaphysics from being the target of debunking to being a crucial part of the machinery of debunking itself. But second it shows how considerations from cognitive science alone cannot be expected to defeat the hardened naive realist.

[13] In Schaffer 2009 I argue that the background metaphysical picture is best understood as structured by the grounding relation, so I would take this to be a picture about what grounds what. But nothing in the main text requires this conception of metaphysics. Fine (2001) introduces, as a second primitive notion alongside grounding, the notion of *in reality*, regimented by a primitive "In reality" operator, and glossed (2001, 25) in terms of a "positive idea of the intrinsic structure of reality." What I speak of as the background metaphysical picture looks to correspond to the propositions that Fine would take as real, though I am not presupposing that one needs to understand this via a new primitive operator.

For the naive realist who would insist that colors, morals, or spirits are part of reality may thereby resist the call to debunk, simply by clinging to their metaphysics and claiming that the relevant intuitions fit reality.

2.1.4 Can Cognitive Science Fly Solo?

Is there any alternative to invoking a metaphysical aspect to debunking (as per *Metaphysics for selectivity)?* In particular, are there any purely internal signs of defective cognition, such that cognitive science could all by itself discern when debunking is called for and when not? Of course the folk theory may itself be incomplete, or incoherent in various respects. (It is an empirical psychological matter whether this is so.) But it may not be. And even if the folk theory is incomplete or incoherent, one may still want to assess various completions of the theory, or various coherent portions of the theory. Is there any way that cognitive science could—on its own—fully assess a complete and coherent (refinement of) folk theory?

It is hard to answer such a question in the abstract, but there is a tempting idea which I think does not work, involving the idea of *hermetic cognition.* Goldman (1993, 123), in discussing the debunking of modal beliefs about essences, says:

> How exactly might cognitive science contribute to this conclusion [that we are not in touch with an extramental modal fact]? It might be in a position (eventually, if not now) to assure us that there is a cognitivist story to be told that explains the existence of our intuitive reactions and their inter-personal uniformity without appeal to any "detection" of extra-mental modal facts. We might then be in the situation of those who reject the inference from religious experience to theism on grounds that the religious experience can be explained purely psychologically without reference to divine sources.

Likewise Goldman (2015, 186)—following Joyce (2006) and Street (2006) in discussing the case of debunking moral beliefs—says:

> Cognitive science can weigh in on this question. It certainly studies perceptual systems (also called "senses") and sometimes discovers new ones. Comparative cognitive science discovered echolocation in bats. Human cognitive neuroscience has discovered many types of *interoceptive* senses in the human brain, i.e., brain mechanisms for monitoring the positions and conditions of one's own bodily organs. . . . There is no comparable discovery of any *moral* sense, whether an exteroceptive sense or an interoceptive sense. This might be interpreted as evidence against the existence of any sense for tracking moral truths.

So one might look for the following internal sign of lack of fit with external reality: the intuition is generated by a cognitive system that is *disconnected from reality,* in the sense of lacking any plausible way of detecting or otherwise tracking the truths it claims to deliver. In other words, the cognitive system lacks the right input profile to render these intuitions suitably connected to the world. The cognitive system may be dismissed as hermetically sealed off from the world, without judgment as to what there might be out in the world.

But first, being hermetic is not *necessary* for debunking. A cognitive system can come equipped with all sorts of sensory transducers and still deliver intuitions that deserve debunking if the affiliated sensory system is sufficiently poor, or if the system itself is doing poor things with its sensory input. Some of the intuitions generated by our folk physics module are presumably of the latter sort. The folk physics module seems strongly connected to perceptual inputs. Only it encodes some false assumptions about reality, and when it generates intuitions solely on this basis, such intuitions deserve debunking.[14] Likewise in the case of color—to be discussed in some detail in section 2.2—the problem is evidently *not* a lack of any sensory input from the world. So metaphysics is still needed to determine when to debunk for cognitive systems that are highly connected to the world.

More interestingly, being hermetic is not *sufficient* for debunking either. A cognitive system can lack any relevant sensory transducers and still deliver intuitions that deserve respecting. Indeed, presumably our capacities for abstract logical and mathematical reasoning are subserved by capacities without any relevantly direct links to perception. (There is no "inner eye" turned upwards to Platonic heaven.) So, barring a radical "debunking" of all of our abstract intuitions in one fell swoop, one needs to allow respectable intuitions severed from any direct perceptual input. How could these intuitions be respectable? There might still be evolutionary pressure, for instance, for these modules to conform to mind-independent mathematical structure, even if these modules are not directly informed by this structure. So there might be some evolutionary pressure towards carrying a sealed-off but still reliable map of mathematical reality.[15]

Indeed—returning to the case of morality—it seems at least *possible* that there is a mind-independent moral structure which typically mandates cooperative familial and social relationships. And there are natural evolutionary pressures that could have led our ancestors to seek cooperative familial and social relationships, and punish those who defected from this moral behavior. In that case there might even be some evolutionary pressure towards carrying a sealed-off but reliable map of "moral reality." (I am not saying that this is a plausible story, and I am not saying

[14] For instance, Kaiser et al. (1986) found that naive subjects—especially between third and sixth grade—tended to predict that a ball emerging from a curved or spiraling tube would continue along a curving trajectory (undergoing curvilinear rather rectilinear motion), presumably due to a naive folk physics of *impetus*.

[15] In this vein Jenkins (2005, 731) suggests that we may carry "a reliable, on-board *Conceptual map* of the structure of the world. And we might well be able to learn about the structure of the world by examining such a map." She sketches an in-principle naturalistic story as to how such a map might be nurtured or implanted by nature (2005, 735–37).

that morality in the end deserves respecting; I am only explaining how the fact that there is no moral sense organ does not suffice to recommend a debunking stance against morality.)

Perhaps the absence of direct and relevant sensory input is still a decent warning sign of intuitions that might best be debunked. I remain neutral. My point is just that it does not yet settle the question of whether or not to debunk, and that metaphysics must still play a role in the end in providing a picture of reality to serve as an external standard of comparison, whether the cognitive system is hermetic or not.

A second tempting idea is to think that there are independently known forms of defective belief-formation. Goldman himself (1979, 9) gives the following as examples of "faulty processes of belief-formation": "confused reasoning, wishful thinking, reliance on emotional attachment, mere hunch or guesswork, and hasty generalization." But it is not obvious that these are uniform cognitive processes, and especially not obvious that any of these forms of belief-formation are to be found in many of the interesting cases such as color (section 2.2). Color just seems directly given in sensation.

Indeed, even in cases where these forms of belief-formation might be thought to apply, one can still question whether these forms of belief-formation are in fact faulty for the given case. For instance, Plantinga (2000, 195–98) suggests, in reply to Freud, that God might have designed us to undergo wishful thinking for the purpose of getting us to believe in God, in which case wishful thinking would be an epistemically proper basis for theistic belief. Likewise, while Singer (2005) and Greene (2008) argue that deontological thinking is based on nonrational emotional responses, Nichols (2014, 747) counters that one still needs "a framework-external specification of why the emotional processes are epistemically defective." (Nichols [2014, 747] also points out that dismissing intuitions when emotionally laden "threatens to be a doomsday argument for normative ethics." Indeed it might be that virtually all human cognition is emotionally laden.)

Perhaps—at least in those cases when one can assimilate the intuition to those of a generally poorly produced sort—one again finds a decent warning sign of an intuition that might best be debunked. My point is just that it does not yet settle the general question of whether or not to debunk in two respects: it does not apply to many crucial cases (for example, color), and it remains possible to allow that the intuition, while being of a generally poorly produced sort, is also of a more specific well-produced sort (for example, Plantinga's idea that wishful thinking might be specifically proper for theistic belief).

Goldman (1992, 53) also introduces the notion of an "autonomous" internal explanation, as the sign of when to debunk in a selective way:

> The [presumption against veridicality] arises only when a feeling or intuition can be given an *autonomous* internal explanation, where an internal explanation is autonomous if it either *excludes* expansion into a fuller explanation citing the metaphysical posit or is *unreceptive* to such expansion.

A system can be autonomous in this sense without being hermetic. Indeed when the nonhermetic folk physics module generates intuitions solely on the basis of false assumptions about the world (for example, intuitions about curvilinear motion on the basis of impetus), the explanation excludes expansion into a fuller explanation citing the metaphysical posit (impetus itself), precisely because there is no such thing as impetus. Likewise if the hermetic folk mathematics module generates intuitions via a sealed-off but reliable map of mathematical reality, the explanation positively invites expansion into a fuller explanation citing the metaphysical posit (mathematical facts), for instance as the facts for which there was evolutionary pressure to conform.

But this shows that Goldman's own criterion of autonomy is not itself internal to cognitive science, but points outwards towards the metaphysical aspect of debunking. By Goldman's own lights, whether a cognitive explanation is receptive to a fuller explanation citing the metaphysical posit depends (inter alia) on whether the posit fits reality. This is not necessarily a criticism of Goldman for he does not explicitly deny the existence of a metaphysical aspect of debunking (he is silent on the matter). In any case I conclude that metaphysics is a required partner to cognitive science for the debunking project. Part of the task of assessing intuitions for debunking is assessing whether they fit reality.

2.2 CASE STUDY: COLOR

I have argued—albeit largely in the abstract—that metaphysics is a required partner to cognitive science for the debunking project. In what remains I focus on rendering the point concrete, by taking up cases. Goldman himself, over the arc of his 1992, 1993, 2007, and 2015, discusses a fairly wide range of cases, including

- the spatial and temporal unity of objects (1987, 539--43; 1992, 62; 1993, 101–8; 2015)
- temporal passage (1992, 53; 2015)
- the individuation of events (2007, 12–18)
- modality and essences (1992, 55–62; 1993, 119–23; 2015)
- causation (1987, 538)
- morality (1987, 538; 2007, 8)
- color (1992, 43, 50; 1993, 114–19; 2007, 5; 2015)
- numbers (1992, 46), and
- deities (1987, 543; 1992, 46; 2015).

Goldman initially sees these cases as generally inviting a (1987, 539) a "revisionary metaphysics" of an antirealist bent, according to which (1987, 538) "certain parts of our conceptual scheme (e.g., space, time, unity, cause, moral value) are really the handiwork of the mind," resulting in some form of (1992, 51) "ontological demotion" to a less realist position than naive common sense begins from. The relevance of cognitive science is in showing how and why the naively realist starting point needs revisionary demotion.

Goldman (2015) comes to draw a distinction between two types of case. In one type of case—which I label "projections"—Goldman takes the phenomenon at issue to lack objective reality, and the intuitions in question to be mere projections of the human mind, which—as Hume ([1777] 1975, 294; cf. Goldman 1987, 542) memorably said of taste—"has a productive faculty, and gilding or staining all natural objects with the colours, borrowed from internal sentiment, raises in a manner a new creation." In a second type of case—which I label "relations"—Goldman takes there to be an objectively real phenomenon but one which must be reinterpreted in a revisionist spirit, as relativized to a cognitive system:

- *Projections*: no objective reality (temporal passage, morality, modality and essences, deities).
- *Relations*: objectively real relations (color, temporal unity of objects, individuation of events).

I primarily focus on the case of color, with an occasional glance at some other cases en route. I focus on color since it is a central example for Goldman and since virtually all of the general morals I want to draw can be drawn from that case. I draw three overall morals:

- Whether the cognitive story supports debunking depends on the background metaphysical picture.
- The detailed cognitive story is sometimes irrelevant to the debunking project.
- There is no metaphysical distinction between projections and relations but rather an epistemic distinction between undermining and rebutting defeaters.

The first bulleted point is the most important moral with respect to my claim about the role of metaphysics in the debunking project. But the second bulleted point makes the complementary claim that the role of cognitive science is sometimes fairly minimal. Sometimes one only needs the (utterly trivial) claim that our cognitive system produces these intuitions, and one needs no further details whatsoever as to how the cognitive system does so. For as long as the cognitive system produces these intuitions, and they do not fit reality (as given by the background metaphysical picture), it already follows that the intuitions are liable to defeat and a consequent loss of evidentiary standing.

2.2.1 Color Cognition

So onto color: in his earlier presentations, Goldman (1993, 117)—drawing on Gleitman's (1981) treatment of hue—says: "This entire picture of our internal color coding system makes it difficult to interpret our color experience, upon reflection, as anything but a highly idiosyncratic artifact of that system." He specifically (1992, 43) picks up on metamerism (the way in which objects with different spectral reflectance distributions can still present the same color appearance, due to the way that human color receptors respond only to cumulative features), saying: "Metamerism shows that the color divisions that humans make are arbitrary or bizarre by the standards of physics. These divisions arc a product of our humanly idiosyncratic visual system."

Later, Goldman (2015)—drawing on Averill's (1992) and Cohen's (2004, 2009) defenses of color relationalism—focuses not on metamerism but instead on differences between the visual systems of normal humans (trichromatic) and pigeons (tetrachromatic). He follows J. Cohen, in opting for what Cohen calls "the ecumenical policy that both sorts of visual systems are right, and that one and the same object can have more than one color property" (2004, 462). The end result is a revisionary construal of color as a relational property of a visual system. As Cohen says: "It is a consequence of this relational construal that one and the same object can be simultaneously green for your visual system and not green for the visual system of the pigeon on your window ledge" (2004, 463).

Color relationalism leads Goldman to a distinction between cases calling for what I have labeled a "projectivist" treatment, and cases calling for what I am labeling a "relational" treatment (with color and temporal unity coming in for the latter). The difference is supposed to be that projectivism is a form of antirealism about the phenomena in question, positing no objective correlate in the world; while relationalism is a form of revisionary realism about the phenomena in question, positing an objective (but relational rather than intrinsic) objective correlate in the world.

Obviously there is much more to be said about color cognition beyond the existence of metamerism and comparative differences with pigeons. But these are the main factors Goldman notes, and will suffice for the purpose of working through a concrete case of potential debunking.

2.2.2 Projections as Relations

I have three comments to offer on Goldman's discussion of color, the first and most minor of which concerns the projectivist/relationalist distinction, or lack thereof.

I simply do not see a real distinction here. As I have commented elsewhere, with respect to projectivist views of causation:

If a causal relation between *c* and *e* is real from perspective *p*, is there not an objective relativized fact that *c* causes *e* relative to *p*? But if perspectival notions can be rendered objective by relativization, then any alleged perspectival truth will come with an objective relativized counterpart. (Schaffer 2010, 848)

So it seems to me that projectivism implies a certain form of relationalism, namely relationalism *relative to the projection*. If some property or relation *F* is indeed projected from perspective *p*, there is the real objective relation of *being projected as F by p*. How could projectivism be true without such an objective backing? So I disagree with Goldman (2015) that there is really a "second template" in relationalism that is distinct from projectivism. If anything, the relational framework seems like a more general and inclusive framework, allowing for relations to perspectives or cognitive systems, but to arbitrary other factors as well.

For instance, consider the case of deities, from a projectivist perspective. If the ancient Greeks projected the wrath of Zeus upon the thunder, then there is a real relation born between the Greeks and the thunder, of *being projected as the wrath of Zeus*. What is the difference between that and the relational treatment of color, involving a real relation between humans and tomatoes, of *being projected as red?*

That is not to say that there can only be a single template for all cases, but only that projections/relations do not furnish distinct templates. This is also not to say that there is no other sort of difference between these cases. Indeed I think that in the cases Goldman labels as "projections" one sees a cognitive system purporting to detect certain features of reality, which look not to feature in the best metaphysical picture; while in the cases labeled as "relations" one sees *conflicting forms of cognition* (either between cognizers as in the color case, or between multiple modules within a single cognizer as Goldman sees the temporal unity case), and one might doubt that either form of cognition is to be preferred.

It seems to me that a relevant difference between the cases that Goldman labels as "projections" and the cases he labels "relations" is not metaphysical but rather *epistemic*, as between undermining and rebutting defeaters. The epistemic structure of the projection cases is that the evidential force of the intuition is undermined by its failure to fit the structure of reality (as given by the background metaphysical picture). The epistemic structure of the relation cases is that the evidential force of the intuition is rebutted by the evidential force of a contrary intuition (from the other conflicting form of cognition). So it seems to me that in both cases we are looking at a demotion from the naive claim that there is an objective feature of reality to the claim that there is merely an objective relation that incorporates the type of cognitive system. The naive claim is just getting debunked in slightly different ways: by being shown not to fit reality (undermining), or by being shown to conflict with another equally compelling intuition (rebutting).

(Note that even in the rebutting cases, metaphysics is still playing a role. For part of the rebutting story is that it would be objectionably arbitrary to favor the one intuition—for example, that produced by the human visual system—over the other—for example, that produced by the pigeon visual system. But the idea that this is arbitrary is itself a metaphysically laden judgment, involving the idea that the background metaphysical picture offers nothing against which to uphold the human or the pigeon perspective as more fitting. Nothing in cognitive science itself tells against the naive-realist-cum-pigeon-chauvinist, who posits real colors in the world as pigeons tend to see them, towards which humans are partly color-blind.)

2.2.3 Do the Cognitive Details Matter?

My second comment on Goldman's treatment of color concerns the extent to which the detailed cognitive story (be it the story of metamerism or the comparative story of human versus pigeon color vision) is really playing a substantial role in the debunking project. As Goldman himself (1993, 114–15) notes, naive realism about color has been in "ontological trouble" at least since Locke's demotion of colors to the status of secondary qualities, as qualities that are merely "imputed" ([1689] 1996, 54) and "no more really in [bodies], than Sickness or Pain is in Manna" (51).[16] For Locke, the ideas of color, sound, taste, and odor do *not* resemble their causes, and so these are the ideas of secondary qualities (unlike motion, in which the idea of motion is thought to resemble real motion, at least given the corpuscularian picture Locke operates with).

What is not clear to me is exactly what (if anything) the various details of metamerism and pigeon vision are adding to the ontological troubles color faces. Was Locke's case for demotion a failure because of his ignorance of pigeons? So long as one has a background metaphysical picture on which color is a secondary quality of some sort or other, coupled with a psychological picture on which the folk naively view colors as intrinsic properties "painted directly onto" the surfaces of objects,[17] isn't the former already enough to debunk the latter? What are the various details of metamerism and pigeon vision adding vis-à-vis the debunking project?

I take the various details of metamerism and pigeon vision to come into the story in rebutting various sophisticated attempts to defend a nonreductive realism about color. For instance, metamerism puts pressure on the idea of identifying color with spectral reflectance distributions, insofar as these map many-one onto perceived color. But of course the nonreductive realist need not think that human color vision is perfect (human cognition is generally fallible of course). So there remains room to respond to the case of metamerism by identifying color with spectral reflectance

[16] Indeed this idea goes back at least to Democritus (Taylor 1999, 9): "By convention sweet and by convention bitter, by convention hot, by convention cold, by convention color; but in reality atoms and void."

[17] Of course it is a psychological claim that people are in fact naive realists about color. See Cohen and Nichols 2010 for some empirical work calling this claim into question.

distribution (or some other natural phenomena) while taking the human visual system to be an imperfect detector of that natural phenomenon.

As things stand I see no need to do any serious cognitive science whatsoever, to worry about the place of color in a physical world. The cognitive details are of course intrinsically fascinating. But from the perspective of the debunking project, it seems that the cognitive details can at most reveal just *how poorly* our color categorizations fit into the background picture of reality. The nonreductive realist, however, can always (plausibly) admit fallibility in cognition. The bulk of the debunking work for color thus comes through the metaphysical idea that our color categorizations fail to fit anything at all in reality.

2.2.4 How and Why Metaphysics Matters

In contrast to the detailed cognitive story which seems largely irrelevant with respect to debunking naive realism about color, the background metaphysical picture does seem deeply relevant. Consider again Locke's distinction between the primary and the secondary qualities. The reason why motion remains a primary quality but color must be demoted comes from Locke's background corpuscularian picture of reality (via Boyle). If Locke had a background physical image of the corpuscles as themselves little colored dots, then color could have been upheld as a primary quality for Locke, in a way that would have fit naive realism.

Or consider Goldman's discussion of metamerism. Metamerism is a phenomenon on which *distinct objective features* (different spectral reflectance distributions, which looked gerrymandered from the perspective of the metaphysical background picture) map to a *common subjective response* (same color appearance). If one insisted on keeping objective intrinsic color in the background picture, one could in principle accept metamerism as either showing a multiple realizability of objective intrinsic color or (perhaps better) showing that the human visual system is an imperfect detector of color for failing to draw some real distinctions.

Or consider Goldman's discussion of the human-pigeon comparison. In this case *different subjective responses* map to *common objective features*. That is, different subjective responses between normal humans and normal pigeons map to common objective features of surfaces and light which are *neutral between* the human and pigeon ways of seeing things. Indeed in this case there is the somewhat tempting option of thinking of the human system as defective relative to the pigeon system.[18] (And thinking of both systems as defective relative to the decachromatic mantis shrimp.) One could be excused for thinking that we are somewhat color-blind compared to pigeons, in just the same way that a human monochromat or dichromat counts as "color-blind" with respect to her trichromatic colleagues. Goldman seems to think that it would be arbitrary or "species-chauvinist" to defer to either humans

[18] This is J. Cohen's (2004, 462) "defer to the pigeons" option.

or pigeons (or mantis shrimp), but that claim encodes an implicit background metaphysical picture on which nature lacks objective intrinsic color which one species might better track. I agree with that background picture but am only trying to bring it into focus and identify its role in the argument that the human-pigeon comparison supports relationalism about color rather than the view that there are objective intrinsic colors, which pigeons merely happen to discern better than humans.

Goldman partially acknowledges this point in passages such as the following (1992, 44): "[W]hile part of the scientific research that fuels the theory of color is essentially research in physics, another part is cognitive science." What I am pointing out is that the research in physics is coming into the picture by supporting a certain background *metaphysical* picture of objective reality, as lacking joints corresponding to color cognition. It is only physics *via a metaphysical interpretation* that fuels the debunking project for color.

Putting these last two sections together: I see little coming from the cognitive science side that contributes to the debunking project for color. The cognitive science side is giving us a picture that is perfectly compatible with naive realism, plus the modest hypothesis that human color vision is highly fallible. (All realistic detector systems are at least to some extent fallible.) What is doing the bulk of the job of debunking naive realism is the background metaphysical picture of a colorless physical reality. Color then can only be a projection of the human mind. But all this can be argued for with just the trivial cognitive claim that our cognitive system produces these color appearances in some way or another.

Putting the discussion of color together (in reverse order), I conclude, as stated earlier:

- Whether the cognitive story supports debunking depends on the background metaphysical picture.
- The detailed cognitive story is sometimes irrelevant to the debunking project.
- There is no metaphysical distinction between projections and relations but rather an epistemic distinction between undermining and rebutting defeaters.

Were color the only case to consider it would be tempting to speak of metaphysics as taking the lead role in the debunking project, and to speak of cognitive science as at best a junior partner. For cases in which cognitive science is playing a more serious role one must look elsewhere.

2.3 CASE STUDIES: TEMPORAL PASSAGE, SPATIAL UNITY

I conclude by considering two further cases, both to see the extent to which the morals drawn from the case of color may generalize, and also to see some different issues arising including respects in which cognitive science is playing a more serious role. The cases that I think are most helpful in these respects are those of temporal passage (the feeling that time "flows") and spatial unity (the way we treat certain collections of objects as themselves individual wholes).

2.3.1 Temporal Passage

I turn to temporal passage, as a case which is both more controversial metaphysically, and illustrates the importance of selective constraints on the debunking project. In his initial presentations, Goldman (1992, 52)—drawing on Horwich (1987)—takes up an idea which "grants that we have a *feeling* of the passage of time but offers a psychological explanation of this feeling to undermine its presumed veridicality." He comments:

> If this psychological explanation is correct, it shows that the feeling of temporal flow could arise from the content of experience alone, without any relevant thing actually flowing. Since the notion of such flow is problematic, the invited inference is to reject the existence of such a "moving now." (1992, 53)

Drawing on Le Poidevin (2007) and Paul (2010b), Goldman (2015) cites experiments in which subjects experience illusions as if an object is moving, or illusions as to the temporal ordering of events. He takes these experiments to illustrate the possibility that the experience of temporal passage might itself be illusory. In the Bayesian framework Goldman uses, this all goes to showing that the probability of the evidence (felt temporal passage) is actually rather high even given antirealism about temporal passage. In a defeater framework, this could be taken to show that the inference from apparent temporal passage to real temporal passage is defeated by how prone we are to suffering nonveridical illusions.

All of the three points made with respect to color extend to this case. First and foremost:

- Whether the cognitive story supports debunking depends on the background metaphysical picture.

If one believes in temporal flow—which I do not but which is a respectable view in the field—then one will take the cognitive story to be a story about an imperfect detector. By far the most eloquent presentation of this idea is given by one of its opponents, namely Williams (1951, 465–66):

> The final motive for the attempt to consummate the fourth dimension of the manifold with the special perfection of passage is the vaguest but the most substantial and incorrigible. It is simply that we *find* passage, that we are immediately and poignantly involved in the jerk and whoosh of process, the felt flow of one moment into the next.

Here is the focus of being. Here is the shore whence the youngster watches the golden mornings swing toward him like serried bright breakers from the ocean of the future. Here is the flood on which the oldster wakes in the night to shudder at its swollen black torrent cascading him into the abyss.

Note that (as Williams indicates by labeling this "the final motive") there are other motives for positing real passage, including motives drawn purely from physics and not from naive feelings of time being like a river or a burning fire or a force that through the green fuse drives the flower. Indeed a more recent and physics-savvy defense of passage is offered by Maudlin, who concludes as follows:

> [I]t is a central aspect of our basic picture of the world that time passes, and that in virtue of that passage things change. And there are no good logical or scientific or philosophical arguments that cast doubt on the passage of time, and there are no impediments to representing, in our present physical theories, that time passes. I draw what ought to be a most uninteresting conclusion, but one that has somehow managed to be philosophically bold: time does pass. (Maudlin 2007, 142; cf. Skow 2011)

Of course anyone—such as Maudlin—who believes in temporal passage as an objective intrinsic feature of reality will allow that we can suffer illusions. Again, all realistic detector systems are prone to error. So from the perspective of the believer in real passage, all that has been said is the utterly trivial claim that our cognitive system produces this sense of passage in some way or another, and the utterly unsurprising claim that the system is fallible. This could have been said of any intuitions whatsoever, including ones that many would wish to uphold, such as that here is one hand and here is another, and that 2 + 3 = 5 (section 2.2).[19] So whether one comes in siding with Williams against passage or with Maudlin for passage is crucial to what one makes of the cognitive story, and whether one thinks of the cognitive system as "gilding and staining" a static manifold, or just as imperfectly detecting some real flow of the temporal river.

Second, and already apparent in the first point:

- The detailed cognitive story is sometimes irrelevant to the debunking project.

[19] Goldman himself (2015, 184) takes the cognitive story to show that "a reasonable metaphysician would substantially revise the likelihood of the occurrence of passage experiences conditional on anti-realism." But this overgeneralizes. We already know that there is some cognitive story to tell for every single case of cognition. Relatedly, Paul puts the point in term of a challenge to the denier of real temporal flow: "[W]ithout the properties of nowness and passage, we'd have no way to account for the features of our temporal experience" (2010b, 337). Her answer to this challenge is that the relevant features of our temporal experience might be illusions. She is surely right about that—indeed all sides should acknowledge the possibility of illusion—but I think she has thereby missed the stronger challenge. The stronger challenge is that the fact that our temporal experience has these features (felt flow) provides *decent albeit fallible evidence* that the world has these features. The real issue is then whether this sort of evidential force might still be debunked.

After all, to the extent that our naive belief in temporal passage is debunked, the only "information" we have psychologically is the absolutely generic information that there is some cognitive process producing the intuitions, and that it can produce error. All the "oomph" of debunking—indeed virtually the entire debate—seems to me to be taking place over the background metaphysical picture.

Third and of least importance for my purposes:

- There is no metaphysical distinction between projections and relations but rather an epistemic distinction between undermining and rebutting defeaters.

In this case one can say that, if human agents project passage onto an objectively static and undirected manifold (as per the perspectivalism of Price 1996), then there is an objective relation of passing relative to a given projected viewpoint. If my "now" includes an apple falling off a table, then the apple is undergoing the jerk and whoosh of free fall relative to my current agential perspective on the manifold.

The case of temporal passage holds additional interest in two respects. First of all the background metaphysical picture at issue is more controversial than with color. Few nowadays accept naive realism for color.[20] But realism about passage is considered very much of a live view.[21] The case thus illustrates—to my mind—the failure of cognitive science to go it alone on any metaphysically controversial issues. To switch to some of Goldman's other controversial examples, consider the status of deities. In both cases, all sides may agree that our cognitive system produces certain theistic intuitions. As Bloom—in an article entitled "Religion Is Natural"—writes:

> One of the most interesting discoveries in the developmental psychology of religion is that the bias towards creationism appears to be cognitively natural. Four-year-olds insist that everything has a purpose, including things like lions ("to go to the zoo") and clouds ("for raining"). When asked to explain why a bunch of rocks are pointy, adults prefer a physical explanation, while children choose functional answers, such as "so that animals could scratch on them when they get itchy." (Bloom 2007, 150)

The theist would presumably regard this naive tendency of the mind as a tendency *to get it right*. That the mind has a tendency to these intuitions should be common ground and cannot possibly count against the theist. If anything it counts for her, insofar as it shows that her views can claim the best fit with folk metaphysics. For

[20] Though see Campbell 1993, Gert 2008, and Allen 2011 for various attempts to defend color primitivism, as well as Byrne and Hilbert 2007 and J. Cohen 2009 (esp. 67–74) for critical responses.

[21] According to Bourget and Chalmers (2014), 26.3% of philosophers favored the static B-theory, 15.5% favored the passage-laden A-theory, while 58.2% gave some other answer. Acceptance of the A-theory was found correlated with identification with Aristotle, and tended to cluster with thinking that teletransportation was death, deontology about morality, and a rejection of switching in Trolley problems.

atheists—such as myself—it is the overwhelming theoretical pressure of a background naturalistic metaphysical picture that drives us to demand a debunking of these naive intuitions.[22]

Goldman (1992, 46) considers the case of the theist who invokes religious experiences rather than naive creationist intuitions, and says:

> The atheist replies that there is an alternate explanation of these phenomena, one that appeals only to psychological and cultural mechanisms. If the latter explanation is more parsimonious, or otherwise provides a "better explanation," then that undermines the God hypothesis.

But this is in serious danger of overgenerating. Consider again my naive perceptual intuition that here is one hand and here is another, which I am presuming should be respected rather than debunked. Yet there is a purely psychological explanation of this belief which casts the visual system as a "debunked" generator of falsidical illusions. This is a very parsimonious explanation, insofar as it is consistent with radically minimal hypotheses about the external world (including the most extreme solipsism). So it really matters whether hands have a good claim to be in the background metaphysical picture.

Or consider the status of numbers. Again all sides should agree with the utter triviality that our cognitive system produces certain numerical intuitions. Goldman (1992, 46–47) considers Chomsky's speculation that "mathematical cognition may be a byproduct of the language faculty" and immediately says: "If this sort of conjecture were confirmed, it could add empirical support to a subjectivist, or quasi-subjectivist, ontology of mathematics." This seems to me again to be at serious risk of overgenerating and "debunking" everything indiscriminately. For presumably every cognitive faculty we have is the product of some evolutionary process, whether directly or by way of a byproduct of some other selective process or just by random chance. If this is empirical support for subjectivism, then the subjectivist is on the verge of a total and unselective "victory."

So the fifth and final point emerging—in line with the general need for selective constraints on debunking (section 2.1.3)—is that the debunking project for temporal passage needs real constraints. It is not enough just to say that our cognitive system produces these intuitions, or even to say that the cognitive system in question is liable to error. For these points apply unselectively to every single cognitive output. Again I say that the test for when to debunk is internal coherence and external fit with the background metaphysical picture. Indeed, in the case of temporal

[22] In this example cognitive science is playing a role in helping determine *what is intuitive in the first place.* As Bloom makes explicit, it is a *discovery* that the cognitive bias towards creationism is so natural and robust (as opposed to a mere culturally local indoctrination process).

passage—just as in the case of color—cognitive science seems to be at best a junior partner in debunking. The main action is in the metaphysics.

2.3.2 Spatial Unity

I move on to spatial unity, as a case which is highly controversial metaphysically, and which illustrates (for the first time) at least two respects in which cognitive science has a main role to play in the debunking project, and is more than a mere understudy to metaphysics in this task. The case of spatial (and temporal) unity is actually the first case Goldman (1987, 541) explicitly discusses, drawing on *gestalt principles* of unification:

> I suggest that Gestalt principles underlie and shape our spatial and cross-temporal "entification" practices, our propensity to view certain sets of spatial elements as parts of one and the same physical object and certain sets of time slices as stages of one and the same continuant.

He gives the case extended discussion in his 1993 (101–8), drawing on gestalt principles as well as Spelke's (1990) work on infant cognition, which he takes to suggest that infants have "an unlearned conception of physical bodies" (1993, 107) which is augmented between six months and two years of age with gestalt principles leading to an adult conception in which:

> We do not readily consider something as a physical body if it lacks cohesion (a pile of leaves), lacks bounds (a drop of water in a pool), or lacks continuity (a row of flashing lights). These may be considered collections of objects or parts of objects, but they are not unitary and independent objects for us. (1993, 108)

Missing from Goldman's discussion, however, is any claim about why these unity intuitions should be debunked rather than respected. Goldman introduces gestalt principles to illustrate "the principles used by the mind to structure the world into units or unities" (1987, 539). So it may be that he is simply *presupposing* that objective reality lacks any principle of unity whatsoever that the principles used by the mind might fit. He may simply be presupposing a metaphysical background picture on which the gestalt principles are not decent indicators of the "real facts" of restricted mereological composition, where the restriction is to pluralities that are sufficiently cohesive, continuous, and bounded. Bur in any case I must reiterate:

- Whether the cognitive story supports debunking depends on the background metaphysical picture.

Thus the central moral of this discussion remains applicable.

What is interestingly different about the case of spatial unity, and makes the case more favorable to Goldman's view than the previous cases I have discussed (color: section 2.2, temporal passage: section 2.3.1), is that cognitive science can actually be very helpful here, in two distinct respects. The first respect in which cognitive science can be helpful comes in revealing what we find intuitive in the first place. As will not surprise those familiar with the metaphysics literature on composition, there is—as Rose and Schaffer (2017, 238) put the point—"wide disagreement among metaphysicians as to what the folk intuit about mereological composition and why they do so, and no empirical discipline to the discussion." Indeed many different metaphysical theories—from universalism (see Thomasson 2007, 183–84) to brutalism (see Markosian 1998, 211)—have been upheld under the banner of fit with folk intuitions. It is an empirical matter—and one that is often not obvious—what is intuitive in the first place, and on this matter cognitive science is needed.[23]

But second, cognitive science may also discover *subtle presuppositions of cognition.* For instance, in the case of folk physics (as mentioned in section 2.1.3) it takes sustained empirical work to reveal the extent to which our physical expectations are generated by something like an impetus theory. Likewise in the case of folk biology it takes sustained empirical work to reveal the role of essentialist assumptions. Thus Atran speaks of the folk idea of a "biological essence" as "an intrinsic . . . teleological agent, which physically . . . causes the biologically relevant parts and properties of a generic species to function and cohere 'for the sake of' the generic species itself" (1998, 550–51).

In the case of mereological composition, Rose and Schaffer (2017, 259–60)—guided by Kelemen's (1999) arguments for "promiscuous teleology" as a default heuristic in cognition[24]—provide an empirical argument that folk intuitions about when composition occurs are driven by teleology, via the principle of "*Teleologically Restricted Composition*: Composition occurs when the plurality has a purpose." If Rose and Schaffer are right about that claim, it opens up a subtle opportunity for debunking, insofar as teleology is not an acceptable part of the background metaphysical picture but is rather—as Hawthorne and Nolan write—"part of a superseded, pre-scientific muddle about how the world works" (2006, 267).

So in this case the detailed cognitive *story* is relevant to the debunking project, in two respects:

[23] Just to be explicit: I—along with Goldman (2015)—classify experimental philosophy as "a sector of cognitive science." Indeed I think that one of the roles for experimental philosophy is to help adjudicate those common philosophical debates over what is intuitive in the first place. For instance, epistemologists disagree as to whether our intuitions about knowledge are sensitive to the practical stakes of the subject. Well-designed experiments can help resolve these disagreements (see Buckwalter and Schaffer 2015).

[24] Kelemen, Rottman, and Seston (2013, 1075) characterize their view as "akin to dual-processing models that characterize early developing intuitions as heuristics that can be increasingly overridden later in development by effortful processing, but which can nevertheless persistently reemerge in cases when intuitions are favored or forced."

- Cognitive science can help reveal what is intuitive in the first place.
- Cognitive science can reveal subtle presuppositions of our cognitive engine.

Of course these subtle presuppositions must still be held up to our background metaphysical picture to determine whether or not they deserve debunking. The crucial point is that cognitive science *can* turn up hidden aspects of our thought as additional matters to hold up against reality. And thus:

- The detailed cognitive story is sometimes irrelevant *but sometimes relevant* to the debunking project.

Cognitive science at its best gives us the *folk metaphysical image*, which must then be compared to *the real metaphysical picture.* The folk metaphysical image might have internal troubles. It might be incomplete or inconsistent. But assuming that it is internally coherent (or can be refined into an internally coherent picture), there remains the question of whether there is external fit with reality. That is a metaphysical assessment.[25]

2.3.3 Conclusions

So is cognitive science relevant to metaphysics? I agree with Goldman that the answer is *yes.* But I have argued—both in general, and through consideration of case studies involving color, temporal passage, and spatial unity—that metaphysics is also a crucial partner within the debunking project, crucial for determining when a given intuition deserves debunking at all. Intuitions (whether in metaphysics or elsewhere) may be assessed for internal coherence or external fit with reality The latter is a metaphysical assessment. Thus in place of Goldman's talk of "the relevance of cognitive science to metaphysics" I find it less misleading to speak of "the joint relevance of cognitive science and metaphysics to the wider project of debunking intuitions."

By working through these case studies I have drawn the following conclusions:

- Whether the cognitive story supports debunking depends on the background metaphysical picture.
- The detailed cognitive story is sometimes irrelevant to the debunking project.

[25] Spatial unity is also a case in which there is no metaphysical distinction between projections and relations but rather an epistemic distinction between undermining and rebutting defeaters. For suppose that the visual system uses gestalt heuristics, while some other module uses teleological considerations. Then one can think of unity as a projection, or as a relation to a mode of thought (a given plurality might be unified relative to gestalt considerations but not relative to teleological considerations).

- There is no metaphysical distinction between projections and relations but rather an epistemic distinction between undermining and rebutting defeaters.

With respect to the second bullet point, I have acknowledged that the detailed cognitive story is sometimes relevant in at least the following two respects:

- Cognitive science can help adjudicate debates about what is intuitive in the first place.
- Cognitive science can reveal subtle presuppositions of our cognitive engine.

I have primarily traced out the following implications for thinking of the debunking project as itself a metaphysically laden project:

- There is no "external vantage-point" from which to debunk metaphysics, nor is there any prospect that cognitive science can take on the debunking project alone.
- Cognitive science alone cannot provide much dialectical leverage against naive realists.
- Metaphysics is an integral part of philosophical methodology, at least as it pertains to the assessment of intuitions for debunking.

And so I conclude that cognitive science and metaphysics are partners in the debunking project, and both deeply relevant to philosophical methodology.[26]

ACKNOWLEDGMENTS

Thanks to Alvin Goldman, Hilary Kornblith, Brian McLaughlin, Michela McSweeney, Shaun Nichols, L. A. Paul, and David Rose.

REFERENCES

Allen, K. (2011). Revelation and the nature of color. *Dialectica 65* (2): 153–76.

Arico, A., Fiala, B., Goldberg, R., and Nichols, S. (2011). The folk psychology of consciousness. *Mind & Language 26* (3): 327–52.

Atran, S. (1998). Folk biology and the anthropology of science: Cognitive universals and cultural particulars. *Behavioral and Brain Sciences 21* (4): 547–609.

Averill, E. W. (1992). The relational nature of color. *Philosophical Review 101* (3): 551–88.

Bennett, K. (2016). There is no special problem with metaphysics. *Philosophical Studies 173* (1): 21–37.

Bloom, P. (2007). Religion is natural. *Developmental Science 10* (1): 147–51.

Bourget, D., and Chalmers, D. (2014). What do philosophers believe? *Philosophical Studies 170* (3): 465–500.

[26] Thanks to Alvin Goldman, Hilary Kornblith, Brian McLaughlin, Michaela McSweeney, Shaun Nichols, L. A. Paul, and David Rose.

Boyer, P. (2001). *Religion Explained.* New York: Basic Books.

Buckwalter, W., and Schaffer, J. (2015). Knowledge, stakes, and mistakes. *Noûs 49* (2): 201–34.

Byrne, A., and Hilbert, D. (2007). Color primitivism. *Erkenutuis 66* (1–2): 73–105.

Campbell, J. (1993). A simple view of color. In J. Haldane and C. Wright, eds., *Reality, Representation, and Projection.* New York: Oxford University Press, 257–68.

Cappelen, H. (2012). *Philosophy without Intuitions.* New York: Oxford University Press.

Cohen, J. (2004). Color properties and color ascriptions: A relationalist manifesto. *Philosophical Review 113* (4): 451–506.

Cohen, J. (2009). *The Red and the Real: An Essay on Color Ontology.* New York: Oxford University Press.

Cohen, J., and Nichols, S. (2010). Colours, colour relationalism and the deliverances of introspection. *Analysis 70* (2): 218–28.

Cohen, S. (1984). Justification and truth. *Philosophical Studios 46* (3): 279–95.

Fine, K. (2001). The question of realism. *Philosophers' Imprint 1*: 1–30.

Fodor, J. (1983). The *Modularity of Mind.* Cambridge, MA: MIT Press.

Gert, J. (2008). What colors could not be: An argument for color primitivism. *Journal of Philosophy 105* (3): 128–57.

Gleitman, H. (1981). *Psychology.* New York: Norton.

Goldman, A. (1979). What is justified belief? In G. Pappas, ed., *Justification and Knowledge.* Dordrecht: Reidel, 1–23.

Goldman, A. (1987). Cognitive science and metaphysics. Journal *of Philosophy 84* (10): 537–44.

Goldman, A. (1992). *Liaisons: Philosophy Meets the Cognitive and Social Sciences.* Cambridge, MA: MIT Press.

Goldman, A. (1993). *Philosophical Applications of Cognitive Science.* Boulder, CO: Westview Press.

Goldman, A. (2007). A program for "naturalizing" metaphysics, with application to the ontology of events. *The Monist 90* (3): 457–79.

Goldman, A. (2015). Naturalizing metaphysics with the help of cognitive science. In K. Bennett and D. Zimmerman, eds., *Oxford Studies in Metaphysics*, vol. 9. New York: Oxford University Press, 171–216.

Greene, J. (2008). The secret joke of Kant's soul. In W. Sinnot-Armstrong, ed., *Moral Psychology*, vol. 3. Cambridge, MA: MIT Press, 35–79.

Grundmann, T. (2009). Reliabilism and the problem of defeaters. *Grazer Philosophische Studien 79* (1): 65–76.

Grundmann, T. (2011). Defeasibility theory. In S. Bernecker and D. Pritchard, eds., *The Routledge Companion to Epistemology.* London: Routledge, 156–166.

Guthrie, S. (1993). *Faces in the Clouds: A New Theory of Religion.* New York: Oxford University Press.

Hawthorne, J., and Nolan, D. (2006). What would teleological causation be? In Hawthorne, *Metaphysical Essays.* Oxford: Clarendon Press, 265–84.

Horwich, P. (1987). *Asymmetries in Time.* Cambridge, MA: MIT Press.

Hume, D. ([1777] 1975). *Enquiries Concerning Human Understanding and Concerning the Principles of Morals.* Ed. P. Nidditch. New York: Oxford University Press.

Jenkins, C. S. (2005). Knowledge of arithmetic. *British Journal for the Philosophy of Science 56* (4): 727–47.

Joyce, R. (2006). *The Evolution of Morality.* Cambridge, MA: MIT Press.

Kahane, G. (2011). Evolutionary debunking arguments. *Noûs 45* (1): 103–25.

Kaiser, M., McCloskey, J. V. L., and Proffitt, D. (1986). Development of intuitive theories of motion: Curvilinear motion in the absence of external forces. *Developmental Psychology 22* (1): 67–71.

Kelemen, D. (1999). The scope of teleological thinking in preschool children. *Cognition 70* (3): 241–72.

Kelemen, D., Rottman, J., and Sesron, R. (2013). Professional physical scientists display tenacious teleological tendencies: Purpose-based reasoning as a cognitive default. *Journal of Experimental Psychology: General 142* (4): 1074–83.

Korman, D. (2009). Eliminativism and the challenge from folk belief. *Noûs 43* (2): 242–64.

Kornblith, H. (1998). The role of intuition in philosophical inquiry: An account with no unnatural ingredients. In M. DePaul and W. Ramsey, eds., *Rethinking Intuition: The Psychology of intuition and Its Role in Philosophical Inquiry.* Lanham, MD: Rowman & Littlefield, 129–41.

Kripke, S. (1980). *Naming and Necessity.* Cambridge, MA: Harvard University Press.

Le Poidevin, R. (2007). *The Images of Time: An Essay on Temporal Representation.* New York: Oxford University Press.

Lewis, D. (1996). Elusive knowledge. *Australasian Journal of Philosophy 74* (4): 549–67.

Locke, J. ([1689] 1996). *An Essay Concerning Human Understanding.* Ed. K. Winkler. New York: Hackett Publishing.

Markosian, N. (1998). Brutal composition. *Philosophical Studies 92* (3): 211–49.

Mason, K. (2010). Debunking arguments and the genealogy of religion and morality. *Philosophy Compass* 5 (9): 770–78.

Maudlin, T. (2007). On the passing of time. In *The Metaphysics within Physics.* Ed. T. Maudlin. New York: Oxford University Press, 104–42.

Moore, G. E. ([1939] 1993). Proof of an external world. In T. Baldwin, ed., *G.E. Moore: Selected Writings.* London: Routledge, 147–70.

Nichols, S. (2014). Process debunking and ethics. *Ethics 124* (4): 727–49.

Paul, L. A. (2010a). A new role for experimental work in metaphysics. *Review of Philosophy and Psychology 1* (3): 461–76.

Paul, L. A. (2010b). Temporal experience. *Journal of Philosophy 107* (7): 333–59.

Paul, L. A. (2012). Metaphysics as modeling: The handmaiden's tale. *Philosophical Studies 160* (1): 1–29.

Plantinga, A. (2000). *Warranted Christian Belief.* New York: Oxford University Press.

Pollock, J. (1987). Defeasible reasoning. *Cognitive Science 11* (4): 481–518.

Price, H. (1996). *Time's Arrow and Archimedes' Point: New Directions for the Physics of Time.* New York: Oxford University Press.

Pust, J. (2012). Intuition. In E. N. Zalta, ed., *The Stanford Encyclopedia of Philosophy.* Summer 2017 ed.

Rose, D., and Schaffer, J. (2017). Folk mereology is teleological. *Noûs 51* (2): 238–70.

Schaffer, J. (2009). On what grounds what. In D. Chalmers, D. Manley, and R. Wasserman, eds., *Metametaphysics.* New York: Oxford University Press, 347–83.

Schaffer, J. (2010). Review of Price and Corry's "Causation, physics, and the constitution of reality." *Mind 119* (475): 844–48.

Singer, P. (2005). Ethics and intuitions. *Journal of Ethics 9* (3–4): 331–52.

Skow, B. (2011). Experience and the passage of time. *Philosophical Perspectives 25* (1): 359–87.

Spelke, E. (1990). Principles of object perception. *Cognitive Science 14* (1): 29–56.

Street, S. (2006). A Darwinian dilemma for realist theories of value. *Philosophical Studies 27* (1): 109–66.

Taylor, C. C. W. (1999). *The Atomists: Leucippus and Democritus.* Toronto: University of Toronto Press.

Thomasson, A. (2007). *Ordinary Objects.* New York: Oxford University Press.

White, R. (2010). You just believe that because . . . *Philosophical Perspectives 24* (1): 573–615.

Williams, D. C. (1951). The myth of passage. *Journal of Philosophy 48* (15): 457–72.

Williamson, T. (2007). *The Philosophy of Philosophy.* Oxford: Blackwell Publishing.

Wittgenstein, L. (1969). *On Certainty.* Ed. G. E. M. Anscombe and G. H. von Wright. Trans. D. Paul and G. E. M. Anscombe. London: Harper & Row.

PART II
ETHICS AND COGNITIVE SCIENCE

3 Moral Metaphysics, Moral Psychology, and the Cognitive Sciences

Peter Railton

Moral philosophy cannot stand entirely apart from substantive questions about mind and language. On the normative side, *ought* implies *can*, so that if morality makes claims upon how we *should* think and act, it must assume that we are actually *capable* of following such recommendations or demands. On the metaethical side, some of the most central debates hinge upon questions about mental representation and architecture—for example, the debate over "internalism" about moral judgment depends on views about the possibility of mental states that are at once cognitive and motivating.

Cognitive science—broadly understood as encompassing psychology, neuroscience, artificial intelligence, evolutionary psychology, game theory, and decision theory—has a good deal to say about the infrastructure in virtue of which we might, or might not, be moral creatures. Recent years have seen a succession of troubling questions launched from cognitive science about the origin, nature, or rationality of moral thought and practice, ranging from questions about the possible evolution of moral motivation, to the relation between skilled motor control and virtue, to the epistemic status of moral intuitions. In many cases, it has been argued that serious rethinking is needed in moral philosophy.

Here we will consider two influential cognitive-science-based challenges of this kind, each with relevance to the *metaphysics* of morality, since both affect the tenability of a *realist* interpretation of moral thought and language, particularly a realism of a *naturalistic* kind. One is based upon highly general considerations in evolutionary psychology, while the other is based upon more recent "dual process" models of the architecture of the human mind. In response to these questions, it will be argued that a fuller understanding of nature and possible origins of human cognition and motivation could help *consolidate*, rather than undermine, a form of naturalistic realism.

To set the stage, we need to take up two preliminary questions: What ideas about the nature of distinctively *moral* thought and practice have figured in these debates? And, how might the interpretation of moral thought and practice, thus understood, intersect with *metaphysical* questions?

3.1 PRELIMINARIES

3.1.1 The Elements of Morality

While disputes about the nature of morality aren't lacking in moral philosophy, there is something closer to agreement on the "truistic" features of common-sense moral discourse and practice. These are the characteristics most moral philosophers seek to accommodate in their accounts—even when their accounts must offer an explanation that does not take the feature entirely at face value. To give up altogether on the task of accommodating these features, philosophers have argued, would be to change the subject from *moral* philosophy. And while some philosophers have argued that we *should* replace morality with a quite different normative guide, most have seen this as a last resort.

Commonsensically, moral evaluations are (i) *cognitive* (spoken of as true or false, or more or less accurate; used as premises in reasoning); (ii) *opinion-independent* (moral evaluation offers a standpoint from which we can in principle challenge any part of received opinion, and genuine moral disagreements can exist across significant differences in point of view); (iii) *nonparochial* (moral evaluations are general, and do not accord special standing to the perspective or interests of any one individual or social group as such); (iv) *authority-independent* (moral evaluations apply even in circumstances where they are not enforced by monitoring and sanctions); (v) *nonhypothetically thought- and action-guiding* (moral considerations afford reasons for deliberation, decision, and action that do not depend upon any "external" incentives or goals); and (vi) *intrinsically concerned with avoiding harms and promoting benefits for others as well as the self*, where "harms" and "benefits" are understood broadly enough to include not only the utilitarian's concern with suffering and well-being, but also the deontologist's concern with treating others with respect, the natural law theorist's concern with rights, and the virtue theorist's concern with eudaimonia and the development of distinctively human excellences.

Conditions (i)–(vi) yield what is sometimes called a *narrow* or *normative* conception of morality. On this conception, a set of categorical principles that singled out one group as the "chosen few" to rule over the rest, or assigned no weight to the interests of those who are not blood relatives, would not count as a *moral* code. Starting off with a narrow conception of morality might seem artificially restrictive, except that the cognitive-science *challenges* to "morality as we ordinarily conceive it" make sense only if we presuppose something like this more demanding, normative conception. For example, if "moral code" were taken to include schemes that favor individuals in direct proportion to genetic relatedness, then evolutionary challenges to the possibility of moral motivation could not get off the ground. And recent claims that psychological and neuroimaging experiments show moral "intuitions" to be traceable to morally irrelevant considerations are interesting only if "morally *relevant* consideration" is understood in a way we take to have actual normative significance.

3.1.2 Metaethics and Metaphysics

We next need to say something about how questions of metaphysics enter into moral philosophy. Several decades ago, it was possible to say that the major metaphysical issue in ethics was *realism* vs. *antirealism* about morality (where "morality" was understood in the narrow or normative sense). However, a curious evolution has taken place in metaethics in the intervening years. Moral realism has gone from being a fringe position (Railton 1986; Nagel 1987; Boyd 1988; Brink 1989) to being the favored self-description for a surprisingly wide array of metaethical positions, naturalistic and nonnaturalistic—including several positions that began their philosophical lives as staunchly antirealist. How has this happened, and how is it relevant to the present discussion?

Consider the following conception of what it is to adopt a realist interpretation of a domain of discourse *D*, whether it be physics, sociology, epistemology, aesthetics, or ethics. On this conception, realist interpretations satisfy at least three conditions: (1) *cognitivism*—the interpretation assigns truth or accuracy conditions to typical thoughts or assertions contained in *D*; (2) *objectivity or independence*—these truth or accuracy conditions are substantially independent of our particular subjective perspectives or opinions; and (3) *verisimilitude*—at least some positive, substantive thoughts or assertions in *D* come out as true or accurate, or nearly so, under the interpretation. Antirealist interpretations, on this conception, deny at least one of (1)–(3).

Expressivists (Ayer 1936; Stevenson 1937) traditionally denied (1) cognitivism, arguing that the cognitive-seeming surface of moral discourse is misleading: the main function of moral thought and language is not to describe the world more or less accurately, but to prescribe or encourage behavior. Thus the mental state expressed by a sincere moral judgment must be *motivating* for anyone who accepts the judgment—otherwise one could accept a moral judgment and feel no pressure to act accordingly. This view has come to be called *motivational judgment internalism* (Darwall 1997). It militates against cognitivism because motivating states have a "world-to-mind" direction of fit (disposing us to *act upon the world* to bring it into line with the content of what we desire), while truth-evaluable states have a "mind-to-world" direction of fit (disposing us to *revise our thinking* in order to bring it into line with the state of the world; see Humberstone 1992; Smith 1994), No single mental state, it is argued, could coherently have *both* directions of fit. The challenge for the expressivist then was to explain how ordinary moral discourse could *behave* cognitively on the surface—e.g., work seamlessly with ordinary cognitive discourse and reasoning (the "Frege-Geach problem")—while being noncognitive at base (Schroeder 2008).

Error theorists, by contrast, accepted at face value the (1) cognitivism and (2) objectivity of ordinary moral discourse, but denied (3) verisimilitude. Any form of evaluation that combined features (i)–(vi) (see section 3.1.1), they argued, would have to attribute to its substantive judgments truth conditions of a kind impossible to meet. Moral evaluation would have to be "objectively prescriptive," that is, motivating for all agents, whatever their subjective motivational states—an impossible recipe. So moral claims must systematically be false, and the problem of finding a replacement becomes urgent (Mackie 1977).

Relativists and *subjectivists*, meanwhile, have tended to agree with error theorists in rejecting the idea that cognition of an independent reality can motivate independently of one's motivational attitudes, but concluded instead that moral judgments must therefore be relativized or indexed to the speaker's or the community's normative commitments or other motivating attitudes (Harman 1975; Dreier, 1990; Wiggins, 1987). Such relativization or indexing thus created a link between one's motivations and one's moral assessments without forgoing (1) cognitivism or (3) verisimilitude—whenever a moral judgment is in place, corresponding motivation will be as well. For relativists and subjectivists, the outstanding problem became to explain the (2) objectivity of morality, e.g., the possibility of genuine moral disagreement about a common subject matter across diverse individuals or communities..

Recently, philosophers in each of these three large clusters of views have migrated from seeing themselves as "antirealist" to claiming the mantle of realism. Thus most expressivists now reject the label "noncognitivism," having come to accept a metaphysically undemanding minimalism about "truth" and "facts" (Ramsey 1927; Horwich 1990). This enables them to say that there is nothing problematic about embracing expressivism while speaking of *the <u>truth</u> that all men are created equal* or *the <u>fact</u> that breaking a promise is wrong*. For the minimalist, such expressions "say the same thing" as the corresponding first-order assertions *all men are created equal* or *breaking a promise is wrong*, since, on their account, our grasp of truth is given by the disquotational schema '"*p*" is true iff *p*' and our grasp of facts is given by 'it is a fact that *p* iff *p*'. For the minimalist, introducing talk of truth or facts into moral discourse adds a device for metalinguistic ascent, not further metaphysical commitment—so there is no reason for the expressivist to resist such talk, or describe her view as "antirealist" about moral truth or facts (though questions of realism may still arise in the form of issues about the determinacy or objectivity of moral judgment, and thus of moral truth or facts). The combination of expressivism with minimalism about truth is sometimes called "quasi-realism" (Blackburn 1993), though of course, for its thorough-going advocates, minimal truth is not "quasi-truth"—it is all the truth there is. In a somewhat similar move, some error-theorists embraced *fictionalism*, and so had the prospect of combining (1) cognitivism with (3) verisimilitude for moral claims, namely, *truth-in-the-fiction* (cf. Joyce 2001).

A number of moral relativists, subjectivists, and "response-dependency" theorists have also claimed to be realists. It was a philosopher's mistake, they argue, to take at face value ordinary talk of moral evaluation as objective and independent of *any* subjective perspective or attitude, or to accept uncritically the appearance of straightforward moral disagreement across cultural divides. Rather, objectivity is an *internal* matter of whether an individual or a community has certain normative concepts or attitudes: e.g., *desirability* is not a matter of what *is* desired, but what would be *appropriate* or *fitting* to desire. Thus, were actual desires to change (if there were no longer a dispreference for authoritarianism), the moral truths would not thereby be altered—rather, this would simply be a form of moral decay on our part. Objectivity in this sense is clearly within the reach of a relativist or subjectivist, so long as the relevant concepts, attitudes, or sentiments are indeed part of the relevant moral practice. Thus the relativist or subjectivist can claim to be a realist after all, once moral claims have been properly relativized (cf. Wiggins 1987, on "anthropocentric properties" or Harman 2015, on "realist relativism").

All in all, it is now self-declared antirealists who sit on the metaethical fringe. This means that our present inquiry into cognitive science and moral metaphysics cannot take the form of asking, simply, whether a realist interpretation of moral discourse is tenable in light of contemporary cognitive science (for a discussion of the current state of the debate, see McPherson 2015). The old issues have not gone away, but now they primarily take the form of a dispute over the sense in which morality can be *objective*, most notably as expressed in terms of the issues about "direction of explanation." This idea is sometimes traced back to a contrast made by the Socratic Plato (Plato, ca. 399–395 BCE, 2002), who challenged Euthyphro (who had come to Athens to lodge a murder charge against his own father) to answer the question (originally expressed in somewhat different terms): "Is justice good because the gods love it, or do the gods love justice because it is good?" In the case of contemporary realisms, the contrast is between a *projective* explanation, according to which the objectivity in moral thought is a matter of how our moral attitudes operate, vs. a *tracking* explanation, according to which at least some of our moral attitudes can be *informatively* explained in terms of detecting what is good or right.

3.2 EXPLANATION AND MORALITY

3.2.1 Two Types of Projectivist Explanations of Moral Thought and Practice

Consider a *moral expressivist* who holds that the principal function of moral discourse is not to describe the world, but to influence people's behavior—in particular, to induce and stabilize *coordination* in people's attitudes, plans, and actions (Gibbard 1994, 2003). Perception tends to coordinate people's views about everyday objects and causal relations in their vicinity because it enables people to *detect* these objects and relations. But morality coordinates attitudes in a different way—the conflicting action tendencies of negative and positive moral emotions require us to seek personal coherence and interpersonal agreement. These creates a role for convergence-seeking moral discourse without needing to posit any ability to detect objective value or rightness.

Or consider a *moral relativist* or *subjectivist*, according to whom the principal function of moral language is to *communicate* our interests and values to others in order to negotiate our differences without resorting to violence, and thus to facilitate the formation of mutually beneficial agreements, relationships, or conventions (Harman 1975, 2015). Here, too, there is no need to posit objective value or rightness—characteristic human needs and interests, operating through an array of contestatory and cooperative social processes, will tend to produce relatively stable points of convergence.

For both the expressivist and the relativist (as well as some error-theorists), the common-sense notion of moral objectivity can be explained in terms of the familiar process of *reification*, in which extant social practices come to be seen as "natural" or "rationally required" or "God-given." This makes possible a projectivist response to the question: Do people come to adopt certain practices because these practices are morally good, or do people come to see certain practices as morally good because they have adopted them?

3.2.2. Two Types of Reason-Tracking Explanations of Moral Thought and Practice

By contrast, those we will call "tracking realists" argue that we *do* have capacities for detecting objective value or rightness, and that these have been used historically to challenge previous social practices reified as "natural"—such as slavery, female subordination, or hereditary autocracy—and to shape new, less unjust social practices. Two elements are necessary for such tracking explanations to be credible (Railton 1986, 2017): the putative facts about objective value must be sufficiently *independent* of extant opinions, and it must be possible to *learn* about these facts through some epistemically legitimate process.

Tracking realist accounts are of two kinds, reflecting differences about the relevant kinds of independence and learning. For example, according to *nonnaturalists*, moral principles are necessary truths on a par with the basic truths of arithmetic or geometry, and, like them, can be known a priori (Moore 1903; Ross 1930; Parfit 2011; Enoch 2011). However, unlike the truths of arithmetic or geometry, they imply the existence of substantive *ought*-claims, and so must be *synthetic*. As a result, they cannot be known by logical or conceptual means alone, and nonnaturalists therefore characteristically posit a distinctive rational capacity, *intuition*, by means of which we grasp them as "self-evident." This self-evidence might not be immediately obvious—a significant amount of analysis, reasoning, and understanding, as well as a well-informed, disinterested perspective, might be needed. But in principle all rational beings can be led to recognize these truths and see their necessity. Such truths would, moreover, meet the conditions of (i) cognitivity, (ii) opinion-independence, (iii) nonparochialism, and (iv) authority-independence. Moreover, since it is self-evident that (for example) suffering is *intrinsically* bad, such truths would include recognition of a pro tanto reason to avoid harm, whosever it might be (vi)—on an *objective* view of things, the reason to avoid another's pain is therefore just as real as the reason to avoid one's own (Nagel 1970, 1987). Since it would be *practically irrational* to recognize the reality and action-relevance of a reason and treat it as having no bearing on what one should do, wholly cognitive rational intuitions will be nonhypothetically action-guiding for rational agents, securing a form of (v), categoricalness (Smith 1994; Korsgaard 1986).

Naturalistic tracking realists typically agree that moral truths are synthetic, and that their supervenience upon the natural is necessary (though metaphysical rather than conceptual), but think that we come to know them by means continuous with a posteriori inquiry generally (for a contrasting approach to supervenience in a form of naturalistic realism, see Jackson 1998). They face a much greater challenge in capturing (v). For while it is relatively clear how we would be able to track and be moved by facts about *instrumental* value or *hypothetical* reasons—learning through experience which acts would or would not advance our aims is commonplace—it is much less obvious how experience could enable us to track *intrinsic* value, or supply *nonhypothetical* reasons for action. Of course, to the extent that our notion of intrinsic value, for example, carries various presuppositions connecting it with empirical psychology—say, that intrinsic goods will be, when we are fully acquainted with them, intrinsically motivating—it will be possible to use experiential evidence to support or challenge intrinsic value attributions. But such observation-based inference will take us only so far. At some point, won't naturalistic tracking realists be forced to rely upon "intuition" in some form as well?

This would be less worrisome if the a posteriori tracking realist had a nonmysterious account of how we come to have such "intuitions." One suggestion has been to develop a parallel with other kinds of "intuition" that appear to be acquired a posteriori, and that involve both descriptive and evaluative elements (Railton 2013, 2017). For example, the most highly skilled athletes, chess players, musicians, comedians, teachers, and social hosts acquire a capacity to spontaneously yet accurately evaluate situations and their prospects and risks, and be moved to act in a way that achieves an appropriate balance of benefits over costs. Pressed to explain how they identified this response, they may be unable to do better than point to the *tacit competence* and *intuitive sense* they have acquired through years of experience.

Are such competencies no more than a tacit sensitivity to odds of success, judged by some external goal, such as winning or amusing or reducing conflict—and so ultimately a species of hypothetical causal knowledge? But the most admired athletes, chess players, musicians, comedians, teachers, and hosts do more than succeed at some externally defined goal. They succeed with a sensitivity to the distinctive values inherent in their activity—of athleticism, strategy, artistry, wit, pedagogy, or sociability. And they do so not with a battery of preprepared routines, but flexibly, adaptively, with insight and originality. The athlete triumphs without playing dirty, the comedian amuses in a manner that bears deeper reflection, the teacher enables his students to gain genuine understanding, the host brings out the best in her guests. For these accomplishments there is no fixed, external standard, only a large number of dimensions of value or disvalue to which the individual must be alive in what he or she notices, feels, thinks, and does. Yet, apart from the dubious idea of a "born genius," it seems we must think that expert capacities to appreciate these values and put them into practice are largely learned via processes that are themselves not mysterious. To apply this model to the moral case, we would need to see how people similarly could acquire skill in being alive to a scheme of values that satisfies (i)–(vi)—i.e., showing abilities in perception, thought, feeling, and action that manifest an appreciation of, and ability to be guided by, the values inherent in morality.

3.3 COGNITIVE-SCIENCE-BASED CHALLENGES TO TRACKING FORMS OF REALISM

3.3.1 Dual-Process Theories of Mental Architecture

This optimistic picture of "moral skill" has been sharply challenged on the basis of "dual-process" accounts of human mental architecture. Dual-process theory has a number of forms (Chaiken and Trope 1999; Evans 2008), but perhaps the predominant picture holds that the human brain comprises two functional "systems": "System 1" is inherited from our intelligent animal ancestors, largely implicit, centered in the affective system, and "designed" by natural selection to solve the kinds of problems our animal kin faced; "System 2" is a more recent evolutionary product that supports explicit, controlled thought, and is centered in cortical regions concerned with cognition and language.

System 1 comes "online" almost immediately in response to sensory inputs, and "automatically" generates responses based on instinct or acquired stimulus-response reflexes, often affectively charged, and employing simple but generally effective "heuristics" that manifest "little understanding of logic and statistics"; System 2 is capable of controlled, statistical, and logical inference, but generally comes "online" downstream from System 1, and thus its inputs are already shaped by automatic affective responses (Haidt 2001, 2007; Kahneman 2011). System 2 is also slow, effortful, and constrained by limitations of conscious attention and working memory, and so most of the work of moment-to-moment adjustment of behavior is done by the automatic responses and heuristics of System 1 (Bargh and Chartrand 1999).

Dual-process theories have had tremendous success in accounting for a wide range of human behaviors, especially those in which actual human thought and decision-making appear to depart systematically from the classical standards of rationality—as extensively cataloged in the "heuristics and biases" literature (Kahneman 2003). Indeed, even in those cases where our responses seem reasonable, they may result from "presets" or "automatic" habits that do not vary in response to normatively relevant information. Evolution is a tinkerer, not an optimizer, it is argued, building from whatever pieces it has at hand and sticking with any simple solutions it has found in the past until forced to change them. For example, some predators are thought to zero in on fleeing prey by using an evolutionarily conserved heuristic that seems to have first appeared in insects, namely, holding their quarry at a fixed angle in the visual field, without any need to form computationally and energetically demanding causal models to calculate trajectories (Olberg et al. 2000; McBeath et al. 1995).

Thus our inherited System 1 capacities cannot be expected to resemble the norms of rational belief-formation or decision theory, rather than a set of "good enough" heuristics. However, since System 1 is the primary means by which sensory information enters into human cognition, conscious processing will inevitably be deeply shaped by it. This might be thought to raise a serious challenge to tracking realism even about epistemic reasons—the best explanation of spontaneous human cognitive responses would appear not to posit attunement to epistemic reasons as such, but rather a collection of "quick and dirty" dispositions under no unified rational guidance, which might at most be "ecologically rational" (that is, largely sufficient for success in the environment of evolutionary adaptation; Gigerenzer et al. 1999). But System 2 reasoning is able to introduce a priori rational norms for revising our opinions in cases where reliance upon System 1 yields conspicuous predictive or practical failures, and thus reintroduce a significant element of epistemic reasons-responsiveness. This capacity to correct unreliable System 1 responses on the basis of reasoned argument and more self-conscious forms of investigation is presumably part of the explanation of the evolution of System 2—we owe *it*, at least in part, to its capacity to track epistemic reasons. Tracking realism about epistemic value thus seems to be part of a good explanation of why we have the perceptual and cognitive capacities we do.

However, the problem seems to be different in the case of moral thought and decision-making (see, for example, Harman and Thomson 1996). In the moral case, what corresponds to the kind of predictive or practical error that could enable us to spot deficiencies in the beliefs generated by System 1, and to introduce principled System 2 corrections? It would seem that both a priori and a posteriori moral thought has no choice but to rely upon "fit" or "failure to fit" with "moral intuitions," with no independent source of guidance.

So-called "trolley problems", hypothetical moral dilemmas involving life-and-death decisions, have figured prominently in empirical research on human moral judgment in part because subjects reliably respond in ways that generate seemingly irreconcilable intuitive conflicts. For example, in "Switch" versions, most subjects say it is "OK" to push a lever to divert a runaway trolley onto a side track where it will strike and kill one workman, if this is necessary to save the lives of five workmen on the main track. Yet in "Footbridge" versions, most subjects say it is "not OK" to push a large man off a pedestrian overpass to block a runaway trolley and save five workmen. The attempt to identify a principle underlying this "asymmetric" response to sacrificing one to save five has been frustrated by the fact that, in "Loop," which *also* involves using a large person's body to block the runaway trolley (though this time on a looping side track), most will say this is "OK." Although the ratios of approval and disapproval vary somewhat from subpopulation to subpopulation, this puzzling overall pattern of predominant verdicts is surprisingly robust across demographic and cultural differences (Gold et al. 2014).

Social psychologist Jonathan Haidt and neuropsychologist Joshua Greene have offered a dual-process explanation of this pattern by saying that when, as in Footbridge, we contemplate using *direct muscular force* to push someone in front of the speeding trolley, this triggers an immediate, negatively valenced "ME-HURT-YOU" emotional heuristic in System 1 "even a chimpanzee would understand" (Greene and Haidt 2002). This pre-empts or swamps the kind of harm-minimizing System 2 reasoning that seems to be at work in Switch and Loop. Early fMRI research on moral judgment offered support for this diagnosis, by finding that trolley and trolley-like dilemmas in which one must use personal muscular force on a sacrificed victim trigger increased activation in brain regions associated with emotion, and lessened activation in regions associated with calculation of costs and benefits (Greene et al. 2001). Thus, rather than thinking that the marked difference in typical intuitive assessments when subjects consider Footbridge vs. Switch or Loop corresponds to responsiveness to underlying moral reasons or principles, it is instead attributable to an "automatic" negative emotional reaction to a fact (the direct use of muscular force) that would seem to have minimal moral relevance at best (Greene et al. 2009). If we are forced to rely upon intuitions in ethics, then, tracking realists are faced with the awkward fact that our intuitive assessments of individual cases exhibit patterns that we cannot, it seems, appeal to sensory evidence or rational principles to explain or justify (see also Crockett 2013; Cushman 2013)

3.3.2 Evolution and Moral Motivation or Cognition

A second group of cognitive science challenges to the would-be tracking realists has a more direct connection with evolutionary theory.

(1) *Moral motivation.* Ever since Darwin, there has been controversy over whether natural selection could have produced genuinely *moral* motivational dispositions.

No one doubts that most humans have motivational dispositions that make them responsive to positive or negative *incentives* to act in line with moral requirements, or concerned with acquiring a good reputation that can yield indirect benefits to the self or kin. Rather, the question concerns *noninstrumental* motivation to act in response to moral considerations as such, (iv), or *intrinsic* concern with harms or benefits to unrelated others, (vi).

Natural selection, it is argued, could not have favored motivational dispositions toward *genetically altruistic behavior*—that is, behavior that is expected to benefit *unrelated* others but imposes net costs upon the self or kin. True, in competition *between* hunter-gatherer bands, bands containing some individuals willing to make sacrifices even for unrelated band members might have an advantage. But *within* a band, individuals who are disposed to harvest the benefits of such sacrifices while not themselves paying the cost will tend to have higher inclusive fitness. At best, it is argued, our inherited intrinsic motivational dispositions should be expected to be *tribal, reciprocal*, and *reputational*—that is, disposed toward displays of apparent altruism in order to attract allies or partners, willing to sacrifice for unrelated others who can be expected to reciprocate, but unwilling to make genuinely altruistic sacrifices (for an array of perspectives, see Katz 2000). Spontaneous human evaluative representations would thus systematically fail to meet criteria of nonparochialism (iii), nonhypotheticalness (iv), or intrinsic concern with harms and benefits to others as such (vi).

(2) *Moral cognition.* Tracking realism also faces evolution-based problems on the cognitive side of mind—is there a naturalistically respectable account of how our moral attitudes could be epistemically responsive to moral value? Recently, for example, a number of philosophers (Harman and Thomson 1996; Street 2006) have argued that, even if our spontaneous evaluative representations were more in line with criteria (i)–(vi), moral value *as such* would be no part of the best explanation of this. Instead, such alignment would be entirely attributable to the nonmoral, evolutionary advantages of being disposed to form such attitudes, and thus they would not count as *tracking* moral value. Here again we see an apparent contrast with the nonnormative case—the best explanation of our prosaically factual belief-forming dispositions would need to include the fact that these dispositions enable us to track the actual facts and properties of the external world—indeed, that is *why* they were advantageous and thus selected for. But moral evaluative dispositions contribute evolutionary benefits through the ways having such *attitudes* shape our behaviors, with no reference to whether these attitudes actually track moral value.

Moreover, the familiar notion that *moral reasoning* is an effective rational counterweight to arational forces shaping moral judgment has been called into question by Mercier and Sperber (2011), who give evolutionary grounds for thinking that our dispositions to engage in, or respond to, declarative reasoning have been shaped by effectiveness in social influence, not by genuinely rational requirements.

3.4 REPLYING TO THE CHALLENGES

How might a defender of a naturalistic realism of the tracking kind respond to these challenges? We will focus in the remainder of this chapter on naturalistic forms of tracking realism, since more rationalist forms of tracking realism may be able to borrow the naturalist's solutions to give the psychological infrastructure for a priori moral understanding. Let us look first at the dual-process account of moral intuition, and then turn to the evolutionary challenge to moral motivation.

3.4.1 Evidence against the (Standard) Dual-Process Account of Moral Judgment

A proper response to the dual-process account would be a massive task, given the extent of the literature and the varieties of dual-process accounts on offer. So here we will restrict ourselves to briefly reviewing three lines of argument: (1) recent research in the cognitive and affective neuroscience of foraging mammals suggests that the characterization of System 1 needs systematic reconsideration; (2) an emerging consensus in the neuroscience and early childhood development suggests that moral judgment involves integrated, domain-general, model-based capacities; (3) a better explanation of the patterns of intuitive judgments in a number of well-known cases—e.g., trolley problems—that have played an important role in motivating dual-process accounts can be given in terms of models of mental architecture reflecting (1) and (2).

(1) *Foraging animals.* According to the dual-process hypothesis, the mammalian brain that underlies "intuitive" thought processes in humans operates on the basis of fast, "automatic," "stimulus-bound" responses involving rough heuristics or simple affective responses, without need for complex computation (Haidt 2001). Yet the last two decades of research on mammalian cognition, affect, and decision-making has provided a remarkably detailed picture of a very different kind of mind. For example, in the case of spatial navigation, microelectrode recording of neural arrays provides strong evidence that rats develop abstract representations of the space they explore, both egocentric (place-based) and non-egocentric (grid-based), that are persistent and independent of current stimulus (Moser et al. 2008). Moreover, these representations are activated when the rat is no longer in the experimental box or maze, repeatedly simulating both trajectories that the rat traveled and trajectories it has constructed by leveraging past experience to "fill in" areas of space not explored (Ji and Wilson 2007; Gupta et al. 2010; Redish 2010). These representations are also used *prospectively* when the rat is exploring a maze (prior to the point of overtraining)—when the animal reaches a choice point, activation in its "cognitive map" shifts forward to mentally explore the possible paths ahead (Redish 2016).

Moreover, the information used to assess these options is statistical in character. In the 1980s, it was realized that rats do not simply acquire associations, but rather use experience selectively to identify the most predictive information in their environment (Rescorla 1988). In the 1990s and 2000s, evidence from single-neuron recording identified how monkeys and other mammals used feedforward-feedback learning to sharply calibrate their expectations to actual frequencies and reward values, keeping separate track of probabilistic and evaluative information and calculating expected value to guide choice (Schultz et al. 1997; Burke and Tobler 2011). By the 2000s and 2010s monkeys were found to form hierarchical representations that permitted them to use higher-order abstract relations, such as patterns, to model their environment and facilitate learning (Wallis et al. 2001), and rats were found to use causal modeling for forward and inverse inferences about events and evidence (Blaisdell et al. 2006; Waldmann et al. 2012).

Far from being limited to simple heuristics or "likes or dislikes," mammalian affect and decision-making reflected abstract modeling of possible outcomes, causal relations, and the calculation of their risk and expected value as a guide to choice (Grabenhorst and Rolls 2008). Indeed, by offering macaques choices among gambles of the kind proposed by Ramsey and by von Neumann and Morgenstern, neural activity in the affective system could be seen to model a diverse array of values in ways that reflect classical risk-inflected cardinal utility functions and transitive preferences (Lak et al. 2014; Stauffer et al. 2014). Contrary to the dual-process hypothesis, then, the mammalian brain inherited from our animal ancestors appears to be, not a collection of "rough and ready," "informationally encapsulated" routines, but an integrated, highly sophisticated statistical learning system "designed" evolutionarily to collect, represent, and utilize causal and evaluative information in ways that exhibit the structure of normative decision theory. At last, neuroscience had the beginnings of an explanation of how foraging animals could reliably accomplish the task of developing optimal foraging strategies in complex environments (Dugatkin 2004)—a skill that extends beyond metabolic gains and losses to include foraging efficiently for information value and optimal selection of mates and cooperation partners (Melis et al. 2006).

But the neuroscience of foraging animals was not finished undoing older ideas of the "animal mind." Optimal foraging requires more than abstract mapping of spatial geometry, causal possibilities, and expected value. This causal-evaluative modeling must dynamically shape attention, perception, cognition, motivation, and motor commands, in contrast to the earlier picture of basic drives and model-free stimulus-response habits (Ma et al. 2011; Sugrue et al. 2004; Matsumoto et al. 2003; Berridge 2008; Liu and Todorov 2007; Tanji et al. 2007). Because model-based control generates structured expectations, it permits *content-focused* learning, since "credit" for fit or failures to fit expectations can be selectively distributed by using the model inversely (Doll et al. 2012).

If the design and operation of System 1 embodies sophisticated probabilistic learning and expected-value-based regulation, what accounts for the well-known statistical blunders of spontaneous human thought previously attributed to System

1 (Kahneman and Tversky 2000)? These failures may have more to do with the information-processing limitations of conscious thought and framing effects of language-dependent System 2—when the classic word problems of the "heuristics and biases" literature are presented to experimental subjects in the form of repeated opportunities to choose, accompanied by feedback, the same blunders do not appear (Friedman 1998; Shanks et al. 2002; Pleskac and Hertwig 2014). Gallistel et al. (2014) found that humans are nearly optimal in detecting probabilities in a dynamic situation, and Behrens et al. (2007) found that humans effectively use reward variance as a measure of uncertainty, and adjust their learning rates accordingly—just as normative statistical theory would recommend. In simulated foraging tasks involving multiple values and risks, humans, like animals, tend to develop optimal strategies (Kolling et al. 2012). (For a general discussion, see Tenenbaum et al. 2011 and Wellman 2014; for a critical discussion, see Marcus and Davis 2013.)

(2) *The infrastructure of moral decision-making.* It would seem, then, that standard dual-process accounts that rely upon a limited picture of the cognitive and affective capacities of System 1 need serious rethinking. What, then, about dual-process accounts of moral judgment?

While much remains unclear about the functional architecture underlying intuitive moral judgment, neuroimaging studies have consistently shown that moral evaluation involves activation of a network that is *domain-general* rather than modular, and that is highly integrated with—indeed, overlaps extensively with—the brain's *default network* (Buckner et al. 2008; Shenhav and Greene, 2010). This network is one of two fundamental modes of brain functioning, the other being focused, task-specific thinking, and the two modes typically alternate spontaneously over time, though they can be coactivated via executive control in goal-directed simulation (Gerlach et al. 2011). What does the default network do? At the experimental level, it has been associated with a distinctive set of related functions: episodic memory, imaginative simulation of future or counterfactual possibilities, "theory of mind" (that is, the attribution of mental states to others, intentional explanations of actions, etc.), and moral judgment. At the theoretical level, it has been hypothesized that the network evolved to enhance our capacity for *prospection*, the simulation and evaluation of alternate courses of action as a way of enhancing efficiency and efficacy in perception, thought, choice, and action (Buckner and Carrol 2006; Seligman et al. 2013; Doll et al. 2015). Humans have a far greater capacity for prospection than our animal ancestors, stretching as far as years and decades, and this perhaps coevolved with our capacity for stable long-term reciprocities and social coordination and cooperation in large-scale groups of unrelated individuals. Moral evaluation thus appears to be an integral function of the neural network best adapted for drawing upon extensive experience to evaluate how acts, motives, and practices affect the lives of those involved—an expression of greater or lesser social-evaluative skill.

A crucial component of this skill with prospection, and with "theory of mind" generally, is *empathy*—which will be understood here as the capacity to spontaneously simulate the experience, thoughts, and aims of others, or of oneself at a distant time or under counterfactual conditions (Decety et al. 2012). Because one's own affective system is used as the "test bed" in empathic simulation, the harms or benefits experienced by others—whether these be kin or strangers, friends or foes, or my own future selves—are represented as such, not mediated by one's own interests. For example, experiencing a mild electric shock, anticipating one, watching another receive one, and imagining another receiving one, activate significantly overlapping parts of the brain (Decety et al., 2016). Normally developing infants of twelve to sixteen months in age spontaneously seek to determine the cause of another's distress, and will initiate helping behavior (within the limits of their abilities) without external reward or encouragement (Zahn-Waxler et al. 1992; Warneken and Tomasello 2006). Evidently for creatures such as ourselves, whose existence and thriving depends upon navigating a social world of mutual dependence and vulnerability, evolution favored a capacity for non-egocentric *evaluative mapping* of the prospects and perils of social space—whether we are among friends, foes, or strangers—to match our inherited capacities for non-egocentric spatial, casual, and temporal mapping (Behrens et al. 2008).

Such evaluations satisfy the moral criteria of being (i) cognitive, (ii) opinion-independent, (iii) nonparochial, and (iv) authority-independent, but what of the criteria (v) and (vi) of supporting nonhypothetical motivation to avoid harms or promote benefits to others? Evaluation guides attention and motivation by default, so that, other things equal, if a social interaction involving others or the self is represented in overall positive terms, it will be preferred, other things equal. Infants typically are readily engaged by stories concerning strangers or fictional characters, are *disturbed* when the protagonist suffers harms, *want* to see her succeed in the end, and *prefer* third parties that help her over those that harm her (for some experimental evidence, see Hamlin and Wynn 2011). Starting from an early age an infant will pay attention to the goals and needs of those around him, even when they are strangers, and exhibit spontaneous motivation to act to promote those goals or meet those needs, even in the absence of any encouragement or external reward (Roth-Hanania et al. 2011; Geangu et al. 2011). Results in social neuroscience suggest that spontaneous acts of helping or cooperating with unrelated others are found *intrinsically* rewarding in adults as well (Rilling et al. 2002). Such a positive attitude toward positive social interactions makes evolutionary sense—a lack of interest in, or a default aversion to, *initiating* potentially positive social interactions would tend to prevent cooperation from getting off the ground. Even tit-for-tat involves a *default* favorable attitude toward cooperation with an unknown other, and motivation to extend an unsecured initial benefit to the other. And the most effective forms of tit-for-tat, even in environments where exploitation is a continuing risk, may be those that are *forgiving*, resetting the default back to positive despite a series of noncooperative responses, as a way of exploring whether cooperation might nonetheless still be possible (Zagorsky 2013).

The default network and the empathic simulations it runs, which underlie individual planning and social exchange alike, seem to be *built* in such a way as to generate evaluations that satisfy criteria (i)–(vi) (e.g., for effects of implicit learning in reducing parochialism, see Dasgupta et al. 2008). Importantly, this is the same system that is active during moral judgment. In contrast to "moral module" pictures, it is a domain-general system that recruits information widely. And in contrast to the standard dual-process picture, it is a highly integrated, cognitively complex system that uses affective coding flexibly to represent value and uncertainty in quantitative terms.

(3) *Behavioral evidence.* If this picture is right, how can we explain the seemingly arbitrary patterns found in intuitive moral judgments? For example, how might the difference in the predominant intuitive response to Footbridge vs. Switch and Loop be a manifestation of an underlying moral-evaluative skill, rather than an emotive response to a morally irrelevant feature—e.g., the use of personal muscular force, in the dual-process explanations offered?

Informal but confidential sampling in an undergraduate classroom (for further discussion, see Railton 2017) provides a hint. When students were asked whether they would trust a roommate more, less, or the same amount if they learned that the roommate had pulled the switch in Switch, some said *more*, some said *less*, but most said *the same amount*, with little resulting net gain or loss of trust. The same pattern emerged for Loop. But for Footbridge, there was a remarkable shift toward distrust. A strong majority expected that they would have *less* trust, and very few expected to have *more* trust—even though over a third had said that the man *should* be pushed in this scenario. Perhaps the persistent asymmetries in trolley problems are due, not to an atavistic or "automatic" emotional response, but to simulating the scenario and modeling the kind of motivational structure it would take to push the man to his death.

Psychological studies suggest that such modeling could reflect a grounding in experience. Subjects who indicate that they would push the man scored higher on psychopathy profile, and lower in empathy, altruism, moral concern, and concern to avoid harm (Bartels and Pizarro 2011; Gao and Tang 2013; Kahane et al. 2015; though see also Conway et al. 2018). And acquired expectations also reflect this, e.g., in trust games people expect those who have pushed to be less desirable partners and less equal in dividing benefits (Everett et al. 2016). Other research suggests that judgments in trolley problems are mediated by models of the character of the agents involved (Uhllmann et al. 2013). Similarly, Sripada found that an attribution of character mediated the seemingly "moralistic" asymmetry in judgments of intentionality found in the "Knobe Effect" (Sripada 2012).

It could well be, then, that the patterns observed in familiar examples of intuitive judgment in trolley problems reflect, rather than contradict, the idea that moral intuitions arise from an acquired competence that generates representations meeting criteria (i)–(vi).

3.4.2 Evolutionary Concerns

Recall that a posteriori tracking realists face challenges from evolutionary theory on at least two fronts: (1) motivational and (2) cognitive. Let us consider responses to these challenges in turn.

(1) *Moral motivation.* Once we have moved beyond the drive-based picture of motivation, and recognized that the allocation of attention and effort is generally regulated by causal-evaluative models, the question becomes, not "Would evolution have given us any altruistic basic drives?" but "Would evolution have given us dispositions to form causal-evaluative representations that reflect criteria (i)–(vi)?" We have already seen contemporary evidence for a positive answer. This congrues with extensive ethnographic observations of surviving hunter-gatherer societies—the closest model we have for the period during which most human evolution took place—indicating that these societies exhibit a common pattern of cooperative, egalitarian distribution in which individuals contribute to the group's subsistence in accord with their ability, but receive subsistence in accord with their needs (Boehm 2012). Since, over the course of a normal life, every individual will pass through periods of nearly complete inability to contribute as well as periods of relatively high ability to contribute, these arrangements are such as would receive a long-term stable positive evaluation from a representational system meeting criteria (i)–(vi). Even (iii), nonparochialism? It is not infrequent for individuals in hunter-gatherer society to move from one group to another—owing to, for example, out-marriage, seasonal variations in hunting or foraging patterns, migration, demographic collapse, social exclusion, etc.—or to have regular trading relations with those in other groups (Marlowe 2003; Apicella et al. 2012). An ability to simulate accurately harms or benefits to those who are not group members would help explain why these movements are possible, and why, as agriculture developed, humans could rapidly adapt to living together in large groups of unrelated individuals.

(2) *Moral cognition and reasoning.* What of the philosophers' argument that such psychological and evolutionary evidence can only show that *having attitudes* with features (i)–(vi) has been evolutionarily beneficial, not that these attitudes have any tendency to *track* moral value? For the realist's tracking "direction of explanation" to be credible, we would need to see moral value actually playing a role in accounting for why these attitudes were selected for.

What would it be to track value? Consider *epistemic* or *evidential value*. What are some of the criteria that would characterize epistemic evaluation? It would be (i) cognitive, (ii) opinion-independent, (iii) nonparochial, (iv) authority-independent, and (v) noninstrumentally thought-guiding (that is, would not require external incentive in order to shape the formation of expectation or belief). Of course, (i)–(v) also characterize distinctively moral evaluation, so we must add (vi*), a concern with truth, and with the weight and bearing of evidence, in seeking information and forming expectations or beliefs. (No doubt (vi*) could be refined, but perhaps it will be relatively uncontroversial.)

Is there evidence that humans are spontaneously disposed to form and be guided in thought and inquiry by evaluations satisfying (i)–(vi*)? One can, it seems, track evidential value without forming beliefs containing epistemic *concepts.* In the first months of life, human infants are keeping track of the conditional probabilities of phonetic sequences, noticing anomalous sequences, and paying greater attention to them—in learning, anomalous sequences have greater information value (Aslin et al. 1998). Once infants have begun to recognize word types, they pay attention to how reliably an adult uses words they recognize to adjust how readily they learn new words from that adult (Koenig and Echols 2003). When infants observe objects drawn from a diverse collection of objects, they show greater surprise when the objects drawn are not representative of the relative frequencies in the collection, and are more likely to attribute the selection to intent (Schulz 2012). As they grow infants develop a steadier sense of appearance vs. reality (Karg et al. 2014), and become willing to learn more readily from an *unfamiliar* adult who displays greater competence than a familiar adult (Sobel and Corriveau 2010; Harris and Corriveau 2011). Infants attend to the epistemic situation of others, displaying implicit competence with false belief task as early as fifteen months, and manifesting spontaneous motivation to volunteer information to unrelated adults when they have seen something relevant to the adult's behavior that the adult has not (Baillargeon et al. 2010). And by three to five years in age, infants draw upon observations of how an adult has treated unrelated others in adjusting how much to trust information given by the adult (Doebel and Koenig 2013).

In these ways and more, it would appear that infant learners manifest a systematic pattern of responses suggesting that their belief-forming dispositions (i) are sensitive to the content of experience; (ii) go beyond immediate experience to make predictive inferences that are especially attentive to expectation-violations, and thus support challenges to their existing opinions (Lane et al. 2014); (iii) give weight to evidence from third-party observation of unrelated individuals that can offset weight given to first-person evidence and evidence from familiar or related individuals; (iv) can use observation to undermine the claims of those with authority over them; (v) involve noninstrumental motivation to acquire and share information; and (vi*) give intrinsic weight to evidence proportional to its quality and quantity. Moreover, spontaneous infant behavior reflects such assessments in just the ways appropriate for epistemic value—shaping what they attend to, notice, infer, and rely upon in action, information-seeking, and information-sharing. Suppose that the *proximate* explanation of why this complex constellation of capacities was selected for is that it tended to promote inclusive fitness; even so, the *distal* explanation of why this would be the case is likely to be that these capacities *attune* the infant's expectations to minimize the discrepancy between strength of expectation and the probability of actual or possible outcomes—that is, to increase *accuracy* in representing the physical and social environment. Recent breakthroughs in artificial intelligence incorporate a similar generic algorithmic design (Le et al. 2012; Mnih et al. 2015) and integrate the learning of a value function to guide expectation and action (Silver et al. 2015). Building an effective epistemic agent from scratch would, it seems, take a similar form.

Turning now to moral value, we have said that evaluations meeting criteria (i)–(vi) qualify as distinctively *moral*. And we have seen a host of evidence that infants spontaneously form action-guiding evaluative representations that meet criteria (i)–(vi), i.e., that are to that extent *attuned* to moral value. As in the case of epistemic value, it does not seem necessary to possess beliefs with distinctively moral conceptual content in order to achieve such attunement. However, one might think that such attunement *does* require evidence of mastery of what is distinctive about moral evaluation, and why. An interesting—if somewhat controversial (Kelly et al. 2007)—body of evidence supporting this claim can be found in the literature on children's *implicit* grasp of the *moral/conventional* distinction. Over decades of study in multiple cultures, Turiel and colleagues have found that, by age three to four, children *behaviorally* manifest mastery of this distinction, and do so in ways that are appropriate from a normative perspective. For example, infants treat moral requirements as applying across all contexts, but conventional rules as context- and authority-dependent. Moreover, within a context they tend to treat moral requirements as more serious, and cite such factors as avoiding harm or unfairness to explain why. Moreover, such infants exhibit motivation to comply with moral requirements without need for external reward or encouragement, and will spontaneously intervene to enforce moral requirements in third-party conflicts (Turiel 2006; Smetana 2006; Vaish et al. 2011). This is arguably the result of largely *tacit* learning, since it seems unlikely that adults across a range of cultures, with varying mixtures of traditional vs. liberal attitudes, would have taught children to exercise their own judgment in deciding whether to obey those in authority (themselves included). It also seems unlikely that all most of the children recognizing and respecting this distinction have been taught such concepts as <convention>, <authority-independence>, <context-independence>, and so on. This, then, would seem to be an example of an intuitive skill in tracking a distinction of fundamental normative importance, the learning of which could draw upon the child's ability to simulate empathically the experiences of those affected by particular rules or rule-violations, and thus grasp the evaluative difference between arbitrary vs. harm-related norms.

As in the case of epistemic value, the *proximal* evolutionary explanation of why such dispositions to form and be guided by non-egocentric representations of the evaluative landscape might have been selected for will turn on the inclusive fitness advantages accruing to those with such dispositions. Much debate exists over possible selection mechanisms in this case, including—in addition to kin selection and direct reciprocity—sexual selection, indirect reciprocity, group selection, and social selection. But evolutionary speculation apart, the contemporary evidence is that such spontaneous dispositions evolved, were conserved, and are present in large portion of the population. Human evolution appears to have favored a default disposition in a wide array of circumstances to initiate cooperation with others, even strangers and even in the absence of reciprocity, whether cooperation takes the form of sincere communication or willingness to help or to prevent harm. Crockett and colleagues, for example, found in a careful study that individuals in contemporary England were typically willing to pay more to prevent a stranger from receiving a mild shock than to prevent a comparable shock to themselves (2014). And while Mercier and Sperber have argued that humans use reasoning—epistemic and moral—strategically (2011), a purely strategic story would not explain why we pay such attention to others' reasoning and give their opinions a leading role in our own thinking. Default communicative *cooperation* has no such difficulty explaining why we listen to, and rely extensively upon, others' words—it injects enough *trust* and *trustworthy information* into the epistemic economy, so to speak, to give each other's words real value. And in the moral case, default cooperation injects enough *trust* and *sensitivity to others' interests* into the moral economy to sustain the force of others' moral judgments.

These default dispositions are by their nature nonparochial. That made it possible for the human epistemic and moral economy to "scale up" to the vast collections of unrelated individuals characteristic of contemporary human life. In a classic work, Nagel (1970) characterized the point of view of moral agents as seeing themselves as "one among many," whose needs and concerns are no less real than one's own. It appears that evolution has built us to be capable of responding to information and value as "one among many" in both the epistemic and the moral domain. While such a standpoint must compete with more parochial self- or group-centered attitudes or interests, and will be more prominent and motivating in some social environments than others, it appears that humans have the ability to create and sustain adaptive social environments (Henrich et al. 2004) in which tracking, sharing, and acting on nonparochial reasons have played a central role in the exponential growth of human knowledge and social combination.

3.5 CONCLUDING REMARKS

We began by considering skills in such domains as athletics, games of strategy, performance, teaching, etc., and noting that the literature on such skills suggests that they depend upon the acquisition of complex causal-evaluative models of situations and their prospects, which permit forward modeling to guide perception and action, and inverse modeling to guide the use of experience for increasingly refined learning. These causal-evaluative models appear to be acquired via general-purpose learning capacities and extensive experience—a mode of operation that also characterizes many of the most effective contemporary artificial intelligence systems for perception, game-playing, speech recognition, translation, and autonomous robotics.

Natural selection for optimal foraging and building from scratch for optimal performance appear to have hit upon similar basic designs.

Moral skill and intelligence can be understood as emerging via similar general-purpose intelligence and learning—essentially integrated with such core functions as episodic memory, simulation and evaluation of future possibilities and counterfactual situations, and theory of mind, and, like them, selected for producing reasonably accurate, action-guiding models within the limitations of our evidence. While humans as we know them are often quarrelsome, self-centered, clannish, and even destructive, these tendencies alone could not explain why *Homo sapiens* came to be the dominant hominid and covered the earth with complex societies. Our Late Pleistocene ancestors appear to have deployed a capacity for cooperativeness, sharing of goods and governance, decency toward strangers, and avoidance of conflicts to give themselves a foothold even in some of the most inhospitable natural environments our planet has to offer (Boehm 2012). The "agricultural revolution" changed things, making possible surplus accumulation and an extended division of labor, but also inequality and remunerative raiding. We *Homo sapiens* subsequently distinguished ourselves both for unprecedentedly large-scale cooperativeness and for unprecedentedly destructive competition. Fortunately, our core capacity to be alive to nonparochial value is so nearly universal, and so reliably developed through experience in the main, that the result is far from a zero sum. And so we carry on.

ACKNOWLEDGEMENTS

I would like to thank my colleagues Allan Gibbard and Chandra Sripada for many helpful conversations, and the editors for very helpful comments on an earlier draft.

REFERENCES

Apicella, C. L., Marlowe, F. W., Fowler, J. H., and Christakis, N. A. (2012). Social networks and cooperation in hunter-gatherers. *Nature 481*: 497–501.

Aslin, R. N., Saffran, J. R., and Newport, E. L. (1998). Computation of conditional probability statistics by 8-month-old infants. *Psychological Science 9*: 321–24.

Ayer, A. J. (1936). *Language, Truth, and Logic*. London: Gollancz.

Baillargeon, R., Scott, R. M., and He, Z. (2010). False-belief understanding in infants. *Trends in Cognitive Sciences 14*: 111–18.

Bargh, J. A., and Chartrand, T. (1999). The Unbearable automaticity of being. *American Psychologist 35*: 462–79.

Bartels, D. M., and Pizarro, D. A. (2011). The mismeasure of morals: Antisocial personality traits predict utilitarian responses to moral dilemmas. *Cognition 121*: 154–61.

Behrens, T. E. J., Hunt, L. T., Woolrich, M. W., and Rushworth, M. F. S. (2008). Associative learning of social value. *Nature 456*: 245–50.

Behrens, T. E. J., Woolrich, M. W., Walton, M. E., and Rushworth, M. F. S. (2007). Learning the value of information in an uncertain world. *Nature Neuroscience 10*: 1214–21.

Blackburn, S. (1998). *Ruling Passions*. Oxford: Clarendon.

Blaisdell, A. P., Sawa, K., Leising, K. J., and Waldmann, M. R. (2006). Causal reasoning in rats. *Science 311*: 1020–22.

Boehm, C. (2012). *Moral Origins: The Evolution of Virtue, Altruism, and Shame.* New York: Basic Books.

Boyd, R. (1988). How to be a moral realist. In G. Sayre-McCord, ed., *Essays in Moral Realism*. Ithaca, NY: Cornell University Press, 307–55.

Brink, D. (1989). *Moral Realism and Its Foundations.* Cambridge: Cambridge University Press.

Buckner, R. L., Andrews-Hanna, J. R., and Schacter, D. L. (2008). The brain's default network: Anatomy, function, and relevance to disease. *New York Academy of Sciences 1124*: 1–38.

Buckner, R. L., and Carrol, D. C. (2006). Self-projection and the brain. *Trends in Cognitive Sciences 11*: 49–57.

Burke, C. J., and Tobler, P. N. (2011). Coding of reward probability and risk by single neurons in animals. *Frontiers in Neuroscience 5* (121): 1–9.

Chaiken, S., and Trope, Y. (1999). *Dual Process Theories in Social Psychology.* New York: Guilford Press.

Conway, P., Goldstein-Greenwood, J., Polacek, D., and Greene, J.D. (2018). Sacrificial utilitarian judgments do reflect concern for the greater good: Clarification via process dissociation and the judgments of philosphers. *Cognition 179*, 241–265.

Crockett, M. (2013). Models of morality. *Trends in Cognitive Sciences 17*: 363–66.

Crockett, M., Kurth-Nelson, Z., Siegel, J. Z., Dayan, P., and Dayan, R. J. (2014). Harm to others outweighs harm to self in moral decision making. *PNAS 111*: 17320–25.

Cushman, F. A. (2013). Action, outcome, and value in a dual-system framework for morality. *Personality and Social Psychology Review 17*: 273–92.

Darwall, S. L. (1997). Reasons, motives, and the demands of morality. In S. L. Darwall A. Gibbard, and P.. Railton, eds., *Moral Discourse and Practice: Some Philosophical Approaches*. Oxford: Oxford University Press, 305–12.

Dasgupta, N., and Rivera, L. M. (2008). When social context matters: The influence of long-term contact and short-term exposure to admired outgroup members on implicit attitudes and behavioral intentions. *Social Cognition 26*: 112–23.

Decety, J., Bartal, I. B.-A., Uzefovsky, F., and Knafo-Noam, A. (2016). Empathy as a driver of prosocial behavior: Highly conserved neurobehavioural mechanisms across species. *Philosophical Transactions of the Royal Society B 371*: 20150077.

Decety, J., Michalsha, K. J., and Kinzler, K. D. (2012). The contribution of emotion and cognition to moral sensitivity: A neurodevelopmental study. *Cerebral Cortex 2*: 209–20.

Doebel, S., and Koenig, M. A. (2013). Children's use of moral behavior in selective trust discrimination versus learning. *Developmental Psychology 49*: 462–69.

Doll, B. D., Duncan, K. D., Shohamy, N. D., and Daw, N. D. (2015). Model-based choices involve prospective neural activity. *Nature Neuroscience 18*: 767–72.

Doll, B. D., Simon, D. A., and Daw, N. D. (2012). The ubiquity of model-based reinforcement learning. *Current Opinions in Neurobiology 22*: 1075–81.

Dreier, J. (1990). Internalism and speaker relativism. *Ethics 101*: 6–26.

Dugatkin, L. A. (2004). *Principles of Animal Behavior*. New York: Norton.

Enoch, D. (2011). *Taking Morality Seriously: A Defense of Robust Realism*. Oxford: Oxford University Press.

Evans, J. St. B. T. (2008). Dual-processing accounts of reasoning, judgment, and social cognition. *Annual Review of Psychology 59*: 255–78.

Everett, J. A. C., Pizarro, D. A., and Crockett, M. J. (2016). Inference of trustworthiness from intuitive moral judgments. *Journal of Experimental Psychology—General 145*: 772–87.

Friedman, D. (1998). Monty Hall's three doors: Construction and deconstruction of a choice anomaly. *American Economic Review 88*: 933–46.

Gao, Y., and Tang, S. (2013). Psychopathic personality and utilitarian moral judgment in college students. *Journal of Criminal Justice 41*: 342–49.

Geangu, E., Benga, O., Stahl, D., and Striano, T. (2011). Individual differences in infants' emotional resonance to a peer in distress: Self-other awareness and emotion regulation. *Social Development 20*: 450–70.

Gerlach, K. D., Spreng, R. N., Gilmore, A. W., and Schacter, D. L. (2011). Solving future problems: Default network and executive activity associated with goal-directed mental simulations. *NeuroImage 55*: 1816–24.

Gibbard, A. (1994). *Wise Choices, Apt Feelings*. Cambridge, MA: Harvard University Press.

Gibbard, A. (2003). *Thinking How to Live*. Cambridge, MA: Harvard University Press.

Gigerenzer, G., Todd, P. M., and ABC Research Group (1999). *Simple Heuristics That Make Us Smart*. New York: Oxford University Press.

Gold, N., Colman, A. M., and Pulford, B. D. (2014). Cultural differences in responses to real-life and hypothetical Trolley problems. *Judgment and Decision Making 9*: 65–76.

Greene, J. D., Cushman, F. A., Stewart, L. E., Lowenberg, K., Nystrom, L. E., and Cohen, J. D. (2009). Pushing moral buttons: The interaction between personal force and intention in moral judgment. *Cognition 111*: 364–71.

Greene, J. D., and Haidt, J. (2002). How (and where) does moral judgment work? *Trends in Cognitive Sciences 6*: 517–23.

Greene, J. D., Sommerville, R. B, Nystrom, L. E., Darley, J. M., and Cohen, J. D. (2001). An fMRI investigation of emotional engagement in moral judgment. *Science 293*: 2015–18.

Gupta, A. S., van der Meer, M. A. A., Touretzky, D. S., and Redish, A. D. (2010). Hippocampal replay is not a simple function of experience. *Neuron 65*: 695–705.

Haidt, J. (2001). The emotional dog and its rational tail: A social intuitionist approach to moral judgment. *Psychological Review 108*: 814–34.

Haidt, J. (2007). The new synthesis in moral psychology. *Science 316*: 998–1002.

Hamlin, J.K. and Wynn, K. (2011). Young infants prefer proscocial to antisocial others. *Cognitive Development 26*, 30-39.

Harman, G. (1975). Moral relativism defended. *Philosophical Review 84*: 3–22.

Harman, G. (2015). Moral relativism is moral realism. *Philosophical Studies 172*: 855–63.

Harman, G., and Thomson, J. J. (1996). *Moral Relativism and Moral Objectivity*. Oxford: Blackwell.

Harris, P. L., and Corriveau, K. H. (2011). Young children's selective trust in informants. *Philosophical Transactions of the Royal Society: B 366*: 1179–87.

Henrich, J., Boyd, R., Bowles, S., Camerer, C., Fehr, E., and Gintis, H. (2004). *The Foundations of Human Sociality: Economic Experiments and Ethnography Evidence from Fifteen Small-Scale Societies*. Oxford: Oxford University Press.

Horwich, P. (1990). *Truth*. Oxford: Blackwell.

Jackson, F. (1998). *From Metaphysics to Ethics: A Defence of Conceptual Analysis*. Oxford: Clarendon.

Ji, D., and Wilson, M. A. (2007). Coordinated memory replay in the visual cortex and hippocampus during sleep. *Nature Neuroscience 10*: 100–107.

Joyce, R. (2001). *The Myth of Morality*. Cambridge: Cambridge University Press.

Kahane, G., Everett, J. A. C., Earp, B. D., Farias, M., and Savulescu, J. (2015). "Utilitarian" judgments in sacrificial moral dilemmas do not reflect impartial concern for the greater good. *Cognition 134*: 193–209.

Kahneman, D. (2003). Maps of bounded rationality: Psychology for behavioral economics. *American Economic Review 93*: 1449–75.

Kahneman, D. (2011). *Thinking, Fast and Slow.* New York: Farrar, Straus and Giroux.
Kahneman, D., and Tversky, A. (2000). *Choice, Value, and Frames.* New York: Russell Sage.
Katz, L. D. (2000). *Evolutionary Origins of Morality: Cross-Disciplinary Perspectives.* Thorventon: Imprint Academic.
Koenig, M. A., and Echols, C. H. (2003). Infants' understanding of false labeling events: The referential roles of words and the speakers who use them. *Cognition 87*: 179–208.
Kolling, N., Behrens, T. E. J., Mars, R. B., and Rushworth, M. F. S. (2012). Neural mechanisms of foraging. *Science 336*: 95–98.
Korsgaard, C. (1986). Skepticism about practical reason. *Journal of Philosophy 83*: 5–25.
Lak, A., Stauffer, W. R., and Schultz, W. (2014). Dopamine prediction error responses integrate subjective value from different reward dimensions. *PNAS 111*: 2343–48.
Lane, J. D., Harris, P. L., Gelman, S. A., and Wellman, H. W. (2014). More than meets the eye: Young children's trust in claims that defy their perceptions. *Developmental Psychology 50*: 865–71.
Le, Q. V., Ranzato, M. A., Monga, R., Devin, M., Chen, K., Corrado, G. S., Dean, J., and Ng, A. Y. (2012). Building high-level features using large scale unsupervised learning. In *Proceedings of the 29th International Congress on Machine Learning.* Edinburgh.
Liu, D., and Todorov, E. (2007). Evidence for flexible sensorimotor strategies predicted by optimal feedback control. *Journal of Neuroscience 27*: 9354–68.
Ma, W. J., Navalpakkam, V., Beck, J. M., van den Berg, R., and Pouget, A. (2011). Behavior and neural basis of near-optimal visual search. *Nature Neuroscience 14*: 783–90.
Mackie, J. L. (1977). *Ethics: Inventing Right and Wrong.* London: Penguin.
Marcus, G. F., and Davis, E. (2013). How robust are probabilistic models of higher-level cognition? *Psychological Science 24*: 2351–60.
Marlowe, F. W. (2003). The mating system of foragers in the standard cross-cultural sample. *Cross-Cultural Research 37*: 282–306.
McBeath, M. K., Shaffer, D. M., and Kaiser, M. K. (1995). How baseball outfielders determine where to run to catch fly balls. *Science 268*: 569–73.
McPherson, T. (2015). What is at stake in debates among normative realists? *Noûs 49*: 123–46.
Mercier, H., and Sperber, D. (2011). Why do humans reason? Arguments for an argumentative theory. *Behavioral and Brain Sciences 34*: 74–111.
Mnih, V., et al. (2015). Human-level control through deep reinforcement learning. *Nature 518*: 529–33.
Moore, G. E. (1903). *Principia Ethica.* Cambridge: Cambridge University Press.
Moser, E. I., Kropff, E., and Moser, M.-B. (2008). Place cells, grid cells, and the brain's spatial representation system. *Annual Review of Neuroscience 31*: 69–89.
Nagel, T. (1970). *The Possibility of Altruism.* New York: Oxford University Press.
Nagel, T. (1987). *The View from Nowhere.* New York: Oxford University Press.
Olberg, R. M., Worthington, A. H., and Venator, K. R. (2000). Prey pursuit and interception in dragonflies. *Journal of Comparative Physiology 186*: 155–62.
Parfit, D. (2011). *On What Matters.* Vol. 2. Ed. S. Scheffler. New York: Oxford University Press.
Plato (ca. 399–395 BCE, 2002). *Euthyphro.* In G. M. A. Gruber and J. S. Cooper, trans. and eds., *Plato: Five Dialogues.* 2nd ed. Indianapolis: Hackett.
Pleskac, T. J., and Hertwig, R. (2014). Ecologically rational choice and the structure of the environment. *Journal of Experimental Psychology: General 143*: 2000–2019.
Railton, P. (1986). Moral realism. *Philosophical Review 95*: 163–207.
Railton, P. (2013). The Affective dog and its rational tale: Intuition and attunement. *Ethics 124*: 813–59.

Railton, P. (2017). Moral learning: Conceptual foundations and normative relevance. *Cognition 167*: 172–90.

Ramsey, F. P. (1927). Facts and propositions. *Aristotelian Society Supplementary Vol. 7*: 153–70.

Rilling, J. K., Gutman, D. A., Zeh, T. R., Pagnoni, G., Berns, G. S., and Kilts, C. D. A neural basis for social cooperation. *Neuron 36*: 395–406.

Rolls, E. T., and Grabenhorst, F. (2008). The orbitofrontal cortex and beyond: From affect to decision-making. *Progress in Neurobiology 86*: 216–44.

Ross, W. D. (1930). *The Right and the Good*. Oxford: Oxford University Press.

Roth-Hanania,R., Davidov, M., and Zhan-Waxler, C. (2011). Empathy development from 8 to 16 months: Early signs of concern for others. *Infant Behavior and Development 34*: 447–58.

Schulz, L. (2012). The origins of inquiry: Inductive inference and exploration in early childhood. *Trends in Cognitive Sciences 16*: 382–89.

Schultz, W., Dayan, P., and Montague, P. R. (1997). A neural substrate of prediction and reward. *Science 275*: 1593–99.

Seligman, M. E. P., Railton, P., Baumeister, R., and Sripada, C. S. (2013). Navigating into the future or driven by the past? *Perspectives in Psychological Science 8*: 119–41.

Shanks, D. R., Tunney, R. J., and McCarthy, J. D. (2002). A re-examination of probability matching and rational choice. *Journal of Behavioral Decision Making 15*: 233–50.

Shenhav, A., and Greene, J.D. (2010). Moral judgments recruit domain-general valuation mechanisms to integrate representations of probability and magnitude. *Neuron 67*, 667–677.

Smetana, J. (2006). Social-Cognitive Domain Theory. In J. Smetana and M. Killen, eds., *Handbook of Moral Development*. Mahwah, NJ: Erlbaum, 119–53.

Smith, M. (1994). *The Moral Problem*. Oxford: Blackwell.

Sobel, D. M., and Corriveau, K. H. (2010). Children monitor individual's expertise for word learning. *Child Development 81*: 669–79.

Sripada, C. S. (2012). Mental state attributions and the side-effect effect. *Journal of Experimental Social Psychology 48*: 232–38.

Stauffer, W. R., Lak, A., and Schultz, W. (2014). Dopamine reward prediction error responses reflect marginal utility. *Current Biology 24*: 2491–500.

Street, S. (2006). A Darwinian dilemma for realist theories of value. *Philosophical Studies 127*: 109–66.

Tanji, J., Shima, K., and Mushiake, H. (2007). Concept-based behavioral planning and the lateral prefrontal cortex. *Trends in Cognitive Sciences 11*: 528–34.

Tenenbaum, J. B., Kemp, C., Griffiths, T. L., and Goodman, N. D. (2011). How to grow a mind: Statistics, structure, and abstraction. *Science 331*: 1279–85.

Tobler, P. N., O'Dougherty, J. P., Dolan, R. J., and Schultz, W. (2007). Reward value coding distinct from risk attitude-related uncertainty coding in human reward systems. *Journal of Physiology 97*: 1621–32.

Turiel, E. (2006). The development of morality. In N. Eisenberg, ed., *Handbook of Child Psychology*, vol. 3. 6th ed. New York: Wiley, 789–857.

Uhllmann, E. L., Zhu, L., and Tannenbaum, D. (2013). When it takes a bad person to do the right thing. *Cognition 126*: 326–34.

Vaish, A., Missana, M., and Tomasello, M. (2011). Three-year-old children intervene in third-party moral transgressions. *British Journal of Developmental Psychology 29*: 124–30.

Waldmann, M. R., Schmid, M., Wong, J., and Blaisdell, A. P. (2012). Rats distinguish absence of events from lack of evidence in contingency learning. *Animal Cognition 15*: 979–90.

Warneken, F., and Tomasello, M. (2006). Altruistic helping in human infants and young chimpanzees. *Science 311*: 1301–3.

Wellman, H. (2014). *Making Minds: How Theory of Mind Develops.* Oxford: Oxford University Press.

Zagorsky, B. M., Reiter, J., Chatterjee, K., and Nowak, M. A. (2013). Forgiver triumphs in alternating prisoner's dilemma. *PLOS-One*, e80814.

Zahn-Waxler, C., Radke-Yarrow, M., Wagner, M., and Chapman, M. (1992). Development of concern for others. *Developmental Psychology 28*: 1038–47.

4 Debunking and Vindicating in Moral Psychology

Shaun Nichols

4.1 INTRODUCTION

One of the most obvious ways cognitive science can inform metaphysics is by revealing the psychological underpinnings of our common-sense metaphysical beliefs. Insofar as we base our actual metaphysics on common-sense metaphysical beliefs, it's important to have a sense of the genesis of those beliefs. For the genesis of a belief bears on the epistemic status of the belief. Knowledge of the psychological process that produces a metaphysical belief can serve to *debunk* our intuitive metaphysical commitments, including ethically relevant metaphysical beliefs (Nichols 2014). In this chapter, I'll briefly set out how these process-debunking arguments work. I'll then provide an instance of such an argument that attempts to debunk the common-sense belief in moral objectivity, drawing on work that suggests that the belief in moral objectivity is produced, in part, by epistemically defective process. In contrast to these genetic debunking arguments, some have recently attempted to provide genetic *vindicating* accounts of lay beliefs, according to which key aspects of moral belief are based on rational processes (e.g., Railton 2014; Nichols et al. 2016). I will elaborate the notion of process vindication in some detail, and apply it to the same test case—the belief in moral objectivism. Recent work indicates that the belief in moral objectivism is partly based on rationally appropriate processes, which points towards a possible vindication for the belief in objectivity.

4.2 PROCESS DEBUNKING

The best-known genetic arguments aim to debunk common beliefs. Freud's treatment of religious belief is canonical here. Freud argues that the pull of religious doctrines is "independent . . . of recognition by reason" ([1927] 1961, 37). People's belief in God does not come from evidence or proof. Instead, Freud argues, people believe in God as a result of an epistemically defective process—wish fulfillment.

People's belief in a benevolent God is the result of their wanting there to be a benevolent God.

While Freud's particular argument has been much criticized (see, e.g., Plantinga 2000), the basic form of a process-debunking argument is perfectly sensible. A process-debunking argument aims to show that an agent's belief that p is generated by a process that does not justify the belief that p, and as a result, the agent is not justified in believing that p. The key here is that the process that generates the belief is an epistemically defective basis for forming that belief. If we determine that the process that led to (and sustains) a person's belief is epistemically defective, that undermines the justificatory status of the belief. Since wish fulfillment is typically an epistemically defective way to form metaphysical beliefs, if Freud were right about why we believe in God, this would undercut the justificatory credentials of our belief in God.

There are numerous ways that a process can be epistemically defective for forming a particular type of belief. And even epistemologists who disagree about what makes belief-formation processes defective often agree about central examples of processes that *are* epistemically defective (e.g., Goldman 1979, 9; Cohen 1984, 282–83). Motivated reasoning (like wish fulfillment) is one process that is typically epistemically defective for forming or sustaining beliefs about the world.[1] There is also a range of familiar inferential fallacies. For instance, believing that a fair coin will come up tails because it just came up heads three times in a row is an epistemically defective basis for that belief. A gambler who thus thinks the fair coin will come up tails is not justified in that belief, since it's based on an epistemically defective form of probabilistic inference. Perhaps the coin will come up tails—after all, there's a 50% chance of that—but the gambler's belief is not justified since the process that led to the belief is an epistemically defective way of formulating probabilistic beliefs.

The actual debunking arguments that get deployed in philosophy are much less straightforward than the gambler's fallacy. This is partly because philosophers are interested in weightier beliefs, like the belief in God. And it's partly because the actual details of the processes are quite unclear. So, to stick with Freud's argument, there is some reason to think that motivation influences religious belief (see, e.g., Nichols 2004; Norenzayan and Hansen 2006), but the details of how the motivation works its influence is unclear. Furthermore, even if motivation plays a role, it is hardly the only factor that generates religious belief. Until we have a much sharper picture of the actual processes that generate metaphysically interesting beliefs (like the belief in God), we must make do with qualified debunking arguments: People are unjustified

[1] Motivation itself can facilitate improved reasoning. People tend to reason better when they are reasoning about an issue that matters to them personally (see, e.g., Schaller 1992). But *motivated reasoning* often has pernicious influences of belief maintenance and change, e.g. leading to biased search strategies (see, e.g., Doris 2015; Kunda 1999).

in believing that p *to the extent* that their belief is based on a process that is epistemically defective for believing that p.

4.3 MORAL OBJECTIVITY AND PROCESS DEBUNKING

4.3.1 Moral Objectivity

Over the last several years, there has considerable work on whether lay people think morality is objective (Goodwin and Darley 2008; Nichols 2004; Nichols and Folds-Bennett 2003; Sarkissian et al. 2011; Wright et al. 2008). For present purposes, the operative notion of objectivity is that the truth conditions for objective claims are independent of culture, of geography, and, more generally, of the attitudes and feelings people have toward the claim (see Shafer-Landau 2003).[2] There are uncontroversial examples of claims that are objective in this sense, including claims about matters of fact (e.g., a chlorine atom has seventeen protons) and matters of logic or math (e.g., $2 * 7 = 14$). The truth of $2 * 7 = 14$ holds independently of anyone's attitudes about that proposition. It follows from the preceding characterization of objectivism that, if two individuals make different judgments about the truth of some objective statement, then at least one of them must be wrong. Thus, if an alien denies that $2 * 7 = 14$, then at least one of us is wrong. We can't make different judgments and both be right about objective matters. This kind of objectivism contrasts starkly with relativism. If one person says, "It's 9:00 a.m." and another says, "It's not 9:00 a.m.," they might both be right because time of day is relative to geography.

Many contemporary philosophers maintain that ordinary people presuppose that morality is objective and not relative (e.g., Joyce 2002, 97; Mackie 1977, 33). Psychologists have investigated this by drawing on the preceding idea that for objective claims, if two people make different judgments about the truth of the claims, at least one of them must be wrong.[3] And several studies confirm that people think that if two individuals disagree about a moral claim, one of them must be wrong. In a critical early study, Goodwin and Darley (2008) presented participants with factual statements (e.g., "The earth is not at the center of the known universe"), conventional statements (e.g., "Calling teachers by their first name, without being given permission to do so, in a school that calls them 'Mr.' or 'Mrs.' is wrong behavior"), taste statements (e.g., "Frank Sinatra was a better singer than is Michael Bolton"), and ethical statements, (e.g., "Robbing a bank in order to pay for an expensive holiday

[2] This kind of objectivism is thus stronger than some views labeled as objectivist (e.g. Smith 1993). I'll discuss some of these issues in section 4.5.2.

[3] This is far from a perfect test of the kind of objectivism defined previously. Indeed, as we will see in section 4.5.2, weaker notions of objectivity would make similar predictions. And even for some cases of conventions, like which side of the road to drive on, it would be natural to say that if two people disagree, one of them must be wrong. Nonetheless, the disagreement measure has turned up some interesting differences in people's attitudes, as we will see subsequently in this section and again in section 4.5.1.

is a morally bad action"). Participants were asked, for each statement, whether a person who disagreed with them over the statement "is surely mistaken" or whether "it is possible that neither you nor the other person is mistaken." To simplify slightly, saying "surely mistaken" was interpreted as an objectivist response (1352).

Goodwin and Darley found that people were much more likely to give objectivist responses for the ethical statements than for the statements about taste or social convention (1352–53). Indeed, the ratings of objectivity for ethical statements was almost as high as for plainly factual statements (Goodwin and Darley 2008, 1354). But why?

4.3.2 Objectivity and Epistemically Defective Processes

Goodwin and Darley don't much speculate on why people respond as objectivists. But philosophers have certainly weighed in on this, and the best-known explanations serve as debunking explanation. In particular, Mackie (1977) proposes a number of nonrational, emotion-based explanations for the belief in objectivity. He offers social, motivational, and affective explanations. I'll focus here on the affective explanation.[4] Mackie suggests that the belief in objectivity derives from the tendency to project our moral attitudes onto the world. A fungus that fills us with disgust, says Mackie, may tempt us to think that the fungus itself has some objective foul property. Similarly, our emotional reactions toward ethical violations may persuade us that moral badness exists as an objective property. This would plausibly be an epistemically defective process for coming to beliefs about the metaphysics of morality.

Consistent with the idea that emotions play a distorting role in objectivity judgments, Cameron et al. (2013) found that participants induced to feel disgust by viewing disgusting pictures were more likely to give objectivist responses as compared to those not induced to feel disgust. Cameron and colleagues used a slightly different objectivity measure. They asked participants to evaluate whether an activity practiced in other cultures (e.g., "Marriages are arranged by the children's parents") is wrong regardless of culture in which it is practiced. Pairing the description with a disgusting picture led to increased objectivity responses. Such an influence is plausibly epistemically defective. Cameron and colleagues make this clear by drawing on the distinction between incidental and integral effects of emotions:

> Integral emotions may contain information that should appropriately influence moral judgments: guilt may signal that you have behaved badly towards others, and anger may signal that others have behaved badly towards you (Frank, 1988). In contrast, incidental emotions are conceptually unrelated to subsequent judgments, and so are ethically irrelevant (Doris and Stich, 2005). Whereas incidental emotions may influence

[4] Rose and Nichols (forthcoming) explore Mackie's motivational explanation.

moral judgments, they are not appropriately cited as evidence in the justification of these judgments. (719)

If you are more objectivist about arranged marriages because you are seeing a revolting picture of worms, then you're being swayed by an epistemically defective process.

Cameron and colleagues' central point in the paper is that these effects of incidental emotion can be eliminated when people carefully differentiate emotions, but their results also suggest a nice process-debunking argument. To put this more explicitly in process-debunking terms, we can say about these experiments:

1. People's belief that a claim is objective depends, to some extent, on incidental emotions (like disgust).
2. Incidental emotions are epistemically defective processes for coming to belief that a claim is objective.
3. To the extent that people's belief that P depends on epistemically defective processes (for coming to believe that P), they are not justified in believing that P.

C. To some extent, people are not justified in their belief that a claim is objective.

This is a decent argument. It's valid and the premises are all plausible. Of course, it's explicitly limited in that it only speaks about the experiments, and one might wonder how much it generalizes outside of the experimental context. But that is a common question for the generalizability of social science. If the findings do generalize widely, then it seems that we might have a conclusion of considerable philosophical interest. Depending on how much of everyday belief in objectivity is driven by incidental emotions, we might come to think that our folk intuitions of moral objectivity should carry no epistemic authority.

4.3.3 Limitations

I focus on the study by Cameron and colleagues because it has a clean experimental design, and it provides some of the most direct evidence for the role of an epistemically defective affective process in judgments of objectivity. One limitation, as we saw, is the familiar issue of generalizability. But there is another much more pressing limitation of the study. Process-debunking arguments are qualified by the extent that the belief is influenced by the defective process. And when we look at the amount of influence of incidental emotions in this study, we find that the extent of influence is, for debunking purposes, trivial. The mean difference in objectivity judgments produced by inducing disgust was only 0.1 on a five-point scale.[5] Thus,

[5] More generally, it turns out that the impact of occurrent emotion on moral judgment is quite weak (e.g., Landy and Goodwin 2015; May 2014).

when we look at the conclusion of the preceding debunking argument, it should be recast thusly: "To some slight extent, people are not justified in their belief that a claim is objective." To be sure we cannot take these results to show that people's belief in moral objectivity is largely based on a defective process. The results simply don't explain much of why people believe in moral objectivity. As a result, they don't do much by way of debunking the belief.

4.4 PROCESS VINDICATION

Most work on genetic arguments has concerned debunking (e.g., Kahane 2011; Wielenberg 2010; Avnur and Scott-Kakures 2015; Street 2006). However, a new strand of work has been exploring the positive side of genesis, using cognitive science to vindicate belief (e.g., Nichols et al. 2016; Railton 2014). In its roughest form, the basic idea of process vindication is just that beliefs are vindicated to the extent that they are based on rational processes. A bit more fully, the process vindication claim runs as follows:

If process Q is a rational basis for coming to believe that P, then insofar as S believes that P as a result of process Q, S is (prima facie pro tanto) rational in believing that P.

The consequent is doubly hedged. Belief formation can draw on a wide variety of evidence and processes. As a result, even if we know that there is *a* rational process that contributes to a belief, it doesn't follow that the belief is all-things-considered rational. For one thing, even if process Q is a rational consideration in favor of P, S might have other evidence against P, so we can only say to the extent that S's belief that P is based on process Q, the belief is pro tanto rational. In addition, the pro tanto contribution of process Q might be defeated (not merely outweighed) by other evidence that S has and thus S's belief is only prima facie (pro tanto) rational. Process vindication arguments, as I've characterized them, are thus modest in that they do not pretend to conclude that the targeted belief is all-things-considered rational, but only that the belief is prima facie pro tanto rational.

To articulate process vindication arguments, two issues need to be explored in more detail: What is a process? and What makes a process rational?

4.4.1 Processes: Computational and Algorithmic

To explain the operative notion of process, we can draw on Marr's influential account of levels of analysis. Marr explains levels of analysis in terms of their guiding questions. The first level is the *computational* level, and its basic questions are "what is the goal of the computation, why is it appropriate, and what is the logic of the strategy by which it can be carried out?" At the second level, the *algorithmic* one, the basic questions are "what is the representation for the input and output, and what is the algorithm for the transformation?" (Marr 1982, 25).

Marr illustrates these levels of analysis with the example of a lowly cash register. At the computational level, the process we find in a cash register is addition, which is a function that takes two numbers as input and produces one number, in ways specified by an arithmetic theory (Marr 22). Addition is *what* the cash register does, and we characterize this performance "as a mapping from one kind of information to another" (24). The computational level also concerns the question, *why* is addition used for this device, rather than, say, division or multiplication. Here, Marr says, we can draw on our intuitions about what is appropriate for the context. In the context of exchanging groceries for money, it seems most appropriate to sum the costs of the items purchased (22–23). To determine the *why* of psychological processes, we might draw on other resources as well, of course. Evolutionary psychologists emphasize adaptationist considerations to explain why a certain process is appropriate for a context (Cosmides and Tooby 1995, 1202).[6]

Marr's second level of analysis involves the actual representations and algorithms of the process (23). The computation of addition can be carried out in different ways. One might use different kinds of symbols to represent numbers, e.g., binary, Arabic, or hash marks. And the kinds of algorithms one uses will be constrained by the kinds of representations in play. With hash marks, one can use concatenation for addition (|||| + ||| = |||||||), but obviously this would be a disaster for Arabic numerals (4 + 3 = 43). Even if the representations are fixed as, say, Arabic numerals, different algorithms can be used to carry out addition. One common algorithm for addition mirrors the "carrying" algorithm people learn in grade school—add the least significant digits, carry if necessary, and repeat. Another algorithm for addition uses "partial sums," separately adding the one place-value column, the 10 place-value column, the 100 place-value column, and then summing these partial sums.

4.4.2 Rationality

Next, we need to say something about what makes a process rational. For present purposes, I'll work within the dominant framework in analytic epistemology, evidentialism. According to evidentialism, S's belief is rational or justified just in case it is supported by S's evidence. In addition to its prominence in analytic epistemology, evidentialism also coheres reasonably well with the notion of rationality

[6] Although Marr doesn't mention it, sometimes the answer to the *why* question will not correspond to what we intuitively find appropriate. Imagine that we observe the cash register receive two inputs—$2 and $3—and it generates $6 as output; then, when another $2 item is input, the output is $12. Eventually it becomes clear that *what* the machine is doing is multiplication. Why it's doing multiplication is not because that's the right process in this context (based on our intuitions or evolution). Perhaps the machine was hacked or perhaps there's a short circuit that remapped + to *.

that anchors discussions in naturalized epistemology. Perhaps the best-known statement of rationality in that literature comes from Ed Stein:

> [T]o be rational is to reason in accordance with principles of reasoning that are based on rules of logic, probability theory and so forth. If the standard picture of reasoning is right, principles of reasoning that are based on such rules are normative principles of reasoning, namely they are the principles we ought to reason in accordance with. (Stein 1996, 4)

On this familiar description, one must reason "in accordance" with the principles of probability theory to be rational.

What exactly is it to reason "in accordance" with the rules of logic and probability theory? Much of the literature on reasoning is conspicuously vague about this (for discussion, see Nichols and Samuels 2017). We can draw on the Marrian distinction of the previous subsection to impose greater specificity.[7] On one construal, for a system to be in accordance with the rules of logic and probability theory is for the process to execute algorithms that encode the rules of logic and probability theory. So, for instance, if modus tollens is a proper rule of logic, and a given process deploys a modus tollens algorithm to transition from (P → Q and ~Q) to ~P, then the process is rational on the algorithmic construal of accordance. In that case, we can say the transition from input to output is *algorithmically rational.*

Another way to understand "accordance" with the rules of logic and probability theory is in terms of the computational level description. On that approach, a process accords with logic and probability theory when *what* the process computes—the input-output profile of the process—corresponds to the function of the relevant logico-probabilistic rules. In that case, we can say that the process is "computationally rational." As we saw previously, the computational level description of a process is neutral about the actual algorithms involved in the transition from input to output.[8]

It will be helpful to introduce a further distinction. A computational theory describes what a process does. When we evaluate whether such a process is rational (whether it accords to the rules of logic and probability theory), we might take a global view of this according to which a process is rational iff the process would respond in ways that conform to logic and probability theory in all cases. A process that has this characteristic can be called *globally computationally rational.* This is a very demanding conception, but one way to determine that a process is globally

[7] This discussion is based on joint work with Richard Samuels.

[8] Note that not all processes characterized at the computational level need conform to logic and probability theory. For instance, the jumping spider has a visual system described at the computational level as having the goal of identifying mates (Marr 1982, 32); but it needn't be the case that the transition from stimulus to perceptual response in the jumping spider conforms to the laws of probability.

computationally rational is by knowing the actual algorithm. For the process of addition, if a system uses either the carrying algorithm or the partial-sum algorithm, it will be a globally computationally rational addition process. Often, however, we won't know whether a given process will produce the correct results across all contexts. As a result, we might take a narrower view and regard a circumscribed set of transitions as rational if those transitions conform to the rules of logic and probability theory, without committing on whether the process will conform to the rules of logic and probability in all cases. Let's call this less demanding notion *local computational rationality*. Local computational rationality will come in degrees, but to see how it can come apart from global computational rationality, consider a trip odometer that only has a three-digit span (for reasons of layout efficiency on dashboards). The odometer adds a unit every mile. However, when the miles reach 999, with the next mile, the odometer produces 000. The odometer doesn't store the fact that this "000" is really "1,000," so the algorithm clearly produces the wrong arithmetic result. The trip odometer is thus not globally computationally rational. However, in many contexts, the transitions of the odometer will conform to the process of addition. When we start a trip and reset the odometer, the odometer will conform to the addition function for hundreds of units. So long as we focus on those (sub-1,000) transitions, we can say that the system is locally computationally rational. So even though the trip odometer is not globally computationally rational, in a wide range of cases it is locally computationally rational.[9]

In sum, a process is algorithmically rational when the actual algorithms encode laws of logic and probability theory. A process is globally computationally rational when the input-output profile conforms to the laws of logic and probability theory across all actual and counterfactual circumstances. And a process is locally computationally rational when some circumscribed input-output profile conforms to the rules of logic and probability theory.[10]

4.4.3 Rational Processes in Cognitive Science

Since the 1970s, one prominent line of thought in psychology holds that people are surprisingly bad at statistical reasoning. For instance, people seem to neglect the prior probabilities when making judgments about likely outcomes (e.g., Kahneman

[9] As noted previously, the trip odometers have a small digit span for reasons of layout efficiency. There is a more general point, though, about trade-off between local constraints, the purpose of the process, the likelihood of the error-triggering input, and so on. When processing resources are scarce and the likelihood of the error-ridden input is low, it might be appropriate for a programmer (or a learner) to opt for a quick and dirty algorithm that will typically be locally computationally rational.

[10] Note that we don't need to qualify a Marrian rational process as pro tanto rational. This is because computational rationality and algorithmic rationality have the pro tanto built in since they are defined narrowly in terms of the function of the process. For example, the algorithm for addition is defined in terms of a restricted class of inputs and outputs and the dedicated transitions between them.

and Tversky 1973). More recently, however, a wide range of cognitive phenomena have been promoted as instances of rational inference, including categorization (Smith et al. 2002; Kemp et al. 2007), word learning (Xu and Tenenbaum 2007), and parsing (Gibson et al. 2013). In addition, over the last decade, work in developmental and cognitive psychology suggests that children actually have an early facility with statistical reasoning. I'll present two sets of findings from this emerging developmental research.

It is normatively appropriate to draw inferences from samples to populations when samples are randomly drawn from the population, but typically not otherwise. To see whether children appreciated this aspect of statistical inference, Xu and Garcia (2008) showed infants a person pulling four red ping-pong balls and one white one from a box without looking in the box. In that case, it's statistically appropriate to infer that the box has mostly red balls. In keeping with this, when infants were then shown the contents of the box, they looked longer when the box contained mostly white balls than when the box contained mostly red balls. Xu and Denison (2009) then investigated whether the random sampling made a difference. At the beginning of the task, the experimenter first shows a preference for selecting red balls over white ones. Then the experimenter selects balls from an opaque container (as in Xu and Garcia), but in one case the experimenter is blindfolded and in the other she has visual access to the contents of the box. They found that the children were more likely to expect that the population would resemble the sample in the blindfolded condition. Building on these findings, Kushnir and colleagues found that children use sampling considerations to draw inferences about preferences. When a puppet took five toy frogs from a population with few frogs, the children tended to think the puppet preferred toy frogs; but children tended not to make this inference when a puppet took five toy frogs from a population that consisted entirely of frogs (Kushnir et al. 2010).

As a second example, consider another feature of good probabilistic reasoning: If you have priors (e.g., you know the percentage of the population with a certain disease) then you should use them (e.g., in making inferences from symptoms to the presence of the disease); furthermore, if you get new information, you should update the priors. Girotto and Gonzalez (2008) explored this in children using a task with chips of different shapes and colors. Figure 4.1 depicts a stylized version of the materials—four black circles, three white squares, and one black square.

In the "prior" task, kids were shown just the square chips and told, "I'm going to put all the chips in this bag. I will shake the bag and I will take a chip from it without looking." Then they were told that if the chip was black they would give a chocolate to Mr. Black (a puppet) and if the chip was white it would go to Mr. White (another puppet). Then they were asked which puppet they chose to be. Grade school children correctly chose the puppet most likely to win the chocolate (by fourth grade, at 91%) (332). Girotto and Gonzalez explored whether the child can also update in this kind of task. The child was shown all eight chips and asked which is more likely

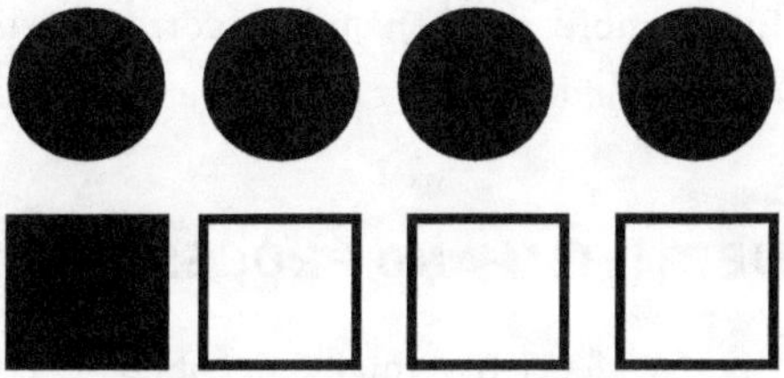

FIGURE 4.1 Features of items in the chips task
Source: Adapted from Girotto and Gonzalez 2008, 328.

to win (with black advantage 5:3). Children tend to say correctly that black is more likely to win. Then all eight chips are put in the bag and the experimenter reaches in. He says, "Ah, listen. I'm touching the chip that I have drawn and now I know something that might help you to win the game. I'm touching the chip that I have in my hand and I feel that it is [a square]" (331). The kids are allowed to revise their judgment, and they tend to answer correctly that now it's more likely that white will win (334). Note that children succeed on these tasks with no training—they produce the correct response immediately. Subsequent work found that children and adults in two preliterate Mayan groups also succeed on this kind of task (Fontanari et al. 2014).

These results, and many others like them, are cited by theorists defending a Bayesian approach to human learning and inference (e.g., Perfors et al. 2011). And the emphasis on the rational nature of the process is quite explicit. Amy Perfors and colleagues write:

> Bayesian probability theory is not simply a set of ad hoc rules useful for manipulating and evaluating statistical information: it is also the set of unique, consistent rules for conducting plausible inference. . . . In essence, it is an extension of deductive logic to the case where propositions have degrees of truth or falsity. . . . Just as formal logic describes a deductively correct way of thinking, Bayesian probability theory describes an inductively correct way of thinking. (Perfors et al. 2011)

It remains unclear whether the child is using a Bayesian algorithm to solve these tasks (see, e.g., Nichols and Samuels 2017). However, it is pretty clear that the inferences in the preceding tasks are, at a minimum, locally computationally rational.[11] That is, the child makes inferences that are appropriate given the evidence presented to her. For instance, in the chips task, the child is given information about the prior distribution of black and white chips, and she correctly determines which chip is more

[11] Indeed, much recent work on Bayesian learning is quite explicitly intended to provide analyses at Marr's computational level, not the algorithmic level (e.g., Denison et al. 2013; Xu and Tenenbaum 2007, 270).

likely to be selected; furthermore, she then updates correctly when new information is provided. The child arrives at the correct belief given the evidence she has.

4.5 MORAL OBJECTIVISM AND PROCESS VINDICATION

In section 4.3, I presented evidence that might contribute to a debunking argument regarding the belief in moral objectivism. I noted that the evidence is far from decisive, and now I will argue that there is also a case to be made for a process vindication of the belief in moral objectivism.

Perhaps the most surprising result in the recent work on moral objectivism is that there is variation in views about objectivism *within the ethical domain* (Goodwin and Darley 2008). People are less likely to give objectivist responses concerning abortion than they are for bank robbery, even though they identify both as *moral* (Wright et al. 2013). This variation in objectivist responses turns out to be correlated with people's estimates of consensus surrounding the issue. When there is high consensus on a moral issue, people are more likely to give objectivist responses concerning that issue; when there is low consensus on an issue, people are more likely to give relativist responses concerning the issue. The relation between consensus and objectivity judgments will provide the basis for a process vindication argument.

4.5.1 Rationality and Consensus

Most of the work on the relation between perceived objectivity and perceived consensus is correlational. But there is a smattering of evidence that at least sometimes perceived consensus drives objectivity judgments (Goodwin and Darley 2012; Ayars and Nichols, forthcoming). For instance, people use consensus information to make inferences about objectivity even for nonmoral cases (like color). When there is high consensus, people tend to think that there is only one right answer, but when there is low consensus, people tend to think that two people who make different judgments can both be right.

What I want to suggest is that using consensus as evidence concerning objectivity is rational. The appeal to consensus as evidence is an ancient idea, prominently used in the common consent (*consensus gentium*) argument for the existence of God (Kelly 2011, 142; Zagzebski 2011, 34), which is found as early as Plato's *Laws*. Cleinias adverts to "the fact that all Hellenes and barbarians believe in [gods]" as an argument for the existence of the gods (*The Laws*, book X). The core thought behind the argument, as Thomas Kelly notes, is that "the fact that theistic belief is widespread among the human population is itself a significant piece of evidence that God exists" (Kelly 2011, 136). The common consent argument moves from a descriptive fact about theistic belief (its ubiquity) to a metaphysical conclusion (God's existence). The idea that consensus often provides evidence is clear from simple examples. When a schoolchild at recess sees lots of kids funneling back into the school, the child can rightly take his peers' behavior as evidence that recess is ending. Similarly, the common consent argument says that the fact that most people believe in God provides evidence for the first-order thesis that God exists.[12] Consensus can also provide evidence, I'll suggest, for second-order theses about whether a true claim is objectively true. To see how consensus applies to second-order theses about objectivity and relativism, let's return to probabilistic inference.

Hypotheses are penalized for flexibility. If we are selecting between two hypotheses that vary in their flexibility, then (provided that we have no prior biases and that the less flexible hypothesis fits the data reasonably well) we should prefer the less flexible hypothesis. But we must also be sensitive to how well the data fit the hypothesis. For illustration, imagine you're a physician trying to determine which disease is present in your community. You're deciding between two hypotheses, each with a prior probability of .5.

H_1: The disease present in the community is M, which typically produces high fever and doesn't produce any other symptoms.

H_2: The disease present in the community is D, which typically produces either high fever or a sore throat, but never both. There is no prior reason to think that a person with D will be more likely to have the fever or the sore throat.

Now imagine that twenty patients have come to see you. Patients 1, 2, 4, 6, 8, 11, 12, 14, 16, 17, and 20 have a high fever and patients 3, 5, 7, 9, 10, 13, 15, 18, 19 have a sore throat. H_1 does a poor job of fitting the data—it has to dismiss almost half of the datapoints as noise. By contrast, H_2 can fit the data perfectly since each patient has one of the possible symptoms produced by disease D. Now imagine instead that among the twenty patients who come to see you, patients 1–3 and 5–20 have a high fever and patient 4 has a sore throat. H_2 can again fit the data perfectly, for one can simply say that patient 4 has the sore throat symptom of D and all the rest of have the fever symptom of D. But at this point, it should seem like things are a bit too easy for H_2. No matter what the distribution of fever/throat symptoms, H_2 can perfectly fit the data. H_2 would fit the data if patient 5 had the sore throat instead of patient 4. H_2 can fit the data if patients 2 and 6 have sore throats, or if 3 and 4 have sore throats. *Any* pattern of symptoms with fevers or sore throats can be fitted by H_2. As a result, H_2 spreads out the probability mass more thinly than does H_1. By contrast, H_1 makes a much less flexible prediction: it predicts lots of high fevers and nothing else. The great flexibility of H_2 can make it a worse explanation of the 19:1 data than

[12] This is related to Condorcet's Jury Theorem, which shows, under some strong assumptions, that a majority opinion will tend to be correct. However, the Jury Theorem, like *consensus gentium*, focuses on first-order claims; I'll argue that consensus also provides evidence for second-order claims concerning objectivism and relativism.

H_1, and this holds despite the fact that H_2 fits the data better.[13] H_1 does have to attribute some of the data, namely patients with sore throats, to noise (e.g., perhaps the sore throats were misreported or were caused by some environmental irritant). But a highly flexible hypothesis that can fit the data perfectly can be less probable than an inflexible hypothesis that has to attribute some data to noise. Another way to put the point is that the flexibility of H_2 means that it can overfit the data, which would be revealed by poor performance in predicting future symptom patterns. We would do well to take this into account when deciding on disease treatment. In hypothesis selection, we want a hypothesis that is sensitive to both fit and flexibility—we seek a hypothesis that fits the data well without being overly flexible (see, e.g., Perfors et al. 2011, 310–12).[14]

Using consensus to decide between objectivist and nonobjectivist hypotheses makes rational sense based on the foregoing considerations about hypothesis selection. Imagine two hypotheses, each with a prior probability of .5, that purport to explain people's judgments about whether P holds:

H_3: There is a single fact about whether P, and this fact (partly) explains the pattern of people's judgments.

H_4: There is no single fact about whether P; rather, whether P holds is relative to context or culture, and this relativity (partly) explains the pattern of people's judgments.

H_3 would correspond to a "naive Bayes" model, which has a single parent node connected to several child nodes (e.g., figures 4.2a and 4.2d). H_4 would correspond to a relativist "multiple fact" model (e.g., figures 4.2b and 4.2c). On the multiple fact model, one parent node might represent that P is a fact (in some contexts), while the other parent node represents that ~P is a fact (in some other contexts). Those who affirm P (represented by the child nodes connected to the P parent node) do so because P is a fact in the relevant context; those who affirm ~P do so because ~P is a fact in the relevant context. Now, if there is low consensus in judgments about whether P holds, the fit between the data and a single fact is poor (figure 4.2a); in that case, appeal to a single fact is a worse explanation of the data since the appeal to multiple facts provides a much better fit (figure 4.2b). When there is high consensus in judging that P, a relativist account can also fit these data (figure 4.2c). But the relativistic model is flexible enough to accommodate *any* pattern of consensus; and, as noted previously, such flexibility is a theoretical cost. As a result, when there is high

[13] Of course, this conclusion will be affected by the likelihood of finding a sore throat among individuals without disease D. So, if it turns out that sore throats are extremely rare except among patients with disease D, then this will offset the penalty to some extent.

[14] Recent work indicates that lay people selecting between hypotheses do seem to penalize hypotheses for flexibility (Blanchard et al. 2018).

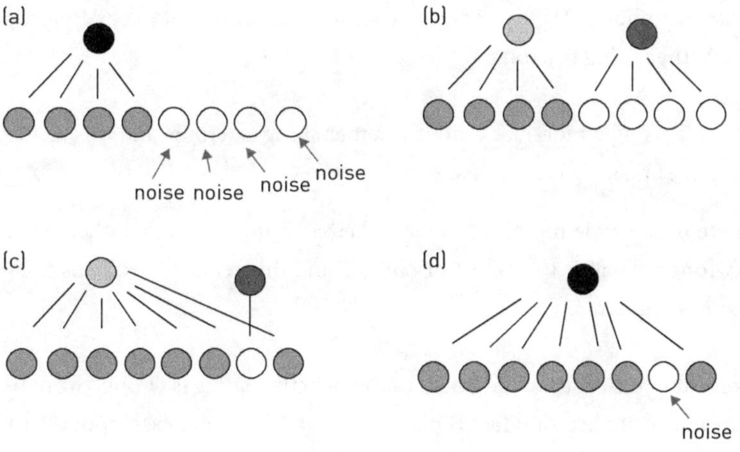

FIGURE 4.2 Objectivist and relativist models for different patterns of consensus

consensus, appealing to a single fact (with a bit of noise) provides a more probable explanation of the data (figure 4.2d).

To make this concrete, imagine a child learning about the seasons. She learns that almost everyone around the world thinks that summer is the hottest season. The consensus surrounding this claim provides reason to think that there is a single fact, viz., that summer is the hottest season. The child also learns that while a large percentage of the global population thinks that July is a summer month, another large percentage thinks that July is not a summer month. The broad divergence of opinion is evidence in favor of the relativist model (figure 4.2b). Importantly, the distribution of opinion here is evidence for two different claims. It is evidence for the second-order thesis that there is no single fact about whether July is a summer month. This parallels the metaethical relativist claim that there is no single fact of the matter. Such a relativist view can be neutral about what facts there are. But the distribution of responses is also evidence for a first-order thesis that in some contexts, July is a summer month while in other contexts, July is not a summer month. This parallels a normative relativist claim that, e.g., polygamy is wrong in some contexts but not in others.[15]

Returning to moral psychology, as we've seen, when perceived consensus surrounding a moral claim is high, people tend to think that the claim is objectively true, and when consensus is low, people tend to think that the claim is only relatively true—that people who make different judgments about the claim could both be right (e.g., Goodwin and Darley 2010, 2012; Wright et al. 2014). As in the preceding case, it is rationally appropriate to use consensus to decide between objectivist and

[15] The first- and second-order claims are related in at least the following way. If in some contexts it's a first-order fact that *July is a summer month* is true and in other contexts it's a first-order fact that *July is a summer month* is false, the second-order thesis follows: there is no single fact about whether July is a summer month.

relativist hypotheses. We can adjust hypotheses H_3 and H_4 to concern judgments about whether φ-ing is wrong:

H_5: There is a single moral fact about whether φ-ing is wrong, and this fact explains the pattern of people's judgments.

H_6: There is no single moral fact about whether φ-ing is wrong; rather, whether φ-ing is wrong is relative to context or culture, and this relativity explains the pattern of people's judgments.

If there is low consensus in judgments about whether φ-ing is wrong, the fit between the data and a single moral fact is poor (figure 4.2a); in that case, appeal to a single moral fact is not a good explanation of the data. The relativist model (H_6) provides a much better fit, as in figure 4.2b. Again, when there is high consensus in judging that φ-ing is wrong, a relativist account can fit these data (figure 4.2c), but that is because of the great flexibility of the relativist model.[16] As a result, when there is high consensus, the objectivist model (H_5) provides a more probable explanation of the data (figure 4.2d).

Thus, if an agent is deciding between an objectivist and a relativist hypothesis, consensus can provide evidence that bears on the decision. If there is high consensus, this favors the objectivist hypothesis since the relativist hypothesis needs to be penalized for its flexibility. But if there is sufficiently low consensus, this favors the relativist model since the objectivist model would need to posit a lot of arbitrary noise.

4.5.2 Substantive Assumptions

Now for the caveats. The foregoing inference depends on strong substantive assumptions. I will discuss three of these.

4.5.2.1 Hypothesis Space

We have assumed that the agent is deciding between two hypotheses, objectivism and relativism, and that the priors for these hypotheses are equal. Thus, the hypothesis space excludes, inter alia, error-theoretic and noncognitivist views on which there are no moral facts.

[16] The relativist can reduce the flexibility of the theory by imposing constraints on how groups can be formed. If the relativist hypothesis groups people purely based on whether they agree with a claim, then the hypothesis will be completely flexible (and accordingly heavily penalized). But a relativist might avoid total flexibility by allowing only natural groupings, e.g., based on community or geography.

In addition, I've been working with an *attitude-independent* notion of objectivism on which objective moral claims describe facts or properties that are independent of anyone's feelings or attitudes about the claims (Finlay 2007, 821). Attitude-independent objectivists can point to a property in the world and maintain that, provided people (or aliens) are sufficiently smart and attentive to the features of the property, there should be convergence on moral claims regarding those properties (for discussion, see Doris and Plakias 2007).[17]

"Objectivism" is sometimes used more capaciously simply for the view that moral claims are made true by moral facts, where those facts might depend on human attitudes (see, e.g., Smith 1993, 399–400).[18] Although I focused on attitude-independent notions of objectivism previously, consensus might also be treated as evidence for attitude-dependent objective facts. One prominent class of attitude-dependent views holds that the objective moral facts concern the attitudes of a fully informed rational individual; for instance, according to Michael Smith, moral facts are facts about what an ideally informed and rational agent would desire, given the desires she has (Smith 1994). Smith himself quite explicitly takes convergence to support such a view: "the empirical fact that moral argument tends to elicit the agreement of our fellows gives us reason to believe that there will be a convergence in our desires under conditions of full rationality. For the best explanation of that tendency is our convergence upon a set of extremely un-obvious a priori moral truths" (1994, 187).

A second class of objectivist attitude-dependent views appeals to *anthropocentric fittingness*. On D'Arms and Jacobson's version of this view, the truth of evaluative judgments depends on the propriety of species-typical affective responses (e.g., D'Arms and Jacobson 2006). They illustrate this with a humbler evaluative concept, *the funny* (see also Shoemaker 2017). On a fittingness account of the funny, what makes something funny involves a species-typical (and perhaps species-specific) response—amusement. Something is funny when amusement is a fitting or appropriate response. Amusement, like many other evaluative reactions, is a pancultural response (Shoemaker 2017, 488; see also D'Arms and Jacobson 2006, 100).

[17] Although this is a natural prediction of objectivism, objectivists sometimes downplay this prediction (e.g., Bloomfield 2001; Shafer-Landau 2003). This reluctance to own the prediction seems largely driven by the desire to evade the argument from disagreement (e.g., Doris and Plakias 2007). To be sure, the prediction of consensus is defeasible—where the moral truths are especially subtle, people might fail to recognize them. But if objectivists maintain that there are moral facts that don't demand great insight or intelligence, we should expect consensus surrounding those facts (see, e.g., Bloomfield 2001, 89).

[18] Smith writes: "we seem to think moral questions have correct answers, that the correct answers are made correct by objective moral facts. . . . The term 'objective' here simply signifies the possibility of a convergence in moral views of the kind just mentioned" (399–400). Finlay calls this more inclusive category "ontological realism" (Finlay 2007, 825) and reserves "objectivism" for attitude-independent forms of ontological realism (2007, 829).

Fittingness theorists haven't explicitly predicted consensus, but it's a natural consequence of the anthropocentric view. Sticking with the funny, if we assume that people are generally accurate at recognizing whether or not amusement is fitting, then if there is high consensus about whether a joke is funny, that is good evidence that amusement is a fitting response to the joke. So if almost everyone thinks a certain joke is funny, and a few people think otherwise, the fittingness view suggests that the outliers are likely exhibiting some kind of performance error—perhaps they don't understand the joke, are in a bad mood, or are subject to some other interfering factor.

Although objectivists of all stripes celebrate high consensus as evidence of objectivism, they are less inclined to use low consensus as evidence of relativism. Even attitude-dependent objectivists tend to resist relativist treatments quite generally, apparently for both the moral and the funny (see, e.g., Shoemaker 2017, 512). However, if an agent is deciding between a relativist and an objectivist model of some evaluative claim, hypothesis selection considerations can provide reason to favor relativism under low consensus. Consider for instance, humor that doesn't inspire wide consensus, like that displayed in the MTV show *Jackass*, which often has extremely disgusting stunts. Many people judge stunts in *Jackass* to be funny; many others don't. Suppose we are deciding between (1) a relativist model which says that *Jackass* is *funny for some people* and not funny for others and (2) an objectivist model which says that *Jackass* really is funny (or really not funny) and those who disagree are mistaken. Given the wide division of opinion on the funniness of *Jackass*, hypothesis selection would favor relativism about the funniness of *Jackass*; otherwise the fit to data is poor (as in figure 4.2a).[19]

4.5.2.2 Tracking Evaluative Properties

I argued that people are being rational in using consensus as evidence for metaethical claims. This depends on a second substantive assumption, viz., that people must think that other people are tolerably good at tracking the (first-order) evaluative facts. If you don't think people are any good at determining whether something is good or funny or wrong, then you shouldn't take their opinions as evidence of whether evaluative claims are objective or relative. It's only if you think that people are reasonably good at determining what is good, funny, or wrong that you should take consensus as evidence.

Is there reason to think that people *do* expect others to be good at determining evaluative properties? For at least some evaluative properties, the answer seems clearly to be yes, as reflected in the widespread use of customer reviews. People consult customer reviews for a huge range of products and services. And there is

[19] More precisely, it would provide prima facie evidence for a pro tanto reason to favor relativism (see section 4.5.3).

evidence that this has an impact on decisions across several areas including hotels (Ye et al. 2009), restaurants (Zhang et al. 2010), and movies (Chintagunta et al. 2010). Presumably, people take the fact that most people liked the movie as indicating that the movie is good.

What about for something closer to moral properties? Work on conformity indicates that people will use other's actions and normative judgments as evidence about the right thing to do. This is found both in experimental games (Bicchieri and Xiao 2009, 198) and in more natural behaviors like littering and volunteering (Cialdini et al. 1991; Cialdini et al 1999). If a subject thinks that others think that one should split evenly on a dictator game, the subject herself is more likely to think that one should split evenly. An even starker result has emerged in developmental psychology. Knowing that members of a certain group typically eat a certain kind of berry leads kids to say it is "not okay" for a member of the group to eat a different kind of berry (Roberts et al. 2016, 10–11).

4.5.2.3 Independence

Thus, in order for consensus to be evidence, people must be good at descrying evaluative properties. A closely related assumption is that the judgments of individuals must be, to some significant extent, independent.[20] It can't be that everyone is just blindly copying the opinion of one guy. The individuals must provide independent evaluative datapoints. Again, for our purposes, the primary question is whether *people* regard the constituents of consensus (or dissensus) as providing independent evaluations. And, at least in some evaluative domains, it's plausible that people tacitly assume independence. Witness the natural concern about whether consumer reviews are faked by companies. If people learned that all the gushing reviews of some indie movie were generated by one industrious fanboy, this would presumably diminish their trust in the evidentiary force of the reviews. It's harder to know whether people also expect others' moral judgments to be largely independent. It is plausible that people do make independent judgments for some morally evaluative properties, like the propriety of guilt or resentment. But it is currently unclear exactly what the contours are of people's assumptions about the independence of those judgments.

4.5.3 Process Vindication for Moral Objectivism

The case for rational metaethics thus depends on some significant assumptions. But with those assumptions in place, an agent is right to use high consensus as evidence of objectivism and low consensus as evidence of relativism. And, as noted, recent work indicates that participants do in fact respond in this way—when presented

[20] Cf. Condorcet's Jury Theorem.

with high consensus regarding a claim, participants move towards objectivism for the claim, and when presented with low consensus, they move towards relativism (Ayars and Nichols, forthcoming). We don't know what algorithm people are using to make these inferences. But given the preceding assumptions, the work on participants' judgments indicates that people are using a process that conforms to the laws of probability.

The vindication here is still muted, however. For flexibility and goodness-of-fit do not exhaust the factors relevant for evaluating a metaethical thesis. For instance, many philosophers take a monistic view of metaethics, holding either that morality is entirely objective or entirely not (but see Gill 2009; Doris and Plakias 2007, 309). Such a monistic commitment would weigh against taking the kind of bifurcated view suggested by considerations of consensus and hypothesis selection. Thus, I am not claiming that consensus gives us an all-things-considered reason to be relativist or objectivist about moral claims. But this is consistent with a process vindication argument, since the conclusion of such an argument is, as noted in section 4.4, importantly qualified. The most we can say is that given the assumptions previously described (section 4.5.2), it is prima facie pro tanto rational for S to believe that it is objectively true that it's wrong to φ on the basis of high consensus that it's wrong to φ. While this argument is qualified in many ways, it is optimistic about the prospect that rational processes play a significant role in generating metaethical beliefs.

4.6 CONCLUSION

The factors that contribute to moral judgment are numerous and varied. Emotions, motivation, heuristics, and rational inference all play a role in moral judgment. This holds for both first-order moral judgment and metaethical judgments. It remains unclear exactly how these factors conspire to generate moral judgment. But as cognitive science presents us with a sharper picture of the processes, we will be in a better position to draw epistemic conclusions about the justificatory status of moral beliefs. A prevailing trend in cognitive science has emphasized the rational shortcomings of humans, which has been recruited in arguments that aim to debunk the process of moral belief formation. Recent work provides a more optimistic view of human cognition, based on statistical learning. I've argued here that this optimistic view of human cognition also plausibly extends to metaethical judgment, and might (partially) vindicate metaethical judgment rather than debunk it.

ACKNOWLEDGMENTS

Thanks to Alisabeth Ayars, Thomas Blanchard, Justin D'Arms, John Doris, Michael Gill, Dan Jacobson, Rachana Kamtekar, Victor Kumar, Jonathan Livengood, Connie Rosati, Richard Samuels, David Shoemaker, David Sobel, Tamler Sommers, and Bas

Van Der Vossen. Research for this essay was supported by Office of Naval Research grant #11492159.

REFERENCES

Avnur, Y., and Scott-Kakures, D. (2015). How irrelevant influences bias belief. *Philosophical Perspectives 29* (1): 7–39.

Ayars, A., and Nichols, S. (2017). Moral empiricism and the bias for act-based rules. *Cognition 167*: 11–24.

Ayars, A., and Nichols, S. (forthcoming). Rational learners and metaethics: Universalism, relativism, and evidence from consensus. *Mind & Language.*

Blanchard, T., Lombrozo, T., and Nichols, S. (2018). Bayesian Occam's razor is a razor of the people. *Cognitive Science 42* (4): 1345–59.

Bloomfield, P. (2001). *Moral Reality*. New York: Oxford University Press.

Cameron, C. D., Payne, B. K., and Doris, J. M. (2013). Morality in high definition: Emotion differentiation calibrates the influence of incidental disgust on moral judgments. *Journal of Experimental Social Psychology 49* (4): 719–25.

Chintagunta, P. K., Gopinath, S., and Venkataraman, S. (2010). The effects of online user reviews on movie box office performance. *Marketing Science 29* (5): 944–57.

Cialdini, R. B., Wosinska, W., Barrett, D. W., Butner, J., and Gornik-Durose, M. (1999). Compliance with a request in two cultures. *Personality and Social Psychology Bulletin 25*: 1242–53.

Cosmides, L., and Tooby, J. (1995). From function to structure. In M. Gazzaniga, ed., *The COGNITIVE neurosciences*. Cambridge, MA: MIT Press, 1199–210.

D'Arms, J., and Jacobson, D. (2006). Anthropocentric constraints on human value. In R. Shafer-Landau, ed., *Oxford Studies in Metaethics*, vol. 1. New York: Oxford University Press, 99–126.

Denison, S., Bonawitz, E., Gopnik, A., and Griffiths, T. L. (2013). Rational variability in children's causal inferences. *Cognition 126* (2): 285–300.

Doris, J. M. (2015). *Talking to Our Selves*. New York: Oxford University Press.

Doris, J. M., and Plakias, A. (2008). How to argue about moral disagreement. In W. Sinnott-Armstrong, ed., *Moral psychology*, vol. 2. Cambridge, MA: MIT Press, 303–31.

Finlay, S. (2007). Four faces of moral realism. *Philosophy Compass 2* (6): 820–49.

Fontanari, L., Gonzalez, M., Vallortigara, G., and Girotto, V. (2014). Probabilistic cognition in two indigenous Mayan groups. *Proceedings of the National Academy of Sciences 111* (48): 17075–80.

Freud, S. ([1927] 1961). *The Future of an Illusion*. Translated by J. Strachey. New York: Norton.

Gibson, E., Bergen, L., and Piantadosi, S. T. (2013). Rational integration of noisy evidence and prior semantic expectations in sentence interpretation. *Proceedings of the National Academy of Sciences 110* (20): 8051–56.

Gill, M. B. (2009). Indeterminacy and variability in meta-ethics. *Philosophical Studies 145* (2): 215–34.

Girotto, V., and Gonzalez, M. (2008). Children's understanding of posterior probability. *Cognition 106* (1): 325–44.

Goldman, A. (1979). What is justified belief? In G. Pappas, ed., *Justification and Knowledge*. Dordrecht: Reidel, 1–23.

Goodwin, G., and Darley, J. (2008). The psychology of meta-ethics. *Cognition 106*: 1339–66.

Goodwin, G., and Darley, J. (2012). Why are some moral beliefs perceived to be more objective than others? *Journal of Experimental Social Psychology 48*: 250–56.

Ichikawa, J. J., and Steup, M. (2016). The analysis of knowledge. In E. N. Zalta, ed., *Stanford Encyclopedia of Philosophy*. Winter 2016 ed.

Kahneman, D., and Tversky, A. (1973). On the psychology of prediction. *Psychological Review 80*: 237–51.

Kelly, T. (2011). Consensus gentium. In K. Clark and R. Vanarragon, eds., *Evidence and Religious Belief*. New York: Oxford University Press, 135–56.

Kemp, C., Perfors, A., and Tenenbaum, J. B. (2007). Learning overhypotheses with hierarchical Bayesian models. *Developmental Science 10* (3): 307–21.

Kunda, Z. (1999). *Social Cognition: Making Sense of People*. Cambridge, MA: MIT Press.

Kushnir, T., Xu, F., and Wellman, H. M. (2010). Young children use statistical sampling to infer the preferences of other people. *Psychological Science 21* (8): 1134–40.

Landy, J. F., and Goodwin, G. P. (2015). Does incidental disgust amplify moral judgment? *Perspectives on Psychological Science 10* (4): 518–36.

Marr, D. (1982). *Vision*. Cambridge, MA: MIT Press.

May, J. (2014). Does disgust influence moral judgment? *Australasian Journal of Philosophy 92* (1): 125–41.

Nichols, S. (2004). Is religion what we want? *Journal of Cognition and Culture 4* (2): 347–71.

Nichols, S. (2014). Process debunking and ethics. *Ethics 124*: 727–49.

Nichols, S., Kumar, S., Lopez, S., Ayars, A., and Chan, H. (2016). Rational learners and moral rules. *Mind & Language 31*: 530–54.

Nichols, S., and Samuels, R. (2017). Bayesian psychology and human rationality. In T. Hung and T. Lane, eds., *Rationality: Constraints and Contexts*. London: Elsevier, 17–35.

Norenzayan, A., and Hansen, I. G. (2006). Belief in supernatural agents in the face of death. *Personality and Social Psychology Bulletin 32* (2): 174–87.

Perfors, A., Tenenbaum, J. B., Griffiths, T. L., and Xu, F. (2011). A tutorial introduction to Bayesian models of cognitive development. *Cognition 120* (3): 302–21.

Plantinga, A. (2000). *Warranted Christian Belief*. Oxford: Oxford University Press.

Railton, P. (2014). The affective dog and its rational tale. *Ethics 124* (4): 813–59.

Rose, D., and Nichols, S. (forthcoming). From punishment to objectivity. *Mind & Language*.

Sarkissian, H., Park, J., Tien, D., Wright, J. C., and Knobe, J. (2011). Folk moral relativism. *Mind & Language 26* (4): 482–505.

Schaller. M. (1992). In-group favoritism and statistical reasoning in social inference. *Journal of Personality and Social Psychology 63*: 61–74.

Shoemaker, D. (2017). Response-dependent responsibility or a funny thing happened on the way to blame. *Philosophical Review 126* (4): 481–527.

Smith, L., Jones, S., Landau, B., Gershkoff-Stowe, L., and Samuelson, L. (2002). Object name learning provides on-the-job training for attention. *Psychological Science 13* (1): 13–19.

Smith, M. (1993). Realism. In P. Singer, ed., *A Companion to Ethics*. Cambridge, MA: Blackwell, 399–410.

Smith, M. (1994). *The Moral Problem*. Oxford: Blackwell.

Stein, E. (1996). *Without Good Reason*. Oxford: Clarendon Press.

Street, S. (2006). A Darwinian dilemma for realist theories of value. *Philosophical Studies 127* (1): 109–66.

Wright, J. C., Grandjean, P. T., and McWhite, C. B. (2013). The meta-ethical grounding of our moral beliefs. *Philosophical Psychology 26* (3): 336–61.

Wright, J. C., McWhite, C., and Grandjean, P. (2014). The cognitive mechanisms of intolerance. *Oxford Studies in Experimental Philosophy 1*: 28–61.

Xu, F., and Denison, S. (2009). Statistical inference and sensitivity to sampling in 11-month-old infants. *Cognition 112* (1): 97–104.

Xu, F., and Garcia, V. (2008). Intuitive statistics by 8-month-old infants. *Proceedings of the National Academy of Sciences 105* (13): 5012–15.

Xu, Fei, and Tenenbaum, J. (2007). Word learning as Bayesian inference. *Psychological Review 114* (2): 245–72.

Ye, Q., Law, R., and Gu, B. (2009). The impact of online user reviews on hotel room sales. *International Journal of Hospitality Management 28* (1): 180–82.

Zagzebski, L. (2011). Epistemic self-trust and the consensus gentium argument. In K. Clark and R. Vanarragon, eds., *Evidence and Religious Belief.* New York: Oxford University Press, 135–56.

Zhang, Z., Ye, Q., Law, R., and Li, Y. (2010). The impact of e-word-of-mouth on the online popularity of restaurants. *International Journal of Hospitality Management 29* (4): 694–700.

PART III
GOD AND COGNITIVE SCIENCE

5 On Perceiving God

Prospects for a Cognitive Science of Religious Experience

Mark Baker and Dean Zimmerman

5.1 INTRODUCTION: A CSR ROAD LESS TRAVELED

In this chapter, we consider possible implications of research in cognitive science for perhaps the grandest metaphysical question of all: does God exist? The most obvious subbranch to consider is the relatively new discipline that calls itself the cognitive science of religion (CSR). CSR uses concepts from cognitive science and evolutionary psychology to explain, among other things, why people from essentially all cultures map ordinary human experiences of living in the world onto distinctively religious beliefs, including the belief that some supernatural agent(s) exist and interact causally with human beings (sometimes called the *consensus gentium*; see, for example, Goldman, this volume).

In practice, the explanations that CSR offers for the *consensus gentium* are often taken to reflect negatively on the possibility that the believed-in supernatural agents actually exist, because the beliefs in question are claimed to be spin-offs of cognitive processes that exist for other purposes. However, we wish to put forward the idea that some key features of a comprehensive account of this important explanandum actually make more sense on the hypothesis that (one or more) God(s) do(es) exist. By "god" we mean a supernatural agent that interacts causally with people. By "supernatural" we mean "has a kind of mental and/or causal powers that no other category of thing (human being, animal, artifact, natural object . . .) could have."[1] (See the end of section 5.3 and section 5.5 for more discussion.) In exploring this idea, we also borrow from a more venerable branch of cognitive science: the study of perception.

[1] Notice that this definition of "supernatural" can be understood as relative to the folk taxonomy of any given culture; it need not be our current scientific notion of what falls outside the scope of natural law. (Thanks to Brian McLaughlin for raising this issue.)

Toward this end, we focus on a weak spot in prominent CSR accounts. Although CSR aspires to account for the naturalness of religious belief in a wide range of cultures studied by anthropologists, it has some weaknesses when it comes to explaining the commonness and naturalness of belief in God in at least one culture of note, namely ours: twenty-first-century overeducated North America. Anthropologist T. R. Luhrmann (2012) has studied this exotic culture, specifically a contemporary branch of evangelical Christianity. She observes that, although standard CSR "describes the way our intuitions evolved and explains why claims about invisible agents seem plausible, and why certain ideas about God are found more often in the world than others," it "does not explain how God *remains* real for modern doubters" (xii)—an assessment that rings true from our own informal observations and discussions with religious believers. Their faith is often supported by religious experiences which happen in contexts very different from what CSR accounts suggest—in their bedrooms, rather than in the forest at dusk—and have very different contents from what CSR accounts might suggest. Similarly, Atran (2002, 195) concludes that "We know next to nothing about the neuro-biology of the vast majority of run-of-the-mill religious experiences and beliefs that sustain most people's faith," an assessment that still seems true today. Existing CSR accounts may thus need to be extended in some interesting ways, including ways that bear on whether they should have an undermining effect on religious belief.

More specifically, we are particularly interested here in a class of religious experiences that seem to those who have them to be perceptions of the presence of a deity. These experiences are perception-like in that they have a similar kind of directness and undeniability to the subject, and they can affect his or her beliefs in a similar way. We consider this subtopic out of intellectual curiosity, but also as people friendly to the possibility that theism is true, and that these religious experiences sometimes are what they seem to be. In focusing on this, we do not claim that current projects in CSR explain nothing, but only that there is a road not (yet) traveled: one that uses the tools of cognitive science to try to understand perception-like experiences that play an important role in generating and sustaining religious belief for adherents of Christianity and other contemporary religions.

In particular, from our theism-friendly perspective, there are at least two cognitive scientific hypotheses that could be extremely useful in accounting for perception-like religious experiences that CSR has neglected: the possibility that the concept of God is a simple one (not composed out of other concepts), and the possibility that there are specialized cognitive mechanisms for mapping special experiential states onto beliefs about God in a subpersonal, computational manner.[2] In contrast,

[2] By saying that the cognitive mechanisms in question output beliefs about God, we are (over) simplifying somewhat: more accurately, we should say that they produce *percepts* about God, or *candidate* beliefs about God, or *tendencies* to believe in God. We don't rule out the possibility that committed atheists might have religious experiences that tempt them to believe in God, but they reject the belief as inconsistent with their other, settled beliefs (see Bering 2011 for actual examples).

most existing CSR tries to do without any religion-specific cognitive resources of this sort, holding that the human mind naturally produces religious beliefs in a more or less "accidental" way, perhaps through a Rube Goldberg-y combination of mental tools properly used for other purposes (see McCauley 2011 for one recent example). We, however, are inclined to deny this deep-held assumption. Anyone who takes seriously the possibility that interactions with God have been a factor relevant to people's survival and reproduction throughout human history should find it plausible that human evolved psychology could include cognitive resources specific to the domain of those interactions.

If indeed domain-specific concepts and computations exist underlying some forms of religious experience, this could have two important consequences. First, it could provide the empirical basis for constructing an argument that the best overall explanation for why humans have these cognitive resources is because people have in fact been in causal contact with supernatural agents, a kind of inference to the best explanation. Second, it could bear positively on the epistemic status of the token beliefs that result from having the right sort of religious experiences. These beliefs may well be the outputs of a cognitive subsystem aimed at truth, operating in the sort of environment for which it was selected, not accidental byproducts of mental tools useful mainly for other purposes.

Our discussion develops as follows. First we call attention to a certain class of modern religious experiences and why they seem not to connect so directly with what CSR offers to explain (sections 5.2 and 5.3). Then we consider what would be involved in analyzing these religious experiences as perceptions (section 5.4). Here we explore a line of thought pioneered by Alston (1991), developing it in more current cognitive scientific terms. This leads us to the likely conclusion that this sort of putative perception depends on specialized cognitive resources. However, we do not find it at all incredible that there would be such resources, if indeed God exists and people's interactions with God have impacted their fitness over evolutionary time (section 5.5). We then close with some brief epistemological reflections (section 5.6).

5.2 MODERN RELIGIOUS EXPERIENCE CHARACTERIZED

Let us consider, then, to what extent recent work in CSR sheds light on how ordinary believers, throughout history and into the present day, have come to believe in God, and how they have sustained that belief. What are the most promising directions to look for cognitive scientific tools that might bear upon this sort of belief acquisition and sustenance, as opposed to that of hunter-gathers?

This could be seen as little different from a person seeing a yellow-bellied sapsucker, but nevertheless not believing in the existence of such birds, for one reason or another. We maintain this simplification throughout. (Thanks to Brian McLaughlin for clarifying this point.)

Religious experience is of course not the only source of religious belief; much religious belief is picked up from testimony, broadly construed. One learns about God from "Granny," as Fodor might put it. And indeed some branches of CSR put heavy weight on the mechanisms of cultural transmission (Boyer 2001; Richerson and Newson 2009, applying the framework of Richerson and Boyd 2005; Wilson 2002; Norenzayan 2013). But this should not obscure the fact that virtually every robust religion that has staying power and an ability to renew itself across generations has included a role for some form of religious experience: either dramatic ones in the form of purported revelations experienced by the founders and made known to others via chains of testimony, or ongoing religious experiences that are meant to be available to all serious adherents, or some combination of the two. Mystic strands and subdivisions have existed in most if not all religions, and while the degree of emphasis on them and the range of people they are taken to be available to (everyone or a special class of priests/monks/saints/shamans) may vary significantly, this element of human experience is generally available for religious leaders to appeal to. For example, spiritual training within both Catholic and Protestant traditions typically includes instruction in how to "live in the Spirit" or to "practice the presence of God." Religious testimony frequently goes beyond "Our founder experienced so-and-so" to include "I experienced so-and-so, and if you do such-and-such you may too." Why does that testimonial gambit work for missionaries and revivalists as well as it does? How well does CSR do at explaining what goes on in both the full-blown religious experiences that are reported by many religious people, and the more subtle divine promptings that figure prominently in the spiritual lives of most of today's more devout Christians, for example?

Let us be more precise about the kind of religious experiences that we are most interested in. We focus on the experiences, great or small, which are described in perception-like terms. People "have a vision," although not necessarily with their eyes, they "hear a voice," although not necessarily with their ears, or they report a vivid sense of God's presence that seems direct and undeniable, even though it does not seem analogous to the deliverances of any of the usual senses. Alston (1991) points to a rich literature of Christian mysticism, emphasizing that many religious experiences are simultaneously experiential, direct, and taken to be of God (14ff). Here is one example from the many that he discusses, from James (1902, 67–68).

> All at once I . . . felt the presence of God—I tell of the thing just as I was conscious of it—as if his goodness and his power were penetrating me altogether. . . . Then, slowly, the ecstasy left my heart; that is, I felt that God had withdrawn the communion which he had granted. . . . At bottom the expression most apt to render what I felt is this: God was present, though invisible; he fell under no one of my senses, yet my consciousness perceived him.

Here is another, which emphasizes the fact that such experiences are not always expected or cultivated:

> But as I turned and was about to take a seat by the fire, I received a mighty baptism of the Holy Ghost. Without any expectation of it, without ever having the thought in my mind that there was such a thing for me, without any recollection that I had ever heard the thing mentioned by any person in the world, the Holy Spirit descended upon me in a manner that seemed to go through me, body and soul. I could feel the impression, like a wave of electricity, going through and through me. Indeed, it seemed to come in waves and waves of liquid love. (James 1902, 250; Alston 1991, 14)

In a similar vein, here is a report of the experience of a Christian woman, Sarah, whom Luhrmann met in an evangelical church in California in the 2000s.

> By the time I met her, Sarah was willing to tell me that when she was praying and she felt God's presence moving through her body or in her mind, she felt extremely close. "When I'm ministering to someone [i.e., praying for that person] and I know that the words that are coming out of me aren't mine, and I know that the pictures that I'm seeing aren't mine, there's a real intense closeness and oneness that you feel." . . . She was able to say to me, when I asked her how she was able to relate to a being she could not see with her eyes, that she did feel as if she "saw" him. "I feel like I do see him and I do see his face and I see his hand on what's going on around me. That is part of the experiential thing." (Luhrmann 2012, 97–98)

And again Sarah says:

> Sometimes when I pray, I see his glory. There's what I call the throne room. . . . You can't really exactly see, but it's being in the presence of the Lord. . . . Sometimes I feel like I'm hearing the prayers that have gone before and the prayers that are going on now and the prayers that are to come, and I just sort of join in the chorus. (Luhrmann 2012, 99)

Such religious experiences are no rare or marginal phenomenon. Surveys show that between 25% and 40% of people polled in America and Britain report having had some kind of religious experience in their lives (Atran 2002, 195; David Yaden, personal communication). They are also found in children as well as adults; see Tamminen 1994 on experiences of God's closeness in Finnish children and adolescents, ages seven to twenty.

Indeed, there seems to be a spectrum of religious experiences to consider: some are more intense, some less; some are rarer, some more frequent; some have content more akin to normal perception, some are more internal or ineffable; some are sought via some kind of spiritual discipline, some come upon one unexpectedly; some come to trained religious experts, some to untutored laity, including unbelievers; some are described more eloquently (even paradoxically), some less so. There seems to be a continuum of experience along all these dimensions. For this work, we focus on the less tutored, more spontaneous, more widespread ends of these spectrums, because here the similarities to ordinary perception are greater, and basic human cognitive machinery is more likely to be at work. However, the fact that there is a broad spectrum of experience in no way strains the analogy we develop with ordinary visual perception: any human being has the cognitive machinery to see a yellow-bellied sapsucker, for example, but we are not surprised that expert birdwatchers might recognize them more readily, see them more often, and describe them more accurately. Indeed, plenty of people will never even see a yellow-bellied sapsucker, for one reason or another. Nevertheless, the expert's experience and the lay person's are akin, built on the same cognitive foundations. We want to consider what those foundations might be, in the religious domain.

5.3 DOES CSR ACCOUNT FOR THIS SORT OF RELIGIOUS EXPERIENCE?

We ask, then, does current CSR help us understand experiences like those of Sarah? Two of CSR's most prominent tools for explaining belief in gods are the (hyperactive) agency detection device (HADD) and the theory of mind module/mechanism (ToM) (Guthrie 1993; Atran 2002; Barrett 2004; Bloom 2009; Bering 2011; Shermer 2011; McCauley 2012). These cognitive tools may help explain how mundane events, like glimpsing movement in one's peripheral vision when walking in the woods at dusk, might be mistaken for sensations of supernatural beings. It is much less clear how they illuminate the workings of a sense of God's presence, as reported by people like those quoted previously. Indeed, visions and "inner voices" with significant religious content can and frequently do occur without the interpretation of environmental sounds or sights as carrying divine messages. For these religious experiences, classic (H)ADD/ToM-type stories seem to miss the mark, in terms of both the context of the experience and the content of the experience.

With regard to context, it is true that modern believers might have a religious experience while walking through the woods, and the sense of motion, hence life, all around them as the wind blows in the trees might well have something to do with it, as (H)ADD predicts. But they are just as likely to have a religious experience when praying with their eyes closed in a quiet room, as many Christians are encouraged to do, where there is no startling stimulus at all.[3] If these potentially intense religious

[3] Of course, some modern religious experiences do involve particular kinds of stimulation, provided intentionally by some kind of ritual, ranging from congregational singing at an American church service to intense African or Melonesian initiation rites. The anthropological wing of CSR has more to say about these induced religious experiences, particularly Whitehouse (2000). We do not deny the importance of such experiences, but concentrate on the other sort, as better candidates for perception of God. Indeed, Greeley (1975) identifies music (49%) and attending group services

experiences are due to hyperactive agency detection, then it is agency detection run utterly amok, completely out of touch with the subject's environment.

With regard to content, it is also hard to see how to extend (H)ADD and ToM from their prehistoric role in evolutionarily significant activities like detecting predators and prey to illuminate the kinder, gentler experiences of "sensing God's presence" that people describe. What do the emotions of being startled by what might be a dangerous jaguar have in common with the experience of having "a wave of electricity, going through and through me . . . com[ing] in waves and waves of liquid love," or the experience of "hearing the prayers that have gone before and the prayers that are going on now and the prayers that are to come?" We see essentially no point of commonality here.[4]

Indeed, we find the (H)ADD-ToM-style theories somewhat incomplete in another way: they are a bit fuzzy about where the concept of God as a powerful incorporeal intentional agent comes from in the first place. Suppose that someone indeed detects a potential agent in the woods, but then looks more closely and does not see a normal agent (e.g., a jaguar). Situations like this are supposed to invite the belief that an *invisible* agent is present—some kind of ghost or spirit. But why is the concept of an invisible agent even available to a person who has no prior experience with such agents? Why do people *ever* draw the inference to a radically new category of being, rather than drawing a whole range of more mundane conclusions, such as "Wow, that jaguar is really well camouflaged" or "There goes another false positive from my overactive agency detector" (see also Murray and Goldberg 2009, 189–93)? We can see why the conclusion that there are spirits in the woods is tempting *if* one already has the concept of a spirit in one's cognitive repertoire, but not how the (H) ADD experience is the source of that concept in the first place.

The obvious potential answer to this challenge is to say that the concept of God is complex, combining independently available concepts of "agent" and "invisible." Indeed, it is plausible to typologize concepts as being either simple or complex, where complex concepts are those that are formed compositionally out of simple concepts by conjunction or other syntactic processes.[5] In addition, concepts are presumably either innate, or they are induced somehow out of experience with instances of the concept. This gives the following rough classification of concepts:

(41%) as two of the most important elicitors of religious experience, but private prayer is right up there with them (48%) (Atran 2002, 171).

[4] We acknowledge that Guthrie's (1993) and Barrett's (2004) use of the (H)ADD is broader than the domain of predation. But we think our point can be generalized to other plausible applications, and the domain of predation is particularly interesting because it is so clear why an "unreliable" system that generates many false positives would be useful in this domain.

[5] There may be other, murkier notions of the complexity of a concept that depend somehow on the intrinsic complexity of the things that the concept applies to, but we have only the precise structural-syntactic notion of complexity in mind here.

(1) Simple and innate
(2) Simple and induced by experience
(3) Complex and innate
(4) Complex and induced by experience

Whether all these types actually exist is controversial. Presumably at least (1) and (4) do exist. Whether (3) exists might be questioned; perhaps no theory makes heavy use of this type. Whether (2) is possible, and if so whether such concepts are common relative to (1), is part of what is at stake in debates between rationalists and empiricists.

Where then would the concept of God plausibly fit into this typology, according to CSR? The strain of CSR research that is most directly relevant to this is Pascal Boyer's (1994, 2001, 2003) influential notion of minimally counterintuitive concepts, adopted also by Barrett (2004) and Atran (2002), among others. Boyer's idea is that supernatural concepts come from recombining features that initially belong to different natural domains of knowledge, so as to create new combinations like "invisible agent." For example, one might import the feature "perceiver" from the cognitive domain of persons into the cognitive domain of plants to get a concept like ebony trees that record what happens around them (a belief of the Uduk people of the Sudan). In a broader sense, Paul Bloom's (2004) notion that beliefs about souls and spirits come from mismatches between our innate categories of "agent" and "physical object" can be seen as a hypothesis of this same general type, building new concepts and beliefs out of resources taken from two different cognitive domains. So too can invocations of Deborah Kelemen's (2004; Kelemen et al. 2005) work on "promiscuous teleology," where certain natural kinds like animals are spontaneously equated with artifacts, with the result that people naturally infer that they have a purpose and a creator-designer.

Indeed, naturalizing forms of CSR which do not make use of the hypothesis that God exists are heavily committed to this view, for the reason already alluded to. Saying that the concept of God is simple and innate would raise the question of why that concept would have occurred in the human mind in the first place, given that it is (some assume) of no direct use. Saying that the concept of God is simple and induced by experience raises the question of how that could happen on the naturalistic hypothesis; it cannot be simply by a form of demonstration, if there is nothing that the concept applies to. Our clearest cases of merely intentional objects are rather obviously constructed out of properties that are otherwise known to exist: a unicorn is a horse with one horn, a griffin is an animal with the front of an eagle and the back of a lion. The same is true for certain things that religious people actually believe in, including Boyer's (2001) recording trees and statues that hear prayer. So standard CSR is heavily committed to "God" being a concept of type (4).

But this strain of research has not made much progress in showing how this particular concept actually fits into the framework. It is not at all obvious that

"God" is a complex concept. Notice that it corresponds to a linguistically simple expression in English and many other languages, consisting of one morpheme rather than several (in contrast to *unicorn* [= "one horn"], *listening statue*, etc.). The word is also learned early in linguistic development, for children who are exposed to it. Boyer (2001, 2003) has proposed an allegedly restrictive and explanatory catalog of the kinds of religious concepts that can easily arise in human minds and cultures, but there is nothing much like the "God" concept to be found in this catalog. The closest he comes (2001, 63) is saying that an omniscient God is [PERSON + special cognitive powers]. But this does not elaborate what those cognitive powers are (involving special perception, knowledge, wisdom, etc.), nor does it even touch on gods' immateriality, or the special powers of causation that they are taken to have. Nor has anyone else adopting this framework filled in this notable gap, so far as we know; for example, Atran's (2002, 98) fifteen-cell typology also has no obvious box for God. Indeed, Barrett (2004, 29) acknowledges that common concepts of God are *massively* counterintuitive, not minimally so, and we think that is true also of the folk concept, even before one adds explicitly taught technical notions like God being triune. Therefore, there is no explanation in these terms of how the concept of God arises. Our best first-pass version is "nonphysical agent having great power and knowledge," but that already consists of at least four basic predicates, and it may well need further elaboration. A minimally counterintuitive analysis of "god" thus seems unlikely, and even a complex conceptual analysis would be challenging, not obviously consistent with the superficial linguistic and developmental evidence.[6]

Summarizing so far, we have considered a case of a modern religious believer praying in a quiet room and sensing something like "God is present to me as loving," and we asked what CSR contributes toward understanding this experience. We find the answer to be almost nothing: it does not elucidate the nature of the concept God that the belief contains (it is not obviously a combination of other concepts), nor the content of the belief (that God is present and loving), nor the context of the belief (in a quiet room, with no unexpected physical stimuli).

There are other important strains in the CSR literature. Some of these, like Norenzayan's (2013), assign a more prominent role to cultural transmission and cultural evolution (see also Richerson and Newson 2009; Wilson 2002; etc.). These accounts create more space for talking about how a complex concept of God might have developed over historical time and been transmitted as a piece of cultural technology, analogous to the development and transmission of complex techniques for making stone axes. One can make a case, then, that cultures with

[6] This argument is a special case of Fodor's classic point that it is hard to take a genuinely reductive approach to *any* concept that corresponds to a simple word in a natural language, but we apply the point only to this very particular case. Responses to Fodor that may work for artifacts of modern technology will not obviously work for this case.

a certain kind of god concept have done better than cultures without it through the vicissitudes of (cultural) evolution; for Norenzayan, this is because belief in a watchful and morally concerned God has value in preventing free-riding and other antisocial behaviors. However, this line of thought still strikes us as incomplete in that it has nothing special to say about how a nascent concept of a powerful immaterial agent got started in the first place, so that it could enter the process of cultural development at all. On this point, Norenzayan (2013, 15–19) simply relies on a standard ToM-type story. Moreover, where the concept of God comes from is the most that one will get from Norenzayan (2013) and similar accounts: they have nothing to offer toward explaining why people have experiences that they take to be caused by being in contact with God. Norenzayan tells us why the priestly class might put carvings with eyes of God around their city, to prime people toward prosocial behavior, but he doesn't tell us why contemporary believers have perception-like experiences of being in the presence of God when no such eyes are present.[7]

It is conceivable that contemporary spiritual disciplines of prayer and "listening" for God's voice do have some tenuous connection to the cognitive mechanisms posited by CSR. Perhaps, as an extension of a very general mental tendency to posit agents when other causes are not obvious, people's intrusive thoughts or sudden alterations in mood are attributed to an invisible agent. But this seems like a far cry from the kind of agency detection that would be useful in ancestral environments, extending it to a range of phenomena that are very different from the circumstances in which it is helpful to posit an agent on scanty input.

We conclude that experiences that purport to be perceptions of God deserve cognitive scientific scrutiny. This then raises questions like the following: What options are there for a cognitive science of religion that takes seriously the widespread reports of perception-like encounters with God, and tries to understand these methods of belief formation in terms of experiential inputs, algorithms, and representational outputs, in the style of cognitive science? How likely is it that the processes in question might constitute something like a "god-faculty," which could be delivering perception of God in some circumstances? And what are the metaphysical and epistemological implications, if a "god-faculty" turns out to be part of our cognitive endowment?

[7] Bulbulia (2009) takes a strong stand that CSR should explain religious experiences as (adaptive) confabulations: people don't really have such experiences, but they think that they do, so as to justify their religious beliefs to themselves. We do not deny that some confabulation happens, particularly in (sub)cultures in which having religious experiences is expected or grants prestige. But the claim that all religious experience is confabulation seems rather incredible to those who have had them, in the same way that Dennett's (1991) claim that we do not really have conscious experiences is to most people.

5.4 RELIGIOUS EXPERIENCE AS PERCEPTION: A NEO-ALSTONIAN APPROACH

In fact, one of the core contributors to CSR, Justin Barrett, is inclined to take some religious experiences at face value, as perceptions of God. Barrett and coauthor Kelly Clark (2010) describe (H)ADD and ToM as together forming a kind of "god-faculty"; and they allow that this "faculty" is probably an evolutionary "spandrel"—selected for survival-enhancing features that had nothing to do with production of beliefs about supernatural beings. However, they claim that this view is consistent with the god-faculty sometimes "operating under optimal conditions for producing reliable religious beliefs. . . . The development of the god-faculty through evolutionary processes prepares one for the acquisition of true religious beliefs when one has genuine religious experiences" (188). Their notion of a faculty is, however, highly malleable; their god-faculty is just whatever mental tools, in whatever circumstances, tend to produce belief in gods. We find perception-like reports of experiencing God to be sufficiently common and important to deserve exploration as a potential god-faculty in a narrower sense than Barrett and Clark have in mind. We are interested in the question of whether there may be an innate type of cognitive processing, defined in terms of inputs and outputs, that promotes belief in gods—a perception-like faculty that might not be a spandrel at all.

What is involved in spelling out cognitive scientifically a view in which certain kinds of religious experiences genuinely constitute perception of God? Though the analysis of perception is fraught with controversy, in at least some central cases of perceiving, the stereotypical early modern account seems to us to be essentially right. After the development of modern science and the widespread rejection of neo-Aristotelian accounts of perception, philosophers and scientists tended to make the following three assumptions about perception. (1) When an external object is perceived by means of some sense modality, there is a family of mental states associated with that mode of sensing. (2) These mental states are caused by perceptible states of the object in a reliable, counterfactual-sustaining pattern (so that some differences in the objects perceived will cause systematic variations in the sensory experiences). (3) Having these sensory experiences tends to cause their subject to believe that an object exists with the corresponding perceptible states.

Like Alston, we believe that the phenomenal aspects of sensory mental states cannot be adequately accounted for in purely representational terms—that is, by positing inner states that represent the world as containing things with the properties objects are perceived to have.[8] On the early modern model, words like *red* and *sweet* are sometimes used to describe an aspect of one's mental states—a phenomenal mode of appearing—and, more often, are used to describe properties of physical

[8] See Alston 1991, 57 n. 46. Alston mentions D. M. Armstrong and G. Pitcher, but we expect he would have been no happier with the views of Dretske (1997) or Tye (2009).

objects that typically appear by virtue of causing mental states of that kind—things that appear under that phenomenal mode. Reductive representationalists, like Dretske and Tye, identify the experience of phenomenal redness or sweetness with the occurrence of a brain state that represents the physical properties of red or sweet objects. There is no room, on their conception, for the idea that a fully systematic and stable swapping of phenomenal feels would result in two subjects with inverted phenomenal experiences but who represent the world as being the same with respect to physical color. We disagree: as Locke claimed,[9] it is possible (logically or metaphysically, at least) for one group of perceivers to experience marigolds as phenomenally yellow, another as phenomenally blue, with neither group misperceiving the colors of flowers. It is a contingent matter which phenomenal experiences are involved in the perception of the physical properties of external objects.[10]

Taking Locke's side in this dispute need not commit us to the irreducibility of "qualia," the phenomenal properties of experiences. There are materialist type-identity theories (Hill 1991) and forms of functionalism that allow for spectrum inversion by identifying differences in qualia with differences in the physical realizers of the qualia role (Shoemaker 1982)—though in fact we prefer a more dualistic account of the relation between the phenomenal properties of experiences and the physical properties of brains and bodies.[11] Adopting the early modern schema for perception need not cast doubt upon the explanatory power of cognitive science. If qualia are real, they must have an effect upon cognition, and they—or representations of them—will show up in cognitive psychology.

Suppose some purported experiences of God's presence involve a family of mental states that are caused by a divine being (in ways that are systematically correlated with properties of the divine being) and that they tend to cause beliefs about the properties of such a being. Such experiences would be good candidates for perceptions of God. We say "good candidates" because the analysis of perception is complex; there is much to be said about the exact way in which properties of the perceived object must be responsible for the experiential states, and about the way these inner states must cause beliefs (or tendencies to have beliefs) about the kind of object in question, in order for the experience to qualify as perception of the object. To constitute a sufficient condition for perception, these conditions have to be fulfilled in "the right way"—there are devious causal paths to be ruled out, and doing so is tricky even when focusing on the clearest cases of perception (see, for example,

[9] Locke [1689] 1975, p. 389; bk. II, chap. xxxii, sec. 15.

[10] Dretske and Tye are able to describe circumstances in which sensory modes are exchanged—e.g., an inversion of a subject's experiences of the spectrum of visible light relative to our experiences (see Dretske 1997, 72; and Tye 2000, 66); but they must say that one of the ways of experiencing light leads to false beliefs about the colors of objects. And that is what Locke denies. For criticism of Locke's argument, see Speaks 2011.

[11] Indeed, we prefer a more fully-fledged dualism of selves and bodies. For some dualistic hypotheses worthy of consideration, see Baker and Goetz 2011. We see no reason to think that dualism should prevent cognitive science from playing an important role in explaining the workings of the mind.

Goldman 1977). Still, if an experience satisfies these conditions, it is well on its way to being perceptual.

Alston (1991) is a model for us of one who claims that religious experiences can count as perceptions of God, on which knowledge of God can be based.[12] Although Alston rejects the stereotypical early modern analysis of perception, his discussion of perceiving God includes this model as a contender and provides resources for those of us who find this kind of view attractive. He asks, in effect: Why think that the conditions necessary for perception are not satisfied by religious believers who take themselves to be aware of God in various ways? Purported experiences of God do potentially fit the early modern perception framework: (1) there are analogues of sensory modes of experience present when subjects claim to be perceiving God, (2) God could well be causally responsible for the sensory experiences in question, and (3) such experiences do seem to generate beliefs about God directly. We discuss these three aspects of the perceptual schema in the next two subsections.

5.4.1 Modes of Experience for Perceiving God

What are the analogues of the sensory modes of experience that are relevant to religious experience as perception of God? At first, this may seem problematic, in that there are no special sense organs that are dedicated to perceiving God, as far as anyone knows. A failure to identify the modes of experience under which God is supposed to appear would make Alston's claims about perceiving God radically incomplete.[13] He thus offers three candidates for the experiential modes under which God might be held to appear—candidates that are suggested by typical descriptions of purported perceptions of God.

First, it is not out of the question that people's normal sense organs can be coopted for perception of God. In full-blown mystical experience, God (or an angel, the Virgin Mary, etc.) is sometimes said to be seen and heard in a literal sense. Even if God is not colored or shaped or a generator of sound, Alston observes that "there is a long tradition that holds that secondary qualities like colors do not really characterize physical substances. Thus it is not inconceivable that God should appear to us as looking bright or sounding a certain way, even though He does not, in His own nature, possess any sensory qualities" (19).

Second, Alston shows that in much of the literature on mystical experience, God's presence is sensed by means of experiences that are said to be *akin* to those generated by the five senses, yet oddly different from them as well. Thus God is seen

[12] Alston himself calls them *mystical* experiences, to distinguish them from a broader class of experiences that could plausibly be deemed religious, like experiences of going to church or of loving one's neighbor.

[13] Plantinga seems to feel these doubts; he retreats from talk of literal perception of God to "something very like perception of God . . . that is epistemically on all fours with perception in that it . . . can be a source of warrant" (Plantinga 2000, 181–82).

or heard in a somewhat more extended sense (e.g., Sarah's experiences of "seeing" God, and "hearing" the prayers of heaven, quoted previously; this seems to be more common than visual or auditory experiences that seem, to the subject, exactly like seeing and hearing). Mystics talk as though there are families of what Alston calls "quasi-sensory" states—states that resemble, but are not quite the same as, the qualia found in auditory, visual, olfactory, and other forms of sensory experience. (Alston champions this model as evidence "that mystical perception involves distinctive, nonaffective phenomenal qualia" [51–54].) In cognitive science terms, we might tentatively think of this as being the effect of mental representations with religious content entering the information-processing stream within (say) the visual system without having the usual sorts of antecedent representations at a lower level of processing. For example, a religious vision might involve something that is present in the V4 area of the visual cortex that does not have an analogue at V1 or V2. This would make sense of a "vision" being like an ordinary visual experience in some ways but not others: for example, what one "sees" might have position and shape but no color or texture. We can perhaps compare this with visual imagery, which is said to make use of some parts of the visual cortex but not others. Then a divine vision would have something like the quasi-perceptual phenomenal character of visual imagery, but it would not be caused by the voluntary use of the visual imagination. Similar scenarios can be imagined within the auditory, tactile, and olfactory cognitive systems. This would give us an interpretation of the different sense-like modes of mystical experience, and how "having a vision" might be different both from literally seeing something and from "hearing God's voice."

Alston also mentions a third candidate for playing the role of a sensory mode of experience that may be involved in some putative perceptions of God. This is the possibility that emotional states are co-opted to play the role of the modes under which the divine appears to us—that certain kinds of feelings of awe, joy, and so on might serve as the modes of God's appearing. These emotions certainly have their characteristic qualia, even though they are not the qualia caused by the eyes, ears, nose, fingers, or tongue. Presumably distinctively emotional qualitative experiences come from proprioceptive sensors inside our bodies detecting associated physiological changes in heart rate, blood pressure, and so on. Although Alston does not consider this mode of experience especially promising (51), we do take it to be promising for at least two reasons: it shows how traditional sense organs need not be involved in something that is still perceptual, and it does justice to much of the experience of God's presence that is reported by ordinary believers; see, for example, James's anonymous informant who "felt the presence of God . . . as if his goodness and his power were penetrating me altogether." These seem to be nicely captured by a model of perceiving God in which God's presence is felt by means of striking changes in affect. So long as the second and third conditions for perception are appropriately satisfied, the change in affect can serve as the subjective side of a veridical perception. According to the early modern model, even in paradigmatic perception there is a wholly subjective aspect to every sensation—for example, being appeared to redly, which can happen in veridical and illusory experiences as of a red object—but, when conditions are right, the experience is nevertheless an experience of an objective reality outside the subject's head. A sign of the directness of perception (at least a necessary condition of it) is that there is no conscious inference from the subjective state (perhaps no conscious noticing of it at all) to the existence of or the properties of the object perceived. Similarly, in the typical experience of God, the existence of the divine Lover is not inferred from a change in one's own mood; rather, God can be presented as loving, for example, by means of an overwhelming sense of one's being loved. We thus see no intrinsic obstacle to the idea that this experience of God is as direct and unmediated as ordinary perceptual awareness of external objects.

We also imagine that these emotional states might be combined and overlaid on one another in various ways that are distinctive of religious experience. For example, the experiences of fear and peace might be triggered by opposite sorts of situations in dealing with the ordinary physical world that evolutionary psychology routinely considers. But they might come together simultaneously in a kind of emotional "chord" in certain types of religious experiences (e.g., John Newton: "Twas grace that taught my heart to fear / and grace my fears relieved," perhaps simultaneously). This could produce a qualitatively different experience that might even be described as a new emotion, say of religious awe, despite being constructed out of familiar elements known to psychology. There could also be interesting combinations of the quasi-sensory states and the emotional states: for example one might "see" a vision of a bright humanoid shining like the sun five feet to one's left (although not with one's eyes) and simultaneously go into an affective state of unusual alertness plus quiescence. The point is that one can imagine a rich variety of religious experiences by combining ingredients like these in different ways. Indeed, the variety of subjective states capable of generating distinctively religious perception-like experiences might approach the variety of subjective states that are associated with a traditional sense mode like vision, which can be used to perceive a complex physical world.

We think that this is important because it provides a possible account of how religious experiences can lead to religious beliefs that have rich propositional content. Religious believers generally go well beyond a simple belief that God (or some supernatural agent) exists: they tend to believe that God is also loving, or holy, or uniquely awesome, or terrible, or some special combination of those qualities. They tend to think that God was angry with them when they did such and so, but forgave them when they offered a sacrifice, or did a good work, or appropriated Christ's atonement by faith. And it is very hard to see how a classical CSR story in terms of the (H)ADD firing when one walks in the woods, or in terms of our cognitive processes inferring an agent from observing certain kinds of functional complexity in nature, could yield beliefs with this sort of specific content. It seems that cognitive processing along those lines could give little more than a bare belief that something else is out there; any more specific "theology" would have to come from somewhere else (from Granny, or the priestly elites). In contrast, the sorts of complex religious experiences we are sketching here could support religious beliefs with richer content. For example, an unusual experience of fear and peace simultaneously could lead to belief in an unusual being who is uniquely holy and loving at the same time. This then could lead to a CSR that does justice to richer systems of both religious experience and religious belief. Granny and the religious elites will no doubt still have their say on how this grist is elaborated into full-blown theologies in different cultures, but they have more grist to work with, and the richness of religious phenomenology provides more opportunities for there to be resonance between official doctrines and the experiences of ordinary believers.[14]

Those who report perception-like experiences of God describe the phenomenal or subjective aspects of their experiences in quite a number of ways. Some experiences have auditory or visual phenomenology that is relatively straightforward, while others leave their subjects struggling to find the right words to identify the sensory mode under which God seems to them to be presented. In ordinary modes of perception there is also a spectrum, from modes of experience in which the subjective vehicle is readily identifiable to ones in which it is extremely hard to describe.

At the one end, for example, is a rheumatic sufferer who experiences a familiar pain as the onset of a drop in atmospheric pressure. The pain is one thing, a very familiar experience that may not always have been associated with the weather, but that has now come to seem a perception-like awareness. As Goldman (1977, 271) puts it, the rheumatic "may not know enough to have beliefs about changes in atmospheric pressure. But he has beliefs like: 'Something is happening that will lead to rain.' The event that satisfies this belief, and also causes his ache, is the drop in atmospheric pressure. Hence, we say that he *perceives* such drops." Though the pain is very different from the drop in atmospheric pressure, once the pain has become a reliable, noninferential source of information about the environment, it becomes natural to describe the experience as a form of perception.

At the other end of the spectrum, there are kinds of perception in which the subjective vehicle—the mode of appearing—is highly elusive; in the extreme case, psychologists begin to talk of "sensationless perception." A famous example is "facial vision":

> Blind people (and sighted people while blindfolded, too, though less reliably) can detect obstacles—walls, chairs, and the like—without having any (conscious) sensation. In fact, they tend to think that they are picking up information somehow through the skin of

[14] Issues of the reliability of religious experience now arise of course: why have religious experiences in different cultural contexts given rise to what seem to be content-rich but contradictory (folk) theologies. We flag the issue as important, but have no space to take it up directly here.

the face (hence "facial vision"), when in truth the information is coming in through the ears as a subtle form of echolocation. (Lyons 2009, 52)

A cognitive process that typically leads to belief in God could be a form of perception whether or not the subjective aspect involved is easy to describe in itself, or easily recognized as distinct from the God purportedly revealed in the experience.

5.4.2 God Causing the Experiences

Continuing to take seriously the reports that a certain class of religious experiences feel like perceptions, we need to say that this belief production happens automatically, at a subpersonal level. "Mystics" (even everyday ones) report a difference between, on the one hand, an experience of feeling loved or feeling an overwhelming sense of peace and *inferring* from this that a loving God exists; and, on the other hand, a strong and direct feeling that one is in the presence of an infinitely loving being. In one case, the focus of attention is on one's own internal states; in the other, it is on the being perceived as outside of oneself, to such an extent that one may hardly even notice one's unusual emotional state—just as in visual perception one may be entirely focused on a basketball and not at all on one's sensation of an orangish round patch in one's visual field. Alston (1991, 20ff) finds this distinction clearly articulated by theorist and practitioner alike. He quotes theologian John Baillie (1962, 88–89): "Faith does not deduce from other realities that are present the existence of God who is not present but absent; rather it is an awareness of the divine presence itself, however hidden beyond the veils of sense" (26). But the distinction between direct and indirect experiences of God can be found in the writings of Christian mystics, such as Angela of Foligno (Alston 1991, 13). In a similar vein, Luhrmann's (2012, 97) interviewee Sarah clearly distinguishes times when she felt God's presence moving through her from times that she did not feel his presence, but still believed in a more conscious cognitive way that God was (omnipresent, and therefore) close to her.

This is a place where some additional cognitive science (CSR) should come in, then. Vision science seeks to give a step-by-step computational account of how a certain pattern of stimulation on the retina is mapped onto the percept of a basketball at such-and-such a location, with the information-processing algorithms performed automatically by the brain. A cognitive-scientifically respectable explanation will account for the fact that we automatically, inevitably perceive an object in the world, not a sensation on the retina. We might then want something analogous from a full cognitive science of religion: a step-by-step computational account of how a certain pattern of input from the proprioceptive receivers of the body (or activity in the higher visual centers, or whatever) produces the output that God is present to the person as loving (or angry, etc.), couched in terms of information-processing algorithms performable by the brain. Anything less than this is arguably failing to live up to the goal of giving a cognitive scientific explanation of the observed facts about religious experience. But we find that CSR as it exists now does not offer us much of anything of this sort. In order to provide a more complete account of modern religion, CSR must take a deeper interest in the role of modern religious experience in generating and sustaining belief in God. And we see no intrinsic reason why it couldn't be developed so as to do so.[15]

How likely is it that a fuller account of the steps from experience to belief would be consistent with the hypothesis that the process constitutes perception of God? Very likely, we think.

In principle, the second and third components of the early modern perception schema should not be so hard for the religious perceiver to satisfy. Alston rightly assumes that the third component is unproblematic: the experiences religious people describe as perceptions of God obviously generate beliefs about God, or at least tendencies to believe (63–64). But he takes the second, causal component to be more problematic:

> It may well be pointed out, as I did earlier, that not every causal contributor to an experience is perceived via the experience. . . . Thus it is not enough that God figure somehow or other in the causes of the experience; He would have to make the right kind of causal contribution. But what is the right kind? (64)

This is a special case of the general problem of determining the "proper stimulus" of a sense modality, a problem Alston thinks cannot be solved by giving a "general answer applicable to all perceptual modalities" (64). We are not so pessimistic.[16] At least one very promising attempt at a general answer implies that there is a close connection between the second and third components in the early modern analysis of perception. It is not a coincidence that objects perceived tend to cause experiences which, in turn, cause beliefs about those kinds of objects. We thus take up the second and third condition together, in the context of discussing the proper stimulus question.

[15] There is, however, at least one practical problem: when studying visual perception it is easy to present a basketball to someone and to withdraw it at will, so as to see what differences in subjective report and brain states are correlated with the change in stimulation. In contrast, when studying perception of God, we cannot present God to the subject, and withdraw God at will, so as to isolate the difference.

[16] We are not, however, as optimistic as Richard Swinburne. Swinburne points out that, if God exists (and sustains everything, and is omnipresent), then "any causal processes at all which bring about my experience will have God among their causes; and any experience of him will be of him as present at a place where he is." From these two facts it is supposed to follow that "if there is a God, any experience which seems to be of God, will be genuine—will be of God" (Swinburne 1979, 270; see also Wainwright 1973). According to Swinburne, so long as God exists, it is impossible for there to be a failure of the appropriateness of the causal chain from God to an experience that seems to reveal God. We think it is not that easy.

In the case of vision, the proper stimulus question can be posed in this way: Why is it that we see surfaces of objects, rather than the light striking the retina, or the light traveling between an object's surface and the retina? The best answer we know of is due to Alvin Goldman; and it appeals to the kinds of beliefs typically caused by the mental states associated with a particular sense modality, and the proportion of those beliefs that are true (Goldman 1977). Thus it is a point of fact that experiences of redness, say, have a tendency to produce largely true beliefs (directly, noninferentially, presumably computationally) about the properties of the surfaces of objects, not about the wavelengths of light striking the eye or traveling between an object and the eye. And it is this fact that explains why red experiences, in normal conditions, are perceptions of the surfaces of objects and not of the properties of the retina or of the light itself—even though the experiences carry as much or more information about these other things.

On Goldman's account, to work out the proper stimulus of a sense, one looks at the relations typically holding between the perceiver and those things that uniquely satisfy beliefs that are generated by (or constituted by) the "percepts" belonging to that sense—whether it be smells, tastes, visual appearances, or some other kind of experience. Things that stand in these (broadly speaking, causal) relations to the experiencer are the *proper* objects of that kind of perceptual experience. Visual experiences tend to produce beliefs that provide a fairly accurate representation of the objects before one's eyes; these objects have surfaces that reflect light and are located in such a way that the reflected light strikes one's eyes; and so when things stand in that relation to our open eyes and generate experiences of color in us, we perceive those things (even if it happens that the usual beliefs do not accompany visual experiences on some occasions—for instance, in skeptical moods, or when distracted).

A notable feature of Goldman's account is that, had experiences of color tended to generate beliefs that were mostly true of *something else* in the causal history of those experiences, then the proper stimulus for perception by means of color would have been those other things. Color experience *could* have been a means of perceiving proximal stimuli—e.g., the light striking the eye—but instead it serves the much more useful function of allowing us to perceive distal objects. In Goldman's example of the rheumatic, a kind of experience that did not originally serve as the vehicle for perceiving anything external comes to be the sense modality for perceiving a drop in atmospheric pressure; among the many causes of the pain, the change in atmospheric pressure is selected as proper stimulus in virtue of most reliably satisfying the beliefs generated by the pain.

Many people come to have experiences in which they take God to be especially present in a perception-like way. In the previous subsection, we surveyed a number of modes of experience that seem, for some people at some times, to mediate the presence of God. What would it take for God to be, in fact, the proper stimulus for such experiences—the object perceived by means of them? If Goldman is right, at least the following necessary conditions must hold: the experiences in question must typically generate beliefs that are uniquely satisfied by God, and there must be a causal relationship between God and the experiences. But what kind of causal relationship would suffice?

Alston considers two ways in which God could satisfy a causal condition on perception. (i) The experiences involved might "occur only because God intentionally presents Himself to the subject's awareness as so-and-so" (64) in a way that "involves some divine activity over and above that which is directed to everything else in creation" (65). On the other hand, (ii) "It may be that God satisfies the causal condition for being *perceivable* in a certain experience just by keeping that experience in existence"—in other words, by exercising nothing more than the kind of concurrence God is supposed to contribute to sustaining everything (65). Like Alston, we do not rule either possibility out of court.

Possibility (i) suggests a more interventionist picture: God occasionally interferes with the workings of our brains (or souls) in order to make his presence felt. This need not be the only account of God's causal activity that would make (i) true however. God could also perform a special, deliberate act of bringing about a religious experience without miraculous interference in the course of nature. Suppose God knew that, by some quirk of our evolved psychology, whenever a certain combination of emotional states occurs in human beings who antecedently have the concept of God, they have an experience that generates belief in God's presence. And then suppose that God providentially arranges the initial conditions of our universe so as to ensure that such experiences are frequently had. These experiences would, we submit, satisfy the conditions for being perceptions of God: a certain mode of experience is caused by the putative object of perception, and these experiences in turn cause true beliefs about their cause. This strategy for providing us with perceptual experience of God ought also to count as "divine activity over and above" God's general concurrence, but activity that does not involve directly fiddling with our neurons. However the mechanism of (i) is to be understood, it provides a model for the causation of religious experiences that easily explains the transient nature of ostensible perceptions of God, and the sense that they were not initiated by the perceiver.

In contrast, possibility (ii) might seem to be a nonstarter.[17] Since God stands in the relevant causal relation to every experience, it is natural to ask—as Alston does—why we do not "perceive God in every experience?" (65). How can (ii) be made compatible with the transient nature of ostensible perceptions of God? Alston proposes a model:

[17] Indeed, it seems so to Michael Levine (1990).

Although God satisfies all the conditions on His side for being perceivable in every experience, there are various obstacles on our side that, most of the time, inhibit that perception. (65)

Although Alston does not say much about the nature of these obstacles, a little reflection on the solution to the proper stimulus problem removes any mystery there might be about how perception of God could be inhibited despite the ubiquity of God's causal contribution—and the inhibition need not come from inattention or cognitive malfunction. Why, for instance, do visual experiences not typically constitute perceptions of God, given God's involvement in causing them? For the same reason they do not typically constitute perception of the light traveling to one's eyes or the state of one's own retina. Visual experiences tend to generate beliefs about surfaces of objects in front of one's eyes, and it is in virtue of this fact that they constitute the internal, subjective side of the perception of physical objects—not the perception of other things causally implicated in generating the experience, including the light two inches before the eyes, the retina, or God.

So long as religious experience reveals things about God that are always true of God (e.g., God is powerful, knows exactly what you are doing, loves you, etc.), there need be no special contribution that God makes on some but not all occasions. Alston was worried that, if God's contribution is no more distinctive than that, then all perception will count as perceiving God. So long as normal visual, auditory, olfactory, and other sense experiences do not regularly generate beliefs about God, the fact that God is causally involved in their production does not transform them into modes of awareness of God. Conversely, so long as a special kind of experience does generate, typically and automatically, beliefs about God, then—since God is, indeed, among the causes of that experience—God is a perfect candidate to be the proper stimulus for a perception with that kind of experience as its sensory mode.

As Goldman's (1977) extended discussion shows, there are complications to be overcome in the full analysis of what it is for something to be the perceptual object of a sense modality. But the basic criteria seem easily satisfiable by God, as the object, and the kinds of experiences described in contemporary spiritual practices, as the mode of sensing.

One need not, then, suppose that God always acts directly upon the mind or brain in a way that bypasses natural processes in order to think that religious experience constitutes perception of God—though we are not particularly skeptical about how often God presents himself by means of some kind of special intervention. The evolutionary story about the development of a "god module" presented in the next section is consistent with many different ways in which such a module could be constructed. Different sorts of sensory, quasi-sensory, or affective states could constitute the experiential side of perceptions of God. And God's role in causing them might involve his setting up special circumstances that trigger such experiences; it might involve direct intervention in our ancestors' brains; but it might involve something much simpler—God's general sustaining of everything in existence.

5.5 THE PROSPECTS FOR EVOLVED DOMAIN-SPECIFIC RELIGIOUS COGNITION: A NEW FABLE IN THE EVOLUTIONARY PSYCHOLOGY STYLE

The hypothesis that God is the perceptual object of some religious experiences strongly suggests that there must be some dedicated cognitive machinery at work in the human mind/brain which a cognitive scientist might look for. To use a controversial term, it strongly suggests that human beings have a god module. This modularity thesis (as we intend it) is basically just the view that there is a *specialized* algorithm to map (say) activity in the proprioceptive receptors onto (candidate) beliefs about the presence of God. What would the alternative be? It would have to be that some other algorithmic process of the human mind, which exists for other reasons, happens to perform this mapping. For example, the same visual processing algorithms that map some round orange patches onto basketball percepts would also map experienced states of arousal plus quiescence onto God percepts. Or auditory processing algorithms would do this, or orientation-in-space algorithms, or whatever. That possibility strikes us as very unlikely, given how different both the inputs and the outputs are. Unless more is said, this would be like inputting lines of Shakespeare into a program designed to calculate compound interest rates and expecting the program to produce a metrical scan of the lines: it is not going to happen. It is extremely rare for a computational system to do anything interesting with an input that does not meet the formal specifications for which it was explicitly designed. We think it would be almost as amazing in the case of religious experience.

Those familiar with the CSR literature may find this conclusion rather far-fetched. After all, pretty much the entire thrust of that literature has been to deny that there is any specialized god-module. And how could such a module have evolved? What would it have been good for in the human ancestral environment?

We would like to propose a novel but conceptually straightforward answer to these questions: a god module could have evolved because it was useful for interacting with God. Of course, this answer takes seriously the possibility that God actually exists, but we are happy to do so. Indeed the conceptual landscape around these questions could look quite different if one is open to the possibility that theism is true. If supernatural beings exist, it is entirely plausible that human beings would have acquired some specialized cognitive apparatus for thinking about them.

To see how this could come about, suppose that we combine theism with standard evolutionary psychological thinking (the intellectual background for CSR). Theism is simply the idea that a transcendent personal God exists, one who can and does act in the world. If God is personal, then one can talk to God (prayer), and if God acts in the world, then God can choose to do what is asked of him (answered prayer). This is true by a very simple characterization of theism.[18] Now suppose that, as *Homo sapiens* was evolving, something entirely natural—a random mutation, an overactive agency detecting device, a wild hunch—caused a hominid to think, "There is a God" and to try praying to that God. For example, he (King David) might pray that he would escape from his enemies, or she (Hannah) might pray that she would finally bear a child. So far, this imagined scenario is not significantly different from standard CSR accounts. The distinctive step in our imagined evolutionary history is that God in fact heard the prayer, liked it, and decided to do something that caused the enemy to turn the other way, or a sperm to find a fertile egg. Obviously, God choosing to answer prayers of this sort increased the biological fitness of the people who prayed. According to the logic of population genetics, this result holds even if God only answered a fairly small percentage of such prayers (less than, for example, the 10% difference that Benson et al.'s [2006] study was designed to detect). Thus a cognitive architecture that, through suitable beliefs and desires, supports praying such prayers could perfectly well have developed, been elaborated, and fixed in the human species by evolutionary mechanisms.

As an aside, we acknowledge that some scientific studies are said to have refuted the theistic hypothesis that God answers prayer. In fact, the literature on this topic is somewhat mixed, some studies finding a positive effect of prayer, others not. The latter group includes Benson et al. (2006), the largest and most carefully designed study, discussed by Dawkins (2006), among others (see also Goldman, this volume). But these studies inevitably make tacit theological assumptions about how prayer works that are suspect. For example, these studies assume that prayer is additive, such that if more people pray for a certain outcome, God is more likely to grant that outcome. They also assume that prayers in which there is no personal connection between those who are praying and the one being prayed for—as in "properly done" double-blind experiments—have the same value to God as prayers in which there is a close personal relationship. But those assumptions about prayer might well be false (we think they are). Indeed, Benson et al. (2006, 941) explicitly note that their study imposed certain constraints on the pray-ers in the cause of standardization, and this forced people to pray quite differently from how they normally would. The Benson et al. study also made no attempt to prevent the subjects in their study from praying for themselves, or from receiving prayer from family, friends, and their religious community, correctly deeming that to do so would have been "unethical and impractical" (942). In fact, 95% of their test subjects said that they believed that people would be praying for them (937). Benson et al. (2006, 942) thus observe that "Our study subjects may have been exposed to a large amount of nonstudy prayer, and this could have made it more difficult to detect the effects of prayer provided by

[18] Here we understand theism as contrasting with deism, the view that God created the world but does not act within it. We also assume that if God acts in the world, God (probably) has acted in similar ways throughout human history. (We thank Joseph Corabi for suggesting these clarifications.)

the [anonymous, distant] intercessors." They conclude that "Private or family prayer is widely believed to influence recovery from illness, and the results of this study do not challenge this belief." Therefore, we do not see any compelling basis in this or similar studies for denying the experience of many that God sometimes answers prayer.

Note also that our evolutionary fable can account not only for why people have distinctive cognitive processes for producing God beliefs from certain subjective experiences, but also for why they have a concept of God at all. In section 5.3, we mentioned that standard CSR is committed to the god concept being a complex one, constructed out of features from other cognitive domains. This view has certain intrinsic weaknesses: no "minimally counterintuitive" version of a god concept has been proposed, it may not fit with the linguistic simplicity of the term and its early appearance in development, and it leaves it somewhat mysterious why (H) ADD experiences would ever have been interpreted in this way. In contrast, within a theistic version of evolutionary psychology, it is entirely possible that God would be a simple innate concept, just as "water" and "human being" and "snake" arguably are. Our account can thus support the representational resources needed to have beliefs about God as well as the computational resources needed to produce beliefs about God from certain kinds of experiences.

Note also that "God" in our fable would not necessarily have to be the omniscient, omnipresent, and omnipotent God of the monotheistic religions to play the role we have sketched, although that is the possibility we have most in mind. He/she/they would just have to be present and potent enough to cause experiential states and to answer a certain percentage of prayers. This account is thus neutral about just what supernatural agent people have acquired the ability to perceive: it could be Fang ghosts or Chinese ancestor spirits rather than the Christian God. This is presumably good, given the range of religious beliefs attested in the world. A Christian thinker might hold that a Fang tribesman has a genuine religious experience caused by the triune God, giving her the (true) belief that a supernatural agent exists, and also the more specific false (?) belief that ghosts exist. Monotheists might even hold that the invisible agent in their environment that most frequently satisfies the only true beliefs generated by their experience is God; and so, according to Goldman's account of what makes an object the proper stimulus of a perceptual modality, God is at least sometimes perceived by means of their experiences—albeit accompanied by much misperception. Their belief is, de re, about God; but it is, de dicto, about ghosts—as in familiar cases of mistaken identity. Conversely, of course, a Fang thinker might hold that a Christian has beliefs about supernatural agents that are, de re, about ghosts of dead people although, de dicto, they are about Jehovah. Choosing between these alternatives will require some combination of a finer analysis of the character and content of religious experiences and considerations of a different sort.

We emphasize that the type of account we have sketched, never considered in the literature, is straightforward evolutionary psychology when it comes to the internal mechanisms by which evolution shapes the human mind/brain. God does not necessarily do anything to intervene directly in the formation of the human mind/brain itself. The view is simply that God was an evolutionarily significant feature of our ancestral environment, just as saber-toothed tigers and diseases were.[19] And any true theist believes that. From this perspective, it is no more incredible that the human mind would have specialized resources for perceiving and knowing God than it is that the mind has specialized resources for perceiving human faces, or natural language sentences, or contaminated food. This then could be part of the conceptual basis for a reformed CSR that can do justice to the widespread existence of religious experiences with the character of perceptions.

5.6 EPISTEMOLOGICAL REFLECTIONS

Finally, let us reflect briefly on the epistemological implications of our discussion for both current CSR and the "reformed" CSR that we have imagined.

First of all, a disclaimer: we do not doubt that CSR, even in its current embryonic state, should have some impact on the assessment of some arguments for the existence of God. The scientists involved in this new field are clever people with provocative and promising ideas; there is a decent chance they are converging upon the right sort of explanation for the ubiquity of belief in spirits, gods, and perhaps even belief in Big Gods. We agree that the plausibility of their explanations diminishes the positive force of the usual form of the *consensus gentium* argument (see Goldman, this volume).[20] We wonder, though, how much damage to this particular reason for believing in God affects the average believer, who may well not base her belief on the *consensus gentium* argument (and may not even know about it). Non-truth-tracking explanations for the widespread belief in gods and spirits will not make a major difference to one's judgments about God's existence for those whose beliefs are not based upon taking a survey of others.

Reaction to CSR has tended to be extreme. Some herald it as poised to strike the death blow to religion; others regard it as faith-friendly, confirming John Calvin's hypothesis that everyone has a built-in knowledge of God's existence. We find ourselves somewhere in the middle, and hoping for more results from a reformed CSR that takes perception-like religious experiences as an explanatory target.

[19] Whether this view violates a notion of "methodological naturalism" that is supposed to be constitutive of not only CSR but scientific endeavor more generally is a complex question. See Schloss 2009 and Plantinga 2009 for some relevant discussion.

[20] We seriously doubt, however, that CSR results will ever transform the *consensus gentium* into an argument against the existence of God, a possibility that Goldman considers near the end of his chapter. The credence assigned to God's existing will always be above .5 as long as widespread belief in God(s) is at least as likely given that God exists as it is given that God does not exist. We think this is a reasonable assumption. The contrary might only be true if God actively intervenes to mislead people to not believe in God—conceivable, to be sure, but not true on most existing theologies.

The work currently flying under the flag "cognitive science of religion" is a subfield of psychology/anthropology in the making, too new to claim great empirical successes or much unanimity among its small band of experts. This has not stopped some of its practitioners (Boyer, Bloom, Bering . . .) and popularizers (Dawkins, Dennett, Hitchens . . .) from drawing consequences from it that are dire for the rationality of contemporary religious belief. According to Jesse Bering, thanks to CSR, scientists finally "have God by the throat"; and Bering, at least, is "not going to stop until one of us is dead."[21] But, as many critics have pointed out, the precise way in which CSR is supposed to undermine belief in God is often left quite vague, or merely insinuated.[22]

When explicitly spelled out, the more extreme debunking arguments based on CSR allege that there are evident defects in the belief-forming processes that yield god-beliefs when they are based on, for example, (H)ADD and ToM firing in the absence of ordinary agents. They then claim that this defect in some way undermines current practices as well, rendering contemporary religious people either unjustified in their belief in God or unjustified once apprised of the relevant parts of CSR.[23]

On the other end of the spectrum from the debunkers is the reaction of Clark and Barrett (2011). They take it as established that belief in God or gods is natural to humans in normal human environments, and they assume that, within a broadly Reidian epistemology, the natural products of human cognition should be considered innocent until proven guilty. Not doing so would open the doors to a much more general skepticism regarding other objects of perception and cognition, and hence it would be self-defeating.

We find ourselves somewhere between these two poles. On the one hand, the claim that current CSR radically undermines the degree of justification that contemporary believers in God have is open to at least two objections. First, it assumes that the CSR accounts are more or less true and complete, whereas we have questioned their completeness here. The origin of the god concept does not seem to us to fit the paradigmatic pattern of minimal counterintuitiveness, and it may well be simple and innate. Moreover, the kinds of perception-like experiences that many believers report do not very obviously fit any pattern predicted by CSR; yet they sustain much ordinary belief, and stand at the origins of many religious revivals and movements. Second, it assumes that epistemologically dubious origins for a cognitive faculty mean that that faculty cannot be used synchronically in reliable ways, which is not necessarily the case (see Goldman, this volume, for very pertinent discussion).[24]

[21] Murray 2009, 169; Bering was quoted in the *Broward–Palm Beach New Times*, March 9, 2006.

[22] Thurow (2013a), for example, complains about the absence of arguments linking evidence from CSR and antitheistic conclusions in Bering (2011). Murray (2009) lodges similar complaints against many antitheistic deployments of CSR.

[23] See, for example, Braddock 2016; Wilkins and Griffiths 2012; Leben 2014; arguments along these lines are articulated and criticized by Murray (2009); van Inwagen (2009); Thurow (2013b); Barrett (2007); Barrett and Clark (2013).

[24] See also Murray 2009; van Inwagen 2009; and Barrett 2007.

On the other hand, the optimism expressed in Clark and Barrett (2011) may need tempering. Intuitively, it seems to matter what the actual mechanism for producing a given religious belief is—whether it is something relatively direct and straightforward, (e.g., a coherent cognitive module), or a highly gerrymandered system, the cognitive equivalent of a Rube Goldberg device (cf. McCauley 2011). And is it really true that we can go no further than accepting all of the "natural" results of our cognition in an uncritical way? (Again, see Goldman, this volume, for relevant discussion.)

Reservations about trust in our "god module" may boil down to a form of the generality problem for reliabilist epistemologies: what class of phenomena should be considered when evaluating whether a given belief is or is not formed by means of a reliable process? Clark and Barrett seem to be typing very broadly: human cognition as a whole is quite reliable, so the religious beliefs it produces are probably reliable too. This is an extremely coarse-grained approach to the categorization of belief-forming methods—arguably, too coarse-grained to capture the differences in justification that correspond to differences in reliability.

The approach we tend to favor on the generality problem is Jack Lyons's (2009): he individuates the faculties in terms of cognitive algorithms. The rough idea is to count all and only the beliefs that are produced by the same algorithmic computation when deciding whether a given faculty is reliable. This is much finer grained than Clark and Barrett's broad-brush Reidian approach, given that the mind contains many algorithms residing in its different (functionally defined) modules—and that seems appropriate. Lyons's approach to the analysis of knowledge and the generality problem implies a fruitful interaction between cognitive science and epistemology, something we should expect.

Now it follows from Lyons's view that the details of the cognitive architecture that produces religious belief matter a good deal. Is it true that the same cognitive algorithm that gives "possible predator nearby," computationally, from the sensation of "movement in my peripheral vision" when walking through the woods gives "Holy Loving God nearby," computationally, from the sensation of "perfect peace and joy" when praying in one's bedroom? Standard CSR accounts hope that it is, without filling out even the first detail in computational-algorithmic terms. In contrast, we have questioned the plausibility of that hypothesis here (while not denying that other instances of religious belief formation may be closer to being of a piece with the deliverances of (H)ADD—Fang villagers' beliefs about ghosts outside their village, perhaps). If the algorithm that produces this sort of religious belief is indeed a different one, then it should stand or fall on its own when evaluating reliability; the fact that (H)ADD arguably has evolutionary value without being very reliable does not come into the calculation. Moreover, it may well be that some algorithms that produce religious belief in fact do little else (if the massive modularity thesis of classical evolutionary psychology is true, this should be expected). In that case, whether a belief in God is reliably formed depends entirely upon whether God exists. Lyons's criterion for determining reliability, plus massive modularity, together make it likely that Plantinga (2000) was right about the question of whether belief in God is justified, at least when it comes to "basic" beliefs (beliefs about God that are not inferred from other beliefs): the de jure question is deeply entangled with the de facto question.

Since we think there is a reasonable chance that God does exist, God's existence seems like a decent candidate explanation of why people would have a simple and innate (or at least easily acquired) concept of God, and why they would take God's presence to be directly felt when experiencing certain quasi-perceptual and affective states. The cognitive machinery for this could perfectly well have developed along evolutionary-psychological lines, if people have in fact interacted with God throughout their evolutionary history. It is then quite possible that this particular "God faculty" is reliable in the sense required for yielding justified beliefs, and that religious mystics—both professional and "every day"—really are perceiving God in a way that gives them knowledge. It also becomes readily imaginable that studying empirically whether there are in fact dedicated computational and representational resources for perceiving God using cognitive and neuroscientific techniques could provide evidence about whether we have been in serious causal contact with God over the course of our evolution, thereby raising our credence in the metaphysical claim that God exists.

ACKNOWLEDGMENTS

For helpful comments on a previous draft of this chapter, we thank Brian McLaughlin, Alvin Goldman, Justin Barrett, Andrew Moon, Roger White, James Jones, Joseph Corabi, the participants in the reading group of the Rutgers Center for the Philosophy of Religion, members of the doctoral seminar at the Central European University's Center for Religious Studies, a graduate seminar at Rutgers, and participants at a symposium at the American Philosophical Association Eastern meeting, January 2017. None of these people necessarily agrees with what we say, and all remaining errors are our own. The authors thank the John Templeton Foundation for generous support of the research that resulted in this chapter.

REFERENCES

Alston, W. (1991). *Perceiving God: The Epistemology of Religious Experience.* Ithaca, NY: Cornell University Press.

Atran, S. (2002). *In Gods We Trust: The Evolutionary Landscape of Religion.* New York: Oxford University Press.

Baillie, J. (1962). *The Sense of the Presence of God.* New York: Scribner's.

Baker, M., and Stewart Goetz, eds. (2011). *The Soul Hypothesis.* New York: Continuum.

Barrett, J. (2004). *Why Would Anyone Believe in God?* Walnut Creek, CA: Alta-Mira Press.

Barrett, J. (2007). Is the spell really broken? Bio-psychological explanations of religion and theistic belief. *Theology and Science* 5: 57–72.

Barrett, J., and Church, I. M. (2013). Should CSR give atheists epistemic assurance? On beer-goggles, BFFs, and skepticism regarding religious beliefs. *The Monist* 96: 311–24.

Benson, H., Dusek, J., et al. (2006). Study of the therapeutic effects of intercessory prayer (STEP) in cardiac bypass patients: A multicenter randomized trial of uncertainty and certainty of receiving intercessory prayer. *American Heart Journal* 151 (4): 933–42.

Bering, J. (2006). The cognitive psychology of belief in the supernatural. *American Scientist* 94: 142–49.

Bering, J. (2011). *The Belief Instinct*. New York: Norton.

Bloom, P. (2004). *Descartes' Baby: How the Science of Child Development Explains What Makes Us Human*. New York: Basic Books.

Bloom, P. (2009). Religious belief as an evolutionary accident. In J. Schloss and M. Murray, eds., *The Believing Primate: Scientific, Philosophical, and Theological Reflections on the Origin of Religion*. Oxford: Oxford University Press, 118–27.

Boyer, P. (1994). *The Naturalness of Religious Ideas*. Berkeley: University of California Press.

Boyer, P. (2001). *Religion Explained*. New York: Basic Books.

Boyer, P. (2003). Religious thought and behaviour as byproducts of brain function. *Trends in Cognitive Sciences* 7 (3): 119–24.

Braddock, M. (2016). Debunking arguments and the cognitive science of religion. *Theology and Science* 14: 268–87.

Bulbulia, J. (2009). Religiosity as mental time travel: Cognitive adaptations for religious behavior. In J. Schloss and M. Murray, eds., *The Believing Primate: Scientific, Philosophical, and Theological Reflections on the Origin of Religion*. Oxford: Oxford University Press, 168–78.

Clark, K., and Barrett, J. (2010). Reformed epistemology and the cognitive science of religion. *Faith and Philosophy* 27: 174–89.

Clark, K., and Barrett, J. (2011). Reidian religious epistemology and the cognitive science of religion. *Journal of the American Academy of Religion* 79: 639–75.

Dawkins, R. (2006). *The God Delusion*. Boston: Houghton Mifflin.

Dretske, F. (1997). *Naturalizing the Mind*. Cambridge, MA: MIT Press.

Goldman, A. (1977). Perceptual objects. *Synthese* 35: 257–84.

Greeley, A. (1975). *The Sociology of the Paranormal*. London: Sage.

Guthrie, S. (1993). *Faces in the Clouds: A New Theory of Religion*. New York: Oxford University Press.

James, W. (1902). *The Varieties of Religious Experience*. New York: Modern Library.

Kelemen, D. (2004). Are children "intuitive theists"? Reasoning about purpose and design in nature. *Psychological Science* 15: 295–301.

Kelemen, D., Callanan, M., Casler, K., and Perez-Granado, D. (2005). Why things happen: Teleological explanation in parent-child conversations. *Developmental Psychology* 41: 251–64.

Leben, D. (2014). When psychology undermines beliefs. *Philosophical Psychology* 27 (3): 328–50.

Levine, M. P. (1990). If there is a God, any experience which seems to be of God, will be genuine. *Religious Studies* 26: 207–17.

Locke, J. ([1689] 1975). *An Essay Concerning Human Understanding*. Ed. P. H. Nidditch. Oxford: Clarendon Press.

Luhrmann, T. M. (2012). *When God Talks Back*. New York: Vintage Books.

Lyons, J. (2009). *Perception and Basic Beliefs: Zombies, Modules, and the Problem of the External World*. New York: Oxford University Press.

McCauley, R. (2011). *Why Religion Is Natural and Science Is Not.* New York: Oxford University Press.

Murray, M. J. (2009). Scientific explanations of religion and the justification of religious belief. In J. Schloss and M. Murray, eds., *The Believing Primate: Scientific, Philosophical, and Theological Reflections on the Origin of Religion.* Oxford: Oxford University Press, 168–78.

Murray, M. J., and Goldberg, A. (2009). Evolutionary accounts of religion: Explaining and explaining away. In J. Schloss and M. Murray, eds., *The Believing Primate: Scientific, Philosophical, and Theological Reflections on the Origin of Religion.* Oxford: Oxford University Press, 179–99.

Norenzayan, A. (2013). *Big Gods: How Religion Transformed Cooperation and Conflict.* Princeton, NJ: Princeton University Press.

Plantinga, A. (2000). *Warranted Christian Belief.* Oxford: Oxford University Press.

Plantinga, A. (2009). Games scientists play. In J. Schloss and M. Murray, eds., *The Believing Primate: Scientific, Philosophical, and Theological Reflections on the Origin of Religion.* Oxford: Oxford University Press, 139–67.

Richerson, P., and Boyd, R. (2005). *Not by Genes Alone: How Culture Transformed Human Evolution.* Chicago: University of Chicago Press.

Richerson, P., and Newson, L. (2009). Is religion adaptive? Yes, no, neutral, but mostly we don't know. In J. Schloss and M. Murray, eds., *The Believing Primate: Scientific, Philosophical, and Theological Reflections on the Origin of Religion.* Oxford: Oxford University Press, 100–117.

Schloss, J. (2009). Introduction: Evolutionary theories of religion – Science unfettered, or naturalism run wild? In J. Schloss and M. Murray, eds., *The Believing Primate: Scientific, Philosophical, and Theological Reflections on the Origin of Religion.* Oxford: Oxford University Press, 1–25.

Schloss, J., and Murray, M. J., eds. (2009). *The Believing Primate: Scientific, Philosophical, and Theological Reflections on the Origin of Religion.* Oxford: Oxford University Press.

Shermer, M. (2011). *The Believing Brain.* New York: Times Books.

Speaks, J. (2011). Spectrum inversion without a difference in representation is impossible. *Philosophical Studies* 156: 339–61.

Swinburne, R. (1979). *The Existence of God.* Oxford: Clarendon Press.

Tamminen, K. (1994). Religious experiences in childhood and adolescence: A viewpoint of religious development between the ages of 7 and 20. *International Journal for the Psychology of Religion* 4: 61–85.

Thurow, J. (2013a). Review: Jesse Bering, *The God Instinct. European Journal for Philosophy of Religion* 5: 196–202.

Thurow, J. (2013b). Does cognitive science show belief in God to be irrational? The epistemic consequences of the cognitive science of religion. *International Journal for the Philosophy of Religion* 74: 77–98.

Tye, M. (2000). *Consciousness, Color, and Content.* Cambridge, MA: MIT Press.

van Inwagen, P. (2009). Explaining belief in the supernatural: Some thoughts on Paul Bloom's "Religious belief as an evolutionary accident". In J. Schloss and M. Murray, eds., *The Believing Primate: Scientific, Philosophical, and Theological Reflections on the Origin of Religion.* Oxford: Oxford University Press, 128–38.

Wainwright, W. (1973). Natural explanations and religious experience. *Ratio* 15: 99–101.

Whitehouse, H. (2000). *Arguments and Icons.* Oxford: Oxford University Press.

Wilkins, J., and Griffiths, P. (2012). Evolutionary debunking arguments in three domains: Fact, value, and religion. In G. Dawes and J. Maclaurin, eds., *A New Science of Religion.* New York: Routledge, 133–46.

Wilson, D. S. (2002). *Darwin's Cathedral.* Chicago: University of Chicago Press.

6 God and Cognitive Science

A Bayesian Approach

Alvin I. Goldman

6.1 INTRODUCTION

The general theme of this volume is the role that cognitive science can play in addressing metaphysical questions. The chapter's specific topic is the role of cognitive science in considering the existence or nonexistence of God. Many different arguments and evidence have been deployed in addressing this age-old question. We begin by looking briefly at a few of the long-standing arguments and methods that have been deployed. We then turn to the prospects of applying cognitive science to the question of a deity, and how such projects might play out.

Among philosophers, several of the most influential arguments for God's existence have been based on purely a priori reasoning. The ontological argument is the most famous example in this genre. Other arguments for God's existence are explicitly empirical, for example, William Paley's "argument from design" (1867). Paley argued that among the (observed) works of nature, high functional complexity is a reliable indicator of design by an intelligent agent. A watch, for example, has great functional complexity, and nobody would think that it would have resulted from anything but an intelligent designer (a watchmaker). Similarly, the world as a whole has great functional complexity; so we should infer that it too is the product of an intelligent designer. Expressed slightly differently, this line of argument is that the *best explanation* of the world's functional complexity is its having been designed by an intelligent designer. In a post-Darwinian age, of course, such an inference is moot. Biological evolution shows that highly complex entities can arise from variation and selective retention. This line of argument, however, does not belong in the category of cognitive science. So it is not the kind of approach that interests us here.[1]

[1] I am not saying, of course, that evolution has no role to play, or contribution to make, within cognitive science. It is just that the science of evolution should not be equated with cognitive science.

Next consider a very ancient argument for the existence of God that has also received attention in recent philosophical literature. This is the *consensus gentium*, or "Common Consent," line of argument, an argument found in Plato, Cicero, Seneca, Calvin, the Cambridge Platonists, Gassendi, and Grotius (see Kelly 2011). This argument can be stated quite simply:

Premise: A majority (or supermajority) of people believe that God exists.

Conclusion: Therefore, God does exist.

The assumption, presumably, is that any opinion held widely by mankind is good evidence for its truth.

Thomas Kelly (2011) discusses this argument with a lot of nuance, but declines to embrace it. He writes: "I do not believe that we are currently in a good position to say anything definitive about the evidentiary value of appeals to common opinion." He offers several reasons for this. First, unless we are told how much these opinions are the products of *independent* belief-formation by the several believers, we cannot start to make any definite inferences about the amount of support conferred by widespread agreement (cf. Goldman 2001). If Jones believes that P and everyone else around him simply bows uncritically to his say-so, the resulting widespread consent may provide nothing like strong evidence for the rest of us to believe that P. Here is a variant of this tale that Kelly offers:

> [S]uppose there is originally no evidence whatever that p is true, but that large numbers of people nevertheless unreasonably come to believe it for bad reasons, or for no reasons at all. If widespread belief in p constitutes genuine evidence that p is true . . . then those who baselessly believe p would seem to have magically brought evidence for p into existence, merely by believing it in the absence of any evidence. Here we seem to have gotten something from nothing.

Here is a further reason Kelly offers for resisting the common consent argument:

> [N]o one thinks that the intellectual case for (e.g.) Islam would be any stronger if birthrates in Muslim countries had been twice as high in past decades as they actually were; nor would the case be any weaker if such birthrates had been significantly lower.

In contrast with Kelly, Linda Zagzebski (2011) offers a hearty endorsement of the *consensus gentium* argument, based on a "self-trust" approach. The crucial steps in her argument are as follows:

(1) Every person must have a general attitude of self-trust in her epistemic faculties as a whole. This trust is . . . shown to be inescapable by philosophical reflection.

(2) The general attitude of self-trust commits us to a general attitude of epistemic trust in the faculties of all other human beings.
(3) The fact that other people believe in God is a prima facie reason to believe that God exists, and the fact that many millions of people constituting a strong supermajority believe . . . that God exists increases my prima facie reason to believe in God myself.

We can agree with Zagzebski that—*for a specifiable class of perceptual* faculties—we are indeed entitled to self-trust. Used for a restricted class of objects, properties of objects, and observational perspectives, our ordinary senses deserve our trust. This is because their outputs are consistent, or replicable. Time and again they lead to concurring judgments. I view a certain tree from many perspectives, under varying light conditions; and my perception of (certain of) its properties remains essentially the same across this diverse set of sightings. This is plausible territory for epistemic self-trust. But there are no analogous faculties that can be applied with similar replicability to supernatural entities like gods. In this kind of territory, self-trust is not warranted by pure philosophical reflection. It isn't entirely clear what Zagzebski is thinking when she argues from the trustworthiness of our faculties "as a whole." She seems to think that since *most* of our (native) faculties are reliable and hence epistemically trustworthy, *all* of our (native) faculties deserve to be trusted. But this is dubious. More importantly, what if we have no native *observational* faculties for believing in gods, because gods are unobservable? Then her unqualified principle of self-trust seems unacceptable (or so its strikes this author). When it is doubtful that we have any native faculty that consistently generates identical judgments with respect to a phenomenon of type X, how can beliefs in category X earn epistemic trust by "philosophical reflection" alone? It might be argued—perhaps partly on the basis of cognitive science—that we *do* have a capacity for perceiving God, although the existence and reliability of this faculty aren't commonsensically detectable in the way that our senses of vision, hearing, touch, etc. are detectable. Then self-trust as applied to gods cannot be shown to be "inescapable by philosophical reflection," as Zagzebski claims.

The foregoing defenses of the *consensus gentium* argument make no use of cognitive science. So even if the *consensus gentium* argument did underwrite belief in God, this would not address the question of how cognitive science bears on theism. The possible role of cognitive science, however, is precisely this chapter's topic. When we return to the *consensus gentium* argument, however, we shall find ways to inject selected findings of cognitive science into that argument. Before turning to this possibility, let us examine other efforts by cognitive scientists to make progress in this territory.

6.2 GOD AND COGNITIVE SCIENCE

During the past twenty years findings from cognitive science have been invoked in the study of religion. Indeed, an entire movement has emerged, the *cognitive science of religion* (CSR), to pursue this issue. How, exactly, can cognitive science contribute to the inquiry? The very idea may seem unpromising. After all, cognitive science is concerned with the workings of the human mind. How can studying the human mind shed light on the reality or unreality of a different kind of being: an immaterial being?[2] We shall approach the subject by first considering lines of argumentation that emerged in CSR. Then we shall examine a line of argumentation not discussed in the CSR literature, and show how it can deploy empirical findings to make headway on the theism question.[3]

In a sense, then, this is an essay as much in epistemology as in metaphysics. We are not here concerned with cognitive science's own methodology. We simply assume that cognitive science has a respectable methodology—or set of methodologies—to establish various facts of human psychology. We then ask how such facts could be utilized to raise or lower the level of epistemic support for God's existence.

6.3 THE "NATURALNESS" OF RELIGIOUS BELIEF

Many contributors to the CSR literature describe their work as showing that belief in God (or gods) is "natural." Pascal Boyer launched this formulation with a book entitled *The Naturalness of Religious Ideas: A Cognitive Theory of Religion* (1994). Paul Bloom's (2007) article, "Religion Is Natural," in the journal *Developmental Science*, echoes the same phrase. What do these writers mean in saying that religious belief is "natural"? At least part of what is meant is that the brain has built-in—or early-maturing—psychological mechanisms that promote interpretations of assorted objects as "agents," i.e., bearers of mental states. And this occurs even when there is scant observable evidence that warrants such an interpretation. Since gods are conceptualized as agents, or persons, this natural tendency readily paves the way for belief in God. Boyer (2001) dubbed this tendency a "hypertrophy of social cognition." More recently it is referred to as a "hypersensitive agency detection device" (HADD) (Barrett 2009).

This trait was first studied by Heider and Simmel (1944), who created a film in which geometrical figures (circles, squares, etc.) moved about in systematic ways, designed to tell a tale. When the movie is shown, observers instinctively describe the figures as if they were people. Subsequent research shows that bounded figures aren't needed to get this effect; it can be obtained even with moving dots (Bloom

[2] There are, of course, various arguments from natural science that purport to identify arguments supporting the existence of God. The so-called fine-tuning argument is a contemporary argument of this sort. But the kind of evidence invoked by this argument—drawing entirely from physics—does not resemble the kinds of evidence one might hope to find in cognitive science. So we leave this type of argument aside.

[3] For the author's previous work on the topic of metaphysics and cognitive science, see Goldman 1987, 1989, 1992, 2007, and especially 2015. A Bayesian approach to these issues was laid out in Goldman 2015.

and Veres 1999). In a similar vein the anthropologist Stewart Guthrie reported in his book *Faces in the Clouds* (1993) that people attribute human characteristics to a striking range of nonhuman entities, including airplanes, automobiles, bags, bells, bicycles, boats, etc. Agency is readily imputed to each of these objects in selected circumstances. This tendency to attribute agency based on minimal cues is an early-emerging trait. Similarly, work by Deborah Kelemen (2004) shows that children have a strong bias toward teleological (purposeful) explanations as opposed to mechanical ones. She calls this "promiscuous teleology." Like the traits cited above, this one could easily promote a tendency to posit unseen, immaterial gods with designs and purposes.

Given these examples of "naturalness," it appears that CSR writers consider beliefs to be "natural" when they are the products of early-maturing psychological tendencies that develop spontaneously and do not depend on reasoning or culture. In the preface to his book *Why Would Anyone Believe in God?* Barrett writes: "[O]nce examined from a scientific perspective, both believers and nonbelievers should appreciate how very natural and almost inevitable" is widespread religious belief (2004, vii). In addition to this sense, however, Barrett's book also suggests a very different sense of "natural." What led him to write the book was a challenge by science-leaning academics to overcome the "utter irrationality of religious belief." He replies: "I do not regard belief in God as strange, loony, or irrational" (2004, 7). Religious belief is "very natural" and "almost inevitable." Thus, "naturalness" here contrasts with looniness or irrationality. It seems to mean "rational," "warranted," or "justified," a respectable *epistemic* status rather than a merely psychological one.

With this in mind, let us distinguish two senses of "natural": a purely psychological sense, which I label *psychological* naturalness (Ψ-naturalness) and *epistemic* naturalness (E-naturalness). Psychological evidence of the sort sketched above may go some distance toward supporting the claim that belief in God is Ψ-natural. But how does it provide support for the view that belief in God is E-natural? How does it show that people who believe in God are *justified* in so believing? As suggested above, Barrett implies that the pertinent psychological findings lend credence to the *rationality* or *justifiedness* of religious belief. However, there is no attempt to defend this thesis in any detail. We are never told what rationality or justification consists in. Nor is it explained how or why the psychological story supports the conclusion that the conditions required for rationality or justification are actually met. The preface of Barrett's book places on the table some core epistemic questions concerning belief in God; but these questions are then dropped for the rest of the book.

6.4 APPEALING TO COGNITIVE SCIENCE TO DEFEND THEISM

In a subsequent paper, however, "Reidian Religious Epistemology and the Cognitive Science of Religion," Clark and Barrett (2011) dive into epistemological issues in some depth, seeking to show that people are *justified* in holding theistic beliefs. They follow epistemological ideas of Alvin Plantinga (1983, 2000) who holds that belief in God is produced immediately—i.e., noninferentially—by a cognitive mechanism present in every human being. Plantinga contends that people can rationally believe in God through the instigation of this faculty without using either *evidence* or *arguments*. Clark and Barrett further argue that such a "god faculty" (or *sensus divinitatis*) is scientifically supported by psychology or cognitive science. What they mean by a "god faculty" remains importantly opaque, however, a theme we shall pursue below. First, however, let us address the purely epistemological matter.

One theory of justification that would allow for justified belief without either evidence or arguments is process reliabilism, a view I have long advocated (see Goldman [1979] 2012). Plantinga's own view is partly derived from that theory, as he acknowledges (Plantinga 1993a, 27–28, 197–205). A (highly) simplified version of process reliabilism holds that a belief is justified if and only if it is produced by a reliable belief-forming process, or sequence of such processes. Clark and Barrett seem to endorse an account of justification (often speaking of "tenability" rather than "justification") that lies in the neighborhood of process reliabilism. This isn't surprising, because, as indicated above, Plantinga incorporates a strong strain of reliabilism in his own "proper function" account of "warrant" (a close correlate of the notion of justification). However, Clark and Barrett do not go down this same path (at least in the paper in question). This introduces problems that require our attention.

A salient feature of process reliabilism is that it does not explicitly invoke "evidence" or "reasoning" in its general formulation. Specifically, in the case of noninferential or "basic" beliefs, justification requires neither evidence possession nor reasoning. This comports with reliabilism. Now let us turn to two questions: (1) Exactly what is a god *faculty*? (2) Can cognitive science be used to establish the existence of a god faculty in an epistemologically sound fashion, i.e., in a way that confers justification on theistic belief?

What does it mean in general to have a cognitive "faculty"? Two possible meanings should be considered:

(F1) Something is an X-faculty if and only if it generates beliefs that affirm the presence or absence of Xs (on a specified class of occasions).

(F2) Something is an X-faculty if and only if it reliably generates beliefs that *accurately* detect the presence or absence of Xs (on a specified class of occasions).

The difference between these senses is that (F2) requires a faculty to be veridical, or accurate, most of the time, whereas (F1) imposes no such accuracy requirement. If humans have a mechanism that frequently generates beliefs in God but there *is* no God, then nobody has a god faculty in the stronger (F2) sense.

Which sense of "god faculty" do Clark and Barrett have in mind? They write, "By 'god-faculty' we mean that the ordinary arrangement and function of cognitive architecture in human minds often produces non-reflective, unreasoned belief in gods" (2011, 14). Nothing is said here about veridicality, accuracy, or truth, so evidently they intend the first—i.e., weak—sense of "faculty." If so, how does belief *justification* arise from a god faculty? Reliabilism would not authorize it. According to reliabilism, a justified belief would issue from a god faculty only if that faculty were reliable, i.e. only if it qualifies as a "faculty" in the (F2) sense.

The next important question is whether cognitive science is well equipped to determine whether there is a god faculty in the (F2) sense. The answer, I submit, is negative. Having a god faculty in the second sense requires (1) a cognitive mechanism that generates beliefs in God, and (2) the actual existence of a God, who makes-true a large proportion of these god beliefs. Some of the beliefs, for example, might have the content "There is an omniscient God." If God did exist and were omniscient, there would be a truth-maker for all of those beliefs, contributing toward a high ratio of true beliefs held by each human believer. But cognitive science has no special research methodology to determine that God exists, or possesses the properties commonly attributed to him. Thus, cognitive science—all by itself—is not in a position to establish, or even strongly support, the satisfaction of condition (2). Hence, it cannot support the existence of a god faculty in the second, i.e., strong, sense of "faculty." Does this mean that cognitive science cannot make *any* contribution to the question of God's existence? Not at all. This chapter's central thesis is that cognitive science *can* make such contributions. But cognitive science cannot make a contribution by directly and independently establishing the *existence* of a God faculty in a sense that entails—or even probabilifies—God's genuine existence. At most it can show that the human mind has propensities that readily lead to *beliefs* in God, an utterly different matter.

Certain writings in the CSR literature may suggest that the relevant researchers *have* established (and even supported) the "stronger," God-entailing kind of thesis. Consider the title of Deborah Kelemen's article "Are Children 'Intuitive Theists'?" (2004). Kelemen answers her question in the affirmative, writing that children are "disposed to view natural phenomena as resulting from nonhuman design" (2004, 295). Readers may construe this as an endorsement of the god faculty idea, which may further be interpreted in the *strong* sense of "faculty." But this would be a misinterpretation. Like other psychologists working in CSR, Kelemen only discusses how people *think,* or *reason.* She is not intimating anything about the truth or accuracy of such thinking or reasoning. (If there is any intimation, it is in the direction of *non*veridicality rather than veridicality.) Similarly, to the extent that HADD is construed as a faculty, the main drift of the psychological findings is to pinpoint *errors* people make in systematically assigning mindedness to nonminded objects (e.g., Heider and Simmel 1944 and Guthrie 1993). This goes nowhere toward showing that people have a god faculty in the (F2) sense, the veridicality-implying sense of the phrase.

Some contributors to the present volume, however, are more sanguine about the prospects of cognitive science lending support to theism. Mark Baker and Dean Zimmerman (this volume) concede that the preexisting CSR literature does little to provide convincing support for theism. But they argue that the tools of cognitive science, along with those of epistemology, can be sharpened so as to make quite a strong case. They are optimistic about the prospects of a god faculty (or module), understood as a faculty of God *perception*. Let us examine the high points of their chapter to see if a stronger case for God perception can be made than the earlier CSR attempts we have just criticized.[4]

Baker and Zimmerman's account of perception runs as follows:

> [In the modern period] philosophers and scientists [have] tended to make the following three assumptions about perception. (1) When an external object is perceived by means of some sense modality, there is a family of mental states associated with that mode of sensing [such that] (2) [t]hese mental states are caused by perceptible states of the object in a reliable, counterfactual-sustaining pattern (so that some differences in the objects perceived will cause systematic variations in the sensory experiences) [and] (3) [h]aving these sensory experiences tends to cause their subject to believe that an object exists with the corresponding perceptible states.

Their discussion then tries to apply this model to the (possible) case of perceiving a God, and to make sure that the whole story of connections between properties of the perceived object and special types of experiences comport with the spirit of contemporary cognitive science (especially a modular form thereof). Their message throughout is that it is not impossible for there to be episodes of people perceiving God. "Not impossible" is my phrase. But it echoes Baker and Zimmerman's terminology, as indicated in the next paragraph.

What is striking for this reader is how weak and modest is their fundamental thesis. What it means to claim is "We thus see *no intrinsic obstacle* [my italics] to the idea that this experience of God is as direct and unmediated as ordinary perceptual awareness of external objects." What they mean to claim is that, *if God exists*, there is no obstacle to his satisfying something like Goldman's (1977) analysis of perception. But all this talk of there being "no obstacle" to God's being a perceptual object carries one hardly any epistemic distance at all toward showing that God actually exists; or showing that it is highly probable that God exists. The argument merely lays out some *possibilities*, with no determinate probability or likelihood of being true. That there is any substantial probability of God existing and being a perceptual object is

[4] There is no space here for a detailed and thoroughgoing examination of the Baker-Zimmerman proposal. Only a few highlights will be hit.

nowhere even claimed, much less demonstrated. So their treatment seems to fare no better than that of Kelly and Barrett.

In addition, there is another central problem for their entire project. If a divine perceptual object is supposed to have properties that cause religious experiences, and if these experiences are supposed to depend counterfactually on the properties of the perceptual object, how can this work unless those properties are *physical* properties? But it is generally agreed that God does not have physical properties. So how, in keeping with the analysis of counterfactual dependence, can God play the role that philosophical accounts of perceptual objecthood require? Moreover, the authors discussed in this section aim to provide a *naturalistic* account of religion. And among cognitive scientists, a naturalistic account would seem to entail a physicalist account. So the idea of hitching theism to cognitive science encounters a serious roadblock from the start, insofar as science is incompatible with the embrace of purely spiritual, nonphysical, objects and properties.

6.5 COGNITIVE SCIENCE AND DEBUNKING ARGUMENTS

Thus far I have been throwing cold water on various attempts to support the existence of God by direct appeal to cognitive science. What about the possibility of challenging the existence of God by appeal to cognitive science? This would qualify as showing the evidential *relevance* of cognitive science to a certain branch of metaphysics. And this, after all, is the unifying theme of this volume. Possible uses of cognitive science to (help) undercut the epistemological status of theism would be genuine examples of cognitive science's relevance. Hence this chapter should not ignore this kind of enterprise. Let us first turn to enterprises of that sort commonly referred to as *debunking* enterprises.[5]

In general usage, to "debunk" something is to disprove or discredit it. In contemporary philosophy, however, the term refers to something more specific. It commonly refers to ways of challenging the existence or metaphysical status of some putative entity (or property, etc.). One way to establish an entity's existence, allegedly, is to show that belief in the targeted entity is usually—or invariably—produced by a suspect type of belief-forming process or etiology, for example, an evolutionary process. Beliefs so produced, allegedly, fail to meet certain epistemological requirements and are therefore unjustified. Showing that they fail to meet those requirements is called "debunking."

Debunking arguments usually employ a three-step line of argument: (1) Establish the *unreliability* of the process or etiology by which the target beliefs are produced. (2) From the unreliability of these processes or etiologies, infer that the beliefs so produced are unjustified. (3) From the unjustifiedness of these beliefs, infer that

[5] Not all debunking enterprises, of course, feature applications of cognitive science. But they are the types of debunking enterprises for present purposes.

the putative referents of these beliefs either don't exist or have a lesser metaphysical status than is commonly assumed.

Guy Kahane (2011) discusses debunking within this kind of framework. He describes the basic structure of a debunking argument as follows:

Causal premise: "S's belief that p is explained by X."

Epistemic premise: "X is an off-track process." Therefore,

Conclusion: "S's belief that p is unjustified." (Kahane 2011, 106)

In formulating the epistemic premise, Kahane uses the term "tracking," borrowed from Robert Nozick (1981). This is slightly suboptimal, however, because "tracking" originally refers to a pair of subjunctive conditionals as the core epistemic criterion rather than reliable belief-forming processes. Moreover, tracking was introduced by Nozick as a theory of *knowledge* rather than *justification*. However, Kahane often uses "off-track processes" interchangeably with "unreliable processes." So he seems to focus on a *process reliabilist* approach to debunking, which is indeed appropriate as an approach to justification. So, as in section 6.4, I shall continue to discuss (doxastic) justifiedness in terms of reliability and unreliability.

In evolutionary debunking arguments, evolution is often said to be incompatible with a reliable production of true belief, because evolution doesn't "aim" at true belief. Thus, Stuart Rachels and Torin Alter write: "[T]he evolutionary process aims at fitness, not truth, wisdom, or rational attitudes" (2005, 314–15). This characterization of evolution is rather misleading, however, and doesn't mesh well with a process reliabilist understanding of "belief-forming processes." "Evolution" (at least in its biological application) refers to a gradual progression of changes in the genetic composition of a biological population over successive generations. So the typical outputs of evolution are genetic changes, not belief states. Of course, such changes can indirectly contribute to the production of beliefs (in suitable organisms). The upshot, however, is that evolution is best conceptualized as a "higher order" process, or metaprocess, that "builds" an assortment of first-order processes which in turn generate actions like running, jumping, feeding, and mating, or states of mind like beliefs, plans, and emotions. With respect to the generation of belief states, there are many "first order" belief-forming processes with different levels of reliability. Even if the accuracy rates of some of these lower-order processes are not very impressive, other lower-level processes may be highly reliable, especially insofar as the beliefs they generate concern potential predators, mates, food sources, and so forth. Thus, assuming that reliabilism is a good standard of belief justifiedness, even animals might be credited with justified beliefs. This is despite the fact that the global upshot of evolution as a whole does not lie primarily in the sphere of belief.[6]

[6] It goes without saying that not all debunking programs invoke evolutionary processes. Nor do they all invoke belief-generating processes or the reliability of such processes. Some approaches

Here is an analogy that may help clarify the foregoing claim. Consider an automated clock-manufacturing system (cf. an auto-assembly plant), which produces an abundance of clocks with varying design properties, some of which yield highly accurate timekeeping and others of which yield highly inaccurate timekeeping. The upshot is that the system as a whole is meta-*un*reliable, but tokens of selected designs are impressively reliable. If we are prepared to treat such clocks as "thinkers," we could consider the time-determinations of these well-designed clocks to be entirely "justified." The timekeeping reliability of individual clocks should neither be downgraded nor ignored simply because the overarching production system is meta-unreliable.

With respect to cognitive skills arising from evolution, consider recent research on animals with quantitative skills, specifically, judgments of numerosity. No animal researcher claims that any animals other than people have a symbolic number system. However, clever experiments have shown that a wide range of nonhuman animals can manage "almost-math" without numbers (Milius 2016, 22). If we credit animals with accurate beliefs about numerosities, why not classify these judgments as "justified"? Yet the evolutionary process that generated these numerically skilled species also probably generated less successful species, ones with poor number skills, which may not have survived very long. Nonetheless, even if the track record of evolution *as a whole* doesn't meet the standard of high reliability, this shouldn't undercut the track record of skilled and highly successful animal species. Nor should their reliably produced numerical judgments be denied "justifiedness."[7]

I turn next to a different worry about debunking. Returning to Kahane's argument scheme, what kinds of things, precisely, are supposed to get debunked? Is it a targeted *belief*, where a belief is a token mental state of an individual? Or is it either (A) a proposition, (B) the truth of a proposition, or (C) an entity, property, or relation putatively referred to by the proposition(s) in question? Kahane's argument scheme regards belief states as the objects of debunkings. Perhaps this is appropriate, because only belief states (and perhaps other doxastic states) are candidates for being epistemically justified or unjustified. But can the unjustifiedness of a belief state be the basis for inferring the falsity of its content, or the nonexistence of objects to which it makes purported reference? Such additional steps seem to be necessary

focus on the "sensitivity" of a belief, which is the appropriate term to use in connection with "tracking." But this chapter makes no attempt to survey all species of debunking, or to evaluate all of them. My main purpose here is to develop my own "substitute" for debunking, highlighting Bayesian reasoning. For a survey of other possible approaches to debunking, see White 2010. Since White is himself a critic of debunking projects, and I too am resisting that family of approaches, I needn't take issue with all of his candidates for debunking, or his criticisms of them. For other discussions of the evolutionary approach to debunking, see Clark-Doane 2012; Vavova 2014; and Bogardus 2016, among others.

[7] The foregoing discussion is partly prompted by a distinction drawn in *Epistemology and Cognition* between first-order and second-order belief-forming processes (Goldman 1986, 21, 51–53, etc.).

to attain a well-founded debunking (beginning with Kahane's initial core). But this is not automatically achieved simply by showing that a belief state lacks justifiedness in virtue of its poor etiology.

Kahane himself is aware of this problem. He writes: "All [that debunking arguments] can show is that someone is not justified in *believing that* p; . . . *on their own they do nothing to show that* p *itself is* false" (2011, 108). That a belief is a product of an unreliable process does not guarantee that its content is false, nor even *probably false*. The upshot is that the conditions presented by Kahane, ostensibly intended to lead to a debunking result, are too *weak* for the job. A belief's being the product of an unreliable process may suffice for that token belief to be unjustified. But no single unjustified belief in P demonstrates P's falsity. Even a hundred unjustified beliefs in P do not demonstrate P's falsity. Suppose many people visit a local cemetery in the darkness of night, and all claim to have glimpsed a bear there. None of them, however, has had more than the slightest glimpse of something vaguely bearlike, and always at a considerable distance. Finally, none of these "witnesses" exchanges notes with the others. Presumably, each of these bear *beliefs* is unjustified; but this does not imply that the bear-in-the-cemetery proposition is false. Does it imply that the presence of a bear has been debunked? No. In general, that a particular existential proposition is widely believed unjustifiedly does not prove its falsity, nor does it debunk the existence of anything posited by that proposition. The existence—and even popularity—of bad methods of belief formation does not preclude the existence of other methods—good methods—for believing it.

The foregoing problems with the standard debunking project are just a sample of reasons that render that project unpromising. Next, however, I shall begin to develop a more promising project that bears some similarity to debunking. It's a project that looks for ways in which science—in our case, cognitive science—can guide our credal assignments with respect to metaphysical issues. In particular, it shows how empirical cognitive science can provide new evidence that can (reasonably) influence a metaphysician's credence toward a metaphysical question; in our case, the existence of God. This comports with a dominant motif of "naturalistic" philosophy.

6.6 APPLYING BAYESIAN METHODS TO REVISIONARY METAPHYSICS

I have said right along that the existence or nonexistence of God is a metaphysical issue. But, as discussed in previous sections, it is also epistemological. It concerns how metaphysicians, or other seekers of the truth, should go about forming and/or revising their beliefs or credences vis-à-vis the existence of God. This kind of epistemic stance should be contrasted with one discussed in connection with the evolutionary debunking project. The question there was how to assess the (doxastic) justifiedness of a previously formed belief. It assumed that such a justificational status hinges on the causal or explanatory history of the belief. So the task was to assess the belief's J-status by considering its actual etiology.

A slightly different kind of epistemic question may also be posed by metaphysicians (and other inquirers, of course). It would commonly take the following form. "The epistemic agent currently has a credence distribution vis-à-vis a specified metaphysical proposition. She now gets new evidence that might be relevant to the problem. How should the new evidence guide her—if at all—in changing or modifying her credences?" This is a *forward-looking,* or *prospective*, question: "What should be my next doxastic step? Given that I have just received new evidence E, how should this influence my upcoming metaphysical credences?" Essentially the same problem could be posed in justificational terms: "What would I be *justified* in doing next?" But this would be a question of *propositional* (ex ante) justification rather than *doxastic* (ex post) justification. And perhaps it would be best expressed in terms of the *rationality* of various possible *changes* in belief.

When it comes to questions of belief change, one naturally turns to Bayesian reasoning. There are many variants of Bayesianism and fierce debates about the best version(s) thereof. Given our space limitations, this is not the place to explore theoretical issues concerning the "proper" or "optimal" form of Bayesianism. However, I shall make some initial remarks to give the reader a feel for how I am approaching these matters. One basic question likely to arise here is whether I shall be working with a subjective or objective form of Bayesian theory. Instead I shall proceed by using standard, elementary Bayesian tools without wading into the subtler issues concerning their interpretation and/or rationale. Readers are invited to construe the examples in terms of their preferred form of Bayesianism. I hope this will bring out the relevant points when seeking to apply cognitive science to the question of God's existence. The central question, then, is: What (in principle) can be learned from cognitive science that might help a metaphysician revise or adjust his/her credences vis-à-vis the proposition "There is/isn't a God"? The cognitive science examples are intended to be "realistic" in the sense that they are not merely invented or contrived, but are "plausible" examples in terms of actual findings by respected cognitive scientists.

Before discussing Bayesianism as applied to religious subject-matter, consider an example from a different domain.[8] You have an emergency and you call to ask a friend if he would lend you his car. He agrees. As you enter it, you wonder whether the car's gas tank is full. Your judgment will depend on what the gas gauge reads. Before even starting the engine, however, your friend calls and says that the car's gas gauge tends to get stuck; its readings cannot be relied upon. You now start the engine and see that the gauge *reads* "full" (= R). What should you infer about the state of the tank? Had you been asked before you received your friend's call—but after you saw the reading on the gauge—the credence you would have assigned to the tank's being

[8] The example is a modified version of one formulated by Roger White (2010).

full was close to 1.0. But now, in light of your friend's new statement, some Bayesian thinking is in order. In light of the new information, you should ask: what is the probability of the gauge reading "full" conditional on the tank's *being* full (P(R/F), and what is the probability of the gauge reading "full" conditional on the tank's *not* being full (P(R/~F)? It is reasonable to set the value of P(R/F) = .50, and perhaps the same value for P(R/~F). You should also consider the *prior* probabilities of the tank's being full versus not full. A reasonable pair of assignments for these priors might be .40 for full and .60 for not full. Finally, how should you incorporate this information to yield a rational assignment for "the tank is full"? Using a variant of Bayes' theorem, you should make the following analysis:

$$P(F/R) = \frac{P(F) \times P(R/F)}{P(F) \times P(R/F) + P(\sim F) \times P(R/\sim F)} = \frac{(.4) \times (.5)}{(.4) \times (.5) + (.6) \times (.5)} + \frac{.20}{.50} = .40. \qquad (6.1)$$

Clearly, the impact of the friend's testimony is substantial. Without his testimony, a reading of "full" would surely invite a much higher posterior probability than .40.

Now turn to an example that concerns the question of God's existence and invokes psychological evidence. We will apply an analogous Bayesian approach there. The two competing hypotheses are "God exists" (= G) and "God does not exist" (= ~G). First suppose that there is tolerably solid prima facie evidence in support of God's existence. Later we shall consider evidence from cognitive science to override that prima facie evidence and substantially reduce its impact. In this fashion we shall see how cognitive science and Bayesian reasoning could jointly weaken rather than strengthen the case for God's existence.

As promised earlier, we now return to the *consensus gentium:* a high incidence of *belief* in God throughout known human history and culture. We do not, however, immediately accept widespread belief in God as superstrong evidence for God's existence. Partly for reasons discussed in section 6.1, the philosophical metaphysician need not—and should not—accept this fact as irrefutable evidence for the existence of God. The evidence should be weighed critically. One way to do this is to reflect on how widespread belief in God might have come about. One possibility is that God indeed exists and does things—e.g., answers prayers—that lead people to believe in his existence. A second possibility—one that is critical to the present line of argument—is that God does not exist but through a set of psychological and/or sociological mechanisms, belief in God becomes widespread. The task of weighing the plausibility and/or likelihood of these scenarios is where cognitive science enters the picture. Bayes' theorem is conveniently deployed here because it supplies an analytical tool that shows how to integrate the indicated possibilities—or "likelihoods"—in ways that comport with standard probability theory.[9]

[9] In a commentary on an earlier draft of this chapter, Roger White interpreted it as assigning Bayesianism a very "limited scope" because it is restricted to "one specific item of potential evidence,"

Although widespread belief in God does not logically entail his existence, it might be good probabilistic evidence for his existence. But grounds for doubt must also be considered. A pivotal move is to notice that even if widespread belief in God is probable assuming that he does exist, it might equally be true that widespread belief in him is probable even assuming he doesn't exist. Moreover, precisely this combination of conditional probabilities might be supported by cognitive science, including some of the early CSR literature mentioned in section 6.3. People tend to overattribute human properties to nonhuman entities. This is well documented by Boyer, Guthrie, Heider and Simmel, and others. A metaphysician who takes on board these psychological findings and appreciates the probabilistic impact of such findings should find it easy to incorporate this kind of information into a Bayesian framework. Let us see how this might go in the context of the *consensus gentium* argument.

Let us clothe these ideas in some numbers, representing possible credences of a metaphysician. A metaphysician might *initially* endorse the following two claims: (A) The conditional probability (or likelihood) of widespread theistic belief, if God existed, is very high, e.g., .90. This is because, if God existed, he would make himself known by answering prayers, issuing moral injunctions, etc. (B) On the other hand, the conditional probability of there being widespread belief in God, if God did *not* exist, is very low, e.g., .10. This is because, if God did not exist, it is unlikely that many people would believe in him. Finally, assume that the metaphysician initially has an "open mind" about God's existence: she assigns the same prior probability—i.e., .50—both to "God exists" and to "God doesn't exist." Now, letting "W" express the proposition that there is widespread belief in God, our metaphysician initially assigns the following four credences or conditional credences:

$$P(G) = .50;\ P(\sim G) = .50;\ P(W/G) = .90;\ \text{and}\ P(W/\sim G) = .10.$$

Suppose that our metaphysician understands the principles of Bayesian reasoning. So she plugs these numbers into a variant of Bayes' theorem, and makes a calculation.

$$P(G/W) = \frac{P(W) \times P(W/G)}{P(G) \times P(W/G) + P(\sim G) \times P(W/\sim G)}$$
$$= \frac{(.50) \times (.90)}{(.50) \times (.90) + (.50) \times (.10)} + \frac{.4500}{.4500 + .0500} = .90. \qquad (6.2)$$

namely *W* (*widespread* belief in God's existence). But no such restriction is intended here. The *consensus gentium* evidence is only a single *illustration* of the kind of evidence on which Bayesian reasoning might operate. I conjecture that many other such examples could be developed, but space limitations preclude extensive development of this claim.

At this juncture, the metaphysician assigns a high posterior probability, i.e., .90, to the proposition that God exists.

Next the metaphysician consults the kind of cognitive science literature described earlier and learns new things that influence her credal assignments. Specifically, her new evidence undercuts one of her initial likelihood assignments. She now has evidence to the effect that widespread belief in God is fairly probable—much more probable than .10—*even* on the assumption that God does not exist. This seems possible in light of what has been learned about the nature of human mentalizing tendencies (as presented in the CSR literature).

Evidence reviewed in section 6.3 could have this upshot. Specifically, it could be evidence linking belief in supernatural beings with an overactive tendency to impute mental states, or agency, to unseen or imagined objects. Widespread belief in gods may not emerge from the actual existence of God but from people's mentalizing tendencies. Given such tendencies, people would posit unseen gods with mental states such as purposes and beliefs, *even if no such gods exist.*

We see, then, how this body of cognitive scientific information could easily lead an informed and open-minded metaphysician to make a new—and much lower—credal assignment to the proposition that God exists, as follows:

$$P(G/W) = \frac{(.50)\times(.90)}{(.50)\times(.90)+(.50)\times(.90)} = \frac{.4500}{.4500+.4500} = .50. \qquad (6.3)$$

A posterior probability assignment of .50—as contrasted with the initial assignment of .90—is a very considerable reduction. If such a reduction is a reasonable upshot of taking cognitive scientific evidence seriously, it is a very significant impact indeed.

6.7 ADDITIONAL RESEARCH MIGHT FURTHER WEAKEN THE CASE FOR THEISM

Let us now turn to a different line of research concerning religious phenomena. This line of research will rationalize the view that the second likelihood in the foregoing argument may well support an even larger revision of the likelihood of widespread belief in "our" kind of God even if God does not exist. When the value of this likelihood is suitably increased, it will call for a (further) reduction in the posterior probability of God's existence, below the level of .50.

The following discussion draws on a body of work from Ara Norenzayan's (2013) book *Big Gods*. Norenzayan is a psychologist who makes extensive use of the history and anthropology of religion as well as contemporary psychology. He takes no explicit position on the existence of God, but lays out historical facts and contemporary experimental results that shed fascinating light on religious tendencies. For present purposes, I assume these findings to be correct and hence potential evidence concerning how religious belief might be generated.

Norenzayan points out that religions have long multiplied and mutated over time, at a worldwide pace of two or three per day. Most religions die out quickly, save for a potent few that have endured and flourished. Most religious people living on the planet today are cultural descendants of just a few religious movements. Strikingly, the surviving religions are parts of large, anonymous, yet cohesive and highly cooperative societies. The challenge to large and anonymous societies is free-riding. Societal punishment of selfish acts is an effective strategy that stabilizes co-operation, but what happens if people fail to punish? Norenzayan's answer pinpoints the role of belief in supernatural beings such as gods, ghosts, and devils. In particular, the "Big Gods" (as he calls them) of major world religions are believed by their faithful to be "watchers" or "monitors" of antisocial behavior, whose (believed) monitoring of everybody's behavior keeps people in check. Gods of earlier religions were not thought to have this monitoring characteristic. The transition to belief in Big Gods, who were reputed to have interventionist inclinations, had a major impact. "Believers who feared these gods cooperated, trusted, and sacrificed for the group much more than believers in morally indifferent gods, or gods lacking omniscience. . . . Through [various] solidarity-promoting mechanisms, religions of Big Gods forged initially anonymous strangers into large, cohesive moral communities tied together with sacred bonds of a common supernatural jurisdiction" (2013, 8–9).

Norenzayan supplements this history with the following facts. Successful religions make vivid use of the theme that their adherents are under the gaze of watchful eyes. "Everywhere in the Hebrew Bible, the New Testament, and the Koran, it is asserted that God sees everything, even and especially when no one [else] is watching. Similarly, 'Buddha Eyes' are displayed everywhere in Buddhist villages in Tibet and Nepal. Sacred objects at the center of towns and villages depict Lord Buddha, also known as the 'Eye of the World,' in Buddhist scripture, as observing the coming and going from high up" (2013, 24).

Another relevant fact presented by Norenzayan concerns people's responsiveness to being watched or observed. This responsiveness is displayed in various social-psychological experiments. "People are so sensitive to social monitoring that even schematic representations of eyes can change behavior in a more prosocial direction. Such hypervigilance is found even when experimental subjects are merely looking at three black dots arranged to look like a schematic face with eyes (in the form of an upside down triangle standing on its pointed tip)" (2013, 21). Participants in an experiment who played the one-shot Dictator game and were exposed to a watching-eye configuration tempered—i.e., moderated—their selfishness (Rigdon et al. 2009).

In short, the kind of God that has psychological meaning or emotional power for most believers is a person-like being that monitors, responds, punishes, and rewards. People prefer a "personal" God to an abstract or impersonal one, a God who is involved in their lives, who can hear their prayers and offer forgiveness and mercy. The more culturally successful Big Gods perform a balancing act that combines transcending human abilities to observe and intervene with enough human mind perception to make them intuitively graspable and emotionally potent (2013, 26–27). As Pascal Boyer (2001) has stressed, these gods have "privileged access" specifically to behaviors with moral consequences. "In principle, God knows everything about Bill, but in the minds of believers, God knows more about whether Bill cheated on his wife than about what color socks Bill chose to put on this morning before going to work" (Norenzayan 2013, 27).

Norenzayan reports an experiment by Benjamin Purzycki and colleagues (2012) that illustrates this. They had religious American students read a series of questions about God. Some questions had no moral relevance, e.g., "Does God know the recipe for Alice's cake?" whereas others had moral information about good deeds, e.g., "Does God know that Ann gives to the homeless?" A third batch had moral information about bad deeds, e.g., "Does God know that John cheats on his taxes?" Purzycki and colleagues found that even among participants who specifically claimed that God is all knowing and all powerful, they responded more quickly to moral than nonmoral questions and even faster to questions about negative rather than positive behavior. Norenzayan sums things up as follows: "[Believers] intuitively treat God as a powerful and omniscient agent who is particularly interested in negative behaviors that have moral consequences for others, precisely what we would expect if God were a supernatural monitor who oversees cooperative and honest interactions among humans" (2013, 28). This is what he means, apparently, in speaking of "Big Gods" in connection with the major religions.

How does this material bear on the existence or nonexistence of a supernatural being? I shall treat Norenzayan's account as *evidence* about the nature of belief in God, at least the type of belief that is extremely widespread. Bearing this in mind lets us revisit the question of whether it would be rational for a metaphysician to posit a God based on this slightly more determinate specification of mankind's God beliefs (more determinate as compared with the loosely specified deity discussed in section 6.7). In the present case, the relevant evidence is widespread belief in *Big Gods* (B).

First, let us fix notation for present purposes.

B = There is widespread belief in Big Gods.

G = A Big God (one or more of them) exists.

B is presumed to obtain as a piece of evidence. Initial conditional probabilities are as follows:

$P(B/G) = .90$

$P(B/{\sim}G) = .10$

Under these specifications, a Bayesian calculation would yield a posterior probability of a Big God existing of .90 (see section 6.7). Does the analysis change if we assume additional scientific evidence that might affect the probability assignments to the likelihoods? Reasons were adduced in section 6.7 to increase the second likelihood number as compared with the initial assumption. Specifically, consider the conditional probability of widespread belief in Big Gods even if no Big God actually exists (P(B/~G). The value of this expression was initially pegged at .10. But Norenzayan's research might support an even higher value for this second conditional probability. (This conditional probability, of course, is not a posterior probability.)

To assess this, however, we first must review all of the evidence. Much of it is focused on the force of cultural inheritance in the spread of religion. Seeking to explain how prosocial religious belief and practices got off the ground and swiftly populated the earth, Norenzayan cites Richerson and Boyd's *Not by Genes Alone* (2005). It is part of the "iron law" of Darwinian evolution that when there is variation and selective retention in any entity capable of some sort of replication, traits that have fitness advantages will spread at the expense of their less fit rivals. This happens in cultural evolution as well as in genetic evolution. In the case of prosocial (Big Gods) religion, there are societal advantages to such religious beliefs. When people imagine being monitored by God, cooperation and trustworthiness are enhanced, thereby encouraging even long-distance commerce in merchant communities, a signal "fitness advantage" (2013, 56–59). Applied to the present issue (the likelihood of widespread belief in Big Gods even if such gods do not exist) there is a clear reason why the emergence and spread of widespread belief of this kind might still have a high probability—even without the actual existence of Big Gods. There is ample reason, then, to boost the assigned likelihood of widespread belief in Big Gods even assuming their nonexistence. Raising that likelihood to .80, for example, would *reduce* the posterior credence for Big God existence as follows:

$$\begin{aligned} P(G/B) &= \frac{P(G)\times P(B/G)}{P(G)\times P(B/G)+P(\sim G)\times P(B/\sim G)} = \frac{(.50)\times(.90)}{(.50)\times(.90)+(.50)\times(.80)} \\ &= \frac{.4500}{.4500+.4000} = .5294. \end{aligned} \tag{6.4}$$

The value .5294 is substantially lower than .90, the value of the posterior probability under the original numbers (cf. section 6.7).

Having revised the value of the second conditional probability—i.e., P(B/~G)—let us now turn to the first conditional probability, P(B/G). This was tentatively set at .90. Is such a high value warranted? For reasons to be explained, I think not. Consider what Norenzayan tells us about early religions:

> A startling fact about the spirits and deities of foraging and hunter gatherer societies is that most of them do not have wide moral concern. . . . [A]nthropologists tell us that in small bands resembling ancestral human groups, the gods may want to be appeased with sacrifices and rituals, although they are typically unconcerned about moral

> transgressions such as theft and exploitation, which preoccupy the Big Gods of major world religions. . . . Religion's early roots did not have a wide moral scope. (2013, 7)

Norenzayan writes as if he were telling us what preoccupied Big Gods themselves (in that period), but, of course, what this material really tells us what *people* in these early groups *imputed* to the gods. It informs us about what they *thought* gods wanted and cared about, namely, sacrifices rather than cooperative or moral behavior. It also tells us, in effect, that those people didn't *believe* in "Big Gods" defined as Norenzayan has done.

Now let us assume for the moment that there are Big Gods, and they are eternal (or atemporal, perhaps). So they existed at the time of those early religions (as well as now). But then we have a situation in which there are/were Big Gods but people didn't much believe in *them*; they only believed in other kinds of gods. Apparently, then, either Big Gods did not care about conveying their real preferences to the faithful, or they were unable to persuade them. In either case, there was a *failure* on the part of existing Big God(s) to generate widespread *belief* in gods of the indicated kind (i.e., Big Gods). This contravenes the idea that there is a strong conditional probability relation between the existence of Big Gods and widespread belief in them. In light of this additional information, the earlier number for the likelihood P(B/G), .90, seems too high.[10] A significantly lower credal assignment seems preferable. Suppose we lower it from .90 to, say, .40. With this change in the example, our Bayesian equation is the following, yielding a posterior value of .333.

$$P(G/B) = \frac{P(G)\times P(B/G)}{P(G)\times P(B/G) + P(\sim G)\times P(B/\sim G)} = \frac{(.50)\times(.40)}{(.50)\times(.40)+(.50)\times(.80)}$$
$$= \frac{.2000}{.2000+.4000} = .333. \qquad (6.5)$$

This analysis yields the reduced odds of Big Gods existing as quite a bit lower than before. This upshot, of course, will be unwelcome to theists, but it is perfectly compatible with Bayesian reasoning, and is generated from a not-unreasonable assignment of numerical values to the inputs. Notice that the example continues to posit *priors* of .50 both for God's existence and for his nonexistence, an assignment quite generous to theists. But the posterior assignment of P(G/B) will nonetheless be rather low.[11]

[10] In this vein we should not forget Hume's ([1779] 1948) discussion of possible inferior deities whose world-making skills are suboptimal. This would include deities who are inept at generating accurate god beliefs in humans. Such possibilities must to be included when assigning a plausible value to P(B/G).

[11] Some philosophers of religion at Western Washington University have objected that the foregoing discussion relies too heavily and exclusively on Norenzayan's religious history and social science to assign the relevant conditional probability, or likelihood. No doubt there is merit in the idea that more literature and complexity could be introduced here. But the present venue does not allow for such explorations. In any case the point of my line of argument is to illustrate the kinds of

In response, theists might take issue with the high value assigned to P(B/~G). After all, theists believe that God created the world, including not only human beings but all psychological capacities and dispositions possessed by humans. From this perspective, if God never existed, humans would not have had those capacities and hence would not have believed in a (Big) God's existence. So it is questionable whether there is *any* possibility of a (Big) God not existing but people believing in a (Big) God nonetheless. According to this line of argument, the assigned value of .90 for P(B/~G) is excessive. Roger White (2010) argues in roughly this fashion as follows:

> Suppose I am justified in believing in the existence of God. . . . Naturally I think that God is ultimately responsible for my theistic belief. Indeed, I'm justified in thinking this if I'm justified in believing that there is a God. So of course I should think that my theistic belief is not at all independent of whether there is a God. If there weren't I wouldn't be here, let alone mistakenly believe in God. (2010, 582)

White correctly describes one possible scenario: a scenario in which he is *justified* in believing in the existence of God. In that scenario, he claims, he would also be justified in thinking that God is ultimately responsible for his theistic belief. (The latter justification doesn't automatically follow, however, from the former justification. Since doxastic justifiedness depends on how the belief is caused, he may or may not be justified in thinking that God is responsible for his theistic belief. Moreover, there is the problematic assumption that God—if he exists—would necessarily be responsible for White's theistic belief.) But the foregoing is only one possible scenario. A different possible scenario is one in which he believes in the existence in God but this belief isn't justified. White doesn't argue against this as a possibility; and it provides space for the sort of increased conditional probability discussed in the foregoing line of argumentation.

I would further argue that metaphysicians—indeed, all philosophers—should pursue their philosophical inquiries with an *open mind*. They should not presume or stipulate particular answers to their philosophical questions "ahead of time." Thus, White isn't entitled to give any special status to the scenario in which he is justified (initially) in believing in God's existence. That would be begging a critical question. So the possibility of God's not existing must be allowed, and the possibility of humans believing in him despite his nonexistence must also be allowed. And it is perfectly appropriate to assign a nonnegligible value to the probability P(B/~G).

discoveries in cognitive science and social history that might generate a rather different picture of where the key likelihood—or range of plausible likelihoods—might lie. The range, I am suggesting, may be quite different than that explored by most philosophers and theorists of religion. The particular numbers I have chosen are just illustrations. (But illustrations can often be helpfully suggestive.)

However, White has also offered an argument (in correspondence) against the assumed likelihoods used in equation 6.5. There it was assumed that widespread belief in God conditional on God's existence is *less likely* than widespread belief in God conditional on God's nonexistence [$P(B/G) = .40 < P(B/{\sim}G) = .80$]. Isn't this a highly artificial scenario rather than a "natural" one? How could God's actual *non*existence be *more* likely (than his existence) to generate widespread human *belief* in him? Why should we pay any attention to this bizarre scenario?

I dispute the suggestion that this is a bizarre or highly artificial combination of assumptions. Suppose that God—if he existed—would be a rather modest being, not given to self-promotion. He would create the universe but then go about his (other) business without fanfare. He wouldn't encourage humans to create stories about him or worship him. On the other hand, if there weren't any God, assorted humans *would nonetheless* become preachers, prophets, or priests who promote religious movements to improve society or promote their own self-interest. Any of these scenarios could give rise to just the pair of likelihoods assumed in equation 6.5. Would this be a bizarre pair of likelihoods, which should be ignored? Not at all.

Consider the following parallel from a different domain, that of ordinary commercial activity in which a product's quality is the analogue of God's existence vs. nonexistence and its popularity among consumers is the analogue of widespread belief in God. It might initially seem clear that a high-quality product would generate wider approval or popularity among consumers than a low-quality product. But this need not be the case. A high-quality product might be assumed to "sell itself"; no need to spend advertising money on it. Contrariwise, a poor-quality product might need intensive advertising to generate a receptive market. Assume that producers proceed on this basis. They give virtually no advertising to their high-quality products and lots of advertising to the poor-quality products. This practice could easily lower the likelihood of widespread popularity for superior products and raise the likelihood of widespread popularity for inferior products. An analogous situation might obtain for widespread belief in God. If informed of these relative likelihoods in the theism case, a thoughtful Bayesian metaphysician should lower her posterior probability assignment for God's existence, as described in connection with equation 6.5.

6.8 COULD COGNITIVE SCIENCE AND BAYESIAN INFERENCE LEND POSITIVE SUPPORT TO THEISM?

Thus far our examples of Bayesian reasoning applied to evidence from cognitive science have tilted toward atheism rather than theism. Is this an essential feature of these analytical tools? Or could the same tools lend equal support to theism? Is Bayesian reasoning a "neutral" tool, *in principle* offering both sides of the dispute equal opportunities for a defense of their ideas? Or is it a biased analytical tool?

There clearly are some ways by which cognitive science might be deployed to buttress the case for theism. To take a trivial example, suppose that cognitive science laboratory, L_1, makes a discovery that supports reduced credence in God's existence. Later another cognitive science laboratory, L_2, conducts a new study that undercuts the findings of L_1, thereby restoring the previous dialectical status quo. This is obviously a case in point, a case in which cognitive science bolsters the case for theism (slightly). I am confident that there are other cases—more interesting cases—in which the same tilt is exemplified. I don't currently have a nice example of the desired sort, but I welcome suggestions from readers.

6.9 CONDITIONALIZATION

I now turn briefly to some technical issues concerning my treatment of Bayesianism. There is an alternative Bayesian tool for analyzing the earlier scenarios. This tool—the rule of *conditionalization*—is a very standard one in the Bayesian kit, and might well seem like the ideal tool for present purposes. Why, then, didn't I use it from the start? The reason is that the conditionalization rule doesn't provide a transparent mention of the two likelihoods that figure prominently in our earlier analyses. Vivid incorporation of these likelihoods, I felt, would help readers "see" what goes on in the cases in question, thereby promoting a better grasp of how cognitive science fits into the story. A downside of this choice, however, is its neglect of certain details that are pretty prominent in Bayesian analyses. So I shall try to cover a few of these details in this section, to bring my overall treatment into closer synch with standard Bayesianism.[12]

Conditionalization is a rule for updating sets of credences, for moving from one credence function at a given time to another credence function at the next time. A credence function at a given time represents all of a person's credences at that time. The rule of conditionalization is intended to instruct a cognizer how to respond to a body of new evidence. (My earlier treatment dealt only with single items of evidence at a time.) Here is one formulation of the rule:

> Assume you begin with a probabilistically coherent credence function. Your new credence function upon receiving total evidence E ought to be $Cr_{new} = Cr_{old}(- / E)$. That is: Take your previous credence for any proposition *P* and replace it with your previous credence in *P given E*; this gives you your new set of credences. (McGrath 2015, 267)

The phrase ". . . given E" brings the relevant likelihood, or conditional probability, into the picture. As one can infer, the results of operating with the conditionalization rule should not diverge in any relevant respects from the style of analysis that appears in this chapter. Mainly, it is more general, and deals with theoretical issues

[12] Thanks to Daniel Rubio for suggesting this addition.

of probabilistic epistemology that needn't concern us here. It is noteworthy, however, that conditionalization is a *commutative* operation. Commutativity implies that conditionalizing on evidence E1 and then on E2 will yield an equivalent result as conditionalizing (at a given time) on "E1 and E2." This assumption was tacitly in the background in our earlier discussion.

ACKNOWLEDGMENTS

The author thanks Dennis Whitcomb and several of his colleagues at Western Washington University for assorted advice and pointers to relevant literature.

REFERENCES

Barrett, J. (2004). *Why Would Anyone Believe in God?* Walnut Creek, CA: Altamira Press.

Barrett, J. (2009). Cognitive science, religion, and theology. In J. Schloss and M. Murrary, eds., *The Believing Primate: Scientific, Philosophical, and Theological Reflections on the Origin of Religion*. Oxford: Oxford University Press, 76–99.

Bloom, P. (2007). Religion is natural. *Developmental Science 10* (1): 147–51.

Bloom, P., and Veres, C. (1999). The perceived intentionality of groups. *Cognition* 71: B1–B9.

Bogardus, T. (2016). Only all naturalists should worry about only one evolutionary debunking argument. *Ethics* 126: 636–61.

Boyer, P. (1994). *The Naturalness of Religious Ideas: A Cognitive Theory of Religion*. Berkeley: University of California Press.

Boyer, P. (2001). *Religion Explained*. New York: Basic Books.

Clark, K. J., and J. Barrett (2011). Reidian religious epistemology and the cognitive science of religion. *Journal of the American Academy of Religion* 79: 1–37.

Clark-Doane, J. (2012). Morality and mathematics: The evolutionary challenge. *Ethics* 122: 313–40.

Goldman, A. I. (1979). What Is justified belief? In G. Pappas, ed., *Justification and Knowledge*. Dordrecht: Reidel. Reprinted in A. I. Goldman (2012), *Reliabilism and Contemporary Epistemology*. New York: Oxford University Press, 29–49.

Goldman, A. I. (1986). *Epistemology and Cognition*. Cambridge, MA: Harvard University Press.

Goldman, A. I. (1987). Cognitive science and metaphysics. *Journal of Philosophy* 84: 537–44.

Goldman, A. I. (1989). Metaphysics, mind, and mental science. *Philosophical Topics* 17: 131–45.

Goldman, A. I. (1992). Cognition and modal metaphysics. In *Liaisons: Philosophy Meets the Cognitive and Social Sciences*. Cambridge, MA: MIT Press, 49–66.

Goldman, A. I. (2007). A program for "naturalizing" metaphysics with application to the ontology of events. *The Monist* 93 (3): 457–79.

Goldman, A. I. (2011). Toward a synthesis of reliabilism and evidentialism? In T. Dougherty, ed., *Evidentialism and Its Discontents*. New York: Oxford University Press. Reprinted in A. Goldman (2012), *Reliabilism and Contemporary Epistemology*. New York: Oxford University Press, 123–50.

Goldman, A. I. (2015). Naturalizing metaphysics with the help of cognitive science. In K. Bennett and D. Zimmerman, eds., *Oxford Studies in Metaphysics*, vol. 9. New York: Oxford University Press, 171–216.

Guthrie, S. (1993). *Faces in the Clouds: A New Theory of Religion*. New York: Oxford University Press.

Heider, F., and Simmel, M. (1944). An experimental study of apparent behavior. *American Journal of Psychology* 57: 243–59.

Hume, D. ([1779] 1948). *Dialogues Concerning Natural Religion*. Ed. Henry Aiken. New York: Hafner.

Kahane, G. (2011). Evolutionary debunking arguments. *Noûs* 45 (1): 103–25.

Kelemen, D. (2004). Are children "intuitive theists"? Reasoning about purpose and design in nature. *Psychological Science* 15: 295–301.

Kelly, T. (2011). Consensus gentium: Reflection on the "common consent" argument for the existence of God. In K. J. Clark and R. J. Van Arragon, eds., *Evidence and Religious Belief*. New York: Oxford University Press, 135–56.

McGrath, M. (2015). Probabilistic epistemology. In A. I. Goldman and M. McGrath, *Epistemology: A Contemporary Introduction*. New York: Oxford University Press, 267–69.

Milius, S. (2016). Animal math: Searching the barnyard and zoo for the evolutionary roots of human number crunching. *Science News* 190 (12): 22–26.

Norenzayan, A. (2013). *Big Gods: How Religion Transformed Cooperation and Conflict*. Princeton, NJ: Princeton University Press.

Paley, W. (1867). *Natural Theology: Or Evidence of the Existence and Attributes of the Deity Collected from the Appearances of Nature*. N.p.: Cross-Reach Publications

Plantinga, A. (1983). Reason and belief in God. In A. Plantinga and N. Walterstorff, eds., *Faith and Rationality*. Notre Dame, IN: University of Notre Dame Press, 16–93.

Plantinga, A. (1993). *Warrant: The Current Debate*. New York: Oxford University Press.

Plantinga, A. (2000). *Warranted Christian Belief*. New York: Oxford University Press.

Purzycki, B. G., Finkel, D. N., Shaver, J., Wales, N., Cohen, A. B., and Sosis, R. (2011). What does God know? Supernatural agents' access to socially strategic and non-strategic information. *Cognitive Science* 36: 846–69.

Rachels, S., and Alter, T. (2005). Nothing matters in survival. *Journal of Ethics* 9: 311–30.

Richerson, P. J., and Boyd, R. (2005). *Not by Genes Alone: How Culture Transformed Human Evolution*. Chicago: University of Chicago Press.

Rigdon, M. L., Ishii, K. Watabe, M., and Kitzyama, S. (2009). Minimal social cues in the dictator game. *Journal of Economic Psychology* 30: 358–67.

Vavova, K. (2014). Debunking evolutionary debunking. In R. Schafer-Landau, ed., *Oxford Studies in Metaethics*, vol. 9. New York: Oxford University Press, 76–101.

White, R. (2010). You just believe that because . . . In J. Hawthorne and J. Turner, eds., *Philosophical Perspectives*, vol. 24. New York: Wiley-Blackwell, 573–615.

Zagzebski, L. (2011). Evidence and religious belief. In K. J. Clark and R. J. van Arragon, eds., *Evidence and Religious Belief*. New York: Oxford University Press, 22–36.

PART IV
MEANING, LINGUISTICS, AND ONTOLOGY

7 Cognitive Psychology and the Metaphysics of Meaning

Mark Johnston and Sarah-Jane Leslie

The philosopher's only resource is the analysis of concepts we already possess.

——*Michael Dummett, The Nature and Future of Philosophy (2010)*

In "Concepts, Analysis, Generics and the Canberra Plan" (Johnston and Leslie 2012), we distinguished the clusters of psychologically real heuristics that govern our use of terms—what cognitive psychologists often call "concepts"—from the philosophical notion of concepts as the meanings of terms, be they public terms, or mental terms in a supposed language of thought. Throughout what follows, meanings are understood as the semantic determinants of the extensions of the terms in question, and hence of the truth-conditions of the sentences that contain those terms.

Both the semantic and psychological notions of concepts are just fine, and not at all in competition, at least when understood as directed at different targets—the phi-concepts and the psi-concepts, as we propose to call them. But they are sometimes connected by an empirically discredited quasi-supernatural theory of our use of terms. The quasi-supernaturalist theory is that use is quite generally guided by *grasp of semantic meaning*, where this involves supposed occurrent causal influences on a speaker's psychological life by things neither physical nor mental, namely extension-determining "senses" or, in our terms, semantic meanings or phi-concepts.

In our view, there is no general explanatory factor that deserves the name "grasp of semantic meaning." It is a misleading *façon de parler*. Aside from the cases where "grasp of semantic meaning" describes explicit knowledge of one or another of the comparatively few statements of any given language made true by meanings, knowledge to the effect that they are in fact so made true—cases such as "vixens are female foxes," "a copse is a thicket of bushes," and the like—the phrase "grasp of semantic meaning" is merely the description of an achievement, i.e. that of acquiring the ability to express one's thoughts by the use of the relevant terms in a given language, and to understand thoughts so expressed. It is not a tenable account of some means by which the ability is realized.

Talk of semantic meaning has its proper home, not in a psychological theory of what guides use, but in the characterization of languages as abstract objects, where each such language is individuated by (*a*) a syntax and (*b*) a potentially infinite list of pairings of sentences of the language with their respective meanings. The potentially infinite lists can be characterized by exploiting a finite class of primitive subsentential expressions (as specified by the syntax), among them the names and simple predicates of the language, which are assigned their own meanings understood as encodings of their systematic contribution to the meanings of the sentences that contain them. If the language is potentially infinite, yet in principle learnable, then there will be a finite list of recursive rules for the combination of the meanings of subsentential and sentential expressions, rules mirroring the syntax of the language, which generate meanings for each of the potential infinity of sentences of the language. As a result of a specification of L's individuation conditions in terms of which meanings it pairs with which subsentential and sentential expressions, we can specify (1) truth-in-L understood as the statement of the truth conditions for the sentences of L and (2) analytic validity-in-L, which concerns just what argument patterns using terms of L are guaranteed to be truth-preserving solely in virtue of the meaning of those L-terms.

The question then becomes: what is it for a population to speak one of these abstractly individuated languages rather than another? In the jargon of the1970s and 1980s: what is the actual language relation?[1] One standard account, which took various forms, was that a population speaks an abstractly specified language L if there are dispositions had by the speakers in the population to use the sentences of L in a variety of conventional speech acts such as assertions, directives, inquiries, etc., whose contents—*what* is asserted, commanded, inquired as to whether, etc.—are propositions with just the meanings paired with those sentences in the specification of the language L. There was an element of proper idealization in the determination of the relevant dispositions; they were to include ordinary speakers' dispositions to correct such usages as a result of deferring to distinguished speakers of the same language, e.g., in culture circles like ours, these include parents, teachers, and those scientifically in the know. So, to take Tyler Burge's familiar example, even someone who does not know that "arthritis" applies only to inflammation in the joints may speak a language in which this is nonetheless a condition on correct use of "arthritis," precisely because of that person's pattern of deference to distinguished speakers of the relevant part of the language, in this case doctors.

From the fact that a population, or group of speakers, achieves enough conventional coherence to speak a given language L, nothing very specific follows about just

[1] See David Lewis (1969, 1975), Stephen Schiffer (1972), Christopher Peacocke (1975), Martin Davies (1981), and Barry Taylor (1982).

how the dispositions to fall in with the conventionally required use of sentences are psychologically realized in the members of the population. After all, conventional conformity of use derives not just from the coincidence of the outputs of individuals' heuristics, but also from massive ongoing deference to the use of distinguished others around us; first our parents, then our teachers, then, in an age of science adoration, to those in the technical know.

In general, from the fact that two speakers at a given time speak the same language abstractly considered nothing very specific follows about what, as an empirical psychological matter, guides those individual speakers of the language in their use of terms in the language. Conformity in conventional use can cover a multitude of individual psychological differences among speakers in respect of the heuristics that generate that conformity of use. Likewise, one's heuristics can change over time, say as one learns more about a topic, without the meanings of one's terms for that topic changing.

In our view, the idea that what generally guides the speakers of L, the language they together speak, in their use of the expressions of L is a *grasp of the meanings* of such expressions, where meanings are then understood as the items that enter into the individuation of L, involves an odd mixing of explanatory levels, the psychological and the individuative. In familiar atheoretical parlance, to "grasp the meaning of a term" is just to know how to correctly use it in a range of not-too-far-out actual and counterfactual circumstances. In its turn, that just involves having some effective heuristics or criteria for telling when one has, and when one has not, something to which the term applies. There is psychological reality behind talk of a speaker grasping such heuristics, but no general psychological reality behind talk of grasp of meaning. Again "grasping meaning" is the description of an achievement, not the description of some queer process by which this achievement is explained.

7.1 WHAT PSYCHOLOGY HAS TO OFFER

Thanks to cognitive psychology, we now know a great deal about just what our heuristics or criteria are like: they frequently take merely generic and prototypical forms; they often are guided by folk-scientific views about explanatory relations among the features of the thing at hand. The employment of such criteria frequently falls far short of even a tacit understanding of universal necessary and sufficient conditions for the application of the term. But that is the very thing which literally grasping the semantic meaning and being guided in one's use by such a grasp would provide. For in the semantic tradition, meaning is understood as determining sentential truth-conditions across all possible worlds, and in that vein the meaning of a subsentential expression—the concept expressed in the philosophical sense—is understood as determining the expression's extension and antiextension in every possible world.

Although the focus and the questions that drive psychological theorizing about concepts significantly deviate from the semantic tradition, this fact is partly obscured by an entrenched pattern of storytelling found in survey articles on the history of the psychology of concept use. Unfortunately, many summaries of the psychological literature on concepts begin with something called the "classical view," according to which subjects actually exploit represented necessary and sufficient conditions when applying most lexical concepts (i.e., a concept that is expressed by a single word). The standard illustration of the classical view is the concept *bachelor*, supposedly composed of the concepts *unmarried* and *male of marriageable age*, such that anything is a bachelor just in case it is an unmarried male of marriageable age. But, of course, not all concepts can be decomposable in this way; there must, on the classical view, be some or other stock of basic concepts, out of which all other concepts are ultimately composed. So "the classical view" encourages a picture of concept learning as combining such basic concepts to form complex ones. Bringing an item under a decomposable concept is then supposed to be a matter of checking whether the item satisfies the necessary and sufficient conditions specified by the decomposition. That is, the psychologist's target known as the "classical view" of the concept bachelor is not just the banal remark that it is true in virtue of meaning that someone is a bachelor if and only if he is an unmarried male of marriageable age. It is the specific empirical thesis that we actually use the concepts *unmarried* and *male of marriageable age* in deciding whether to count something as a bachelor. It is a thesis about the criteria we actually use, not, or not just, a thesis about the meanings of our terms.

Since the 1970s, the classical view, understood as a thesis about the criteria we actually use, has been quite roundly rejected. Much of the reason for its rejection has to do with the discovery by Eleanor Rosch and her colleagues of so-called typicality effects (e.g., Rosch 1973, 1978; Rosch and Mervis 1975). For many categories, some members of a category are perceived as being more typical examples of the category than others, and it turns out that how typical a category member is actually predicts a very wide range of experimental results. For example, people are quicker to categorize typical members, and are more confident and consistent in their categorization of typical members. When learning a novel concept, people learn to categorize the typical members first, and they learn the concept faster when presented with typical members in the learning phase. There are myriad other effects of typicality on language learning and use, on reasoning, and so on so forth. (For some very helpful reviews, see Laurence and Margolis 1999; Murphy 2002; Smith and Medin 1981.)

Hypothesizing that people are guided by, i.e. represent and exploit, necessary and sufficient conditions will not explain typicality effects. Knowing that something is a bachelor just in case it is unmarried and a male of marriageable age does not give any information about what makes for the typical (James Bond) versus the atypical bachelor (Pope Francis). The bachelors are all alike in respect of being unmarried males of marriageable age. Something else has to be posited to explain typicality effects; but once this something else is recognized there may be no empirical reason to also posit the *actual exploitation* of known necessary and sufficient conditions. The classical view has fallen into disrepute because many investigators believe that precisely this has turned out to be true.

Consider, for example, an experiment conducted by Jerry Fodor and his collaborators, in which they asked whether one could find any differences in processing time that would indicate that one concept is composed in part by another. If, e.g., the concept *murder* is composed in part by the concept *kill* (as has been claimed), then it should take longer to process *murder* than *kill*, since processing the former involves processing the latter as a proper part. However, this prediction of the classical view is not borne out (Fodor et al. 1980).

Subsequently, Armstrong et al. (1983) found some typicality effects even for concepts such as *odd number*. This suggests that typicality effects are compatible with representing necessary and sufficient conditions. However, the crucial point is that in the case of concepts such as *odd number* we have independent reason for supposing that we represent necessary and sufficient conditions—namely that we (i.e., typical competent adult users of the term "odd number") can articulate what they are. There is no corresponding case to be made for the majority of concepts. Thus while the in-principle compatibility point concerning typicality effects and necessary and sufficient conditions is illustrated by Armstrong et al.'s findings, this in itself does not alter the fact that, for most concepts, we simply have no reason to suppose that ordinary, competent adults generally represent and exploit necessary and sufficient conditions suited to specify meanings. Indeed, like the Fodor result, the implication of Armstrong et al.'s findings is that even when speakers do know the semantic decomposition of a concept they are not exclusively using that decomposition as their heuristic for applying the related term.

Seeking to explain typicality effects, many psychologists were led to the prototype theory of the heuristics that make up our concepts. Prototypes can be understood as statistical functions over properties, which assign weights to features based on how likely a category member is to have that feature, or conversely, based on how likely something with that feature is to be a category member. There are a number of different proposals that fall under the heading of the prototype theory (see Murphy 2002, for an extensive review), but they all characteristically appeal to features that are in some way statistically related to category membership. For example, the prototype for *dog* might include features such as *barks, has four legs, has a tail, wears a collar*, and so on. These features are not candidates to figure in universal necessary and sufficient conditions since not all dogs have these features; an unfortunate creature can still be a dog even if it has three legs, no tail, no collar, and no bark. However, the basic idea behind prototype theory as illustrated by the concept *dog* is that if one is confronted with an animal and wishes to determine whether or not it is a dog, one will use this animal's features, or lack thereof, in a complex subpersonal calculation based on the weights of the various features in the prototype of *dog*. (The details of this calculation differ a great deal depending on the particular version of the theory.)

The weight that a feature receives in the prototype is generally taken to be determined by two sorts of statistical facts, namely the *prevalence* of the property among *dogs* (so *barks, has a tail, and so on*, would receive high weights since most dogs have these features), and/or the *cue validity* of the property; that is, how likely it is that something with that feature is a dog. Thus even though, perhaps, most dogs don't wear collars, the probability of something being a dog if it wears a collar is high, so *wears a collar* might receive a significant weight in the prototype. The more highly weighted features an individual has, the more typical an exemplar of the kind it will be. Prototype theory thus places typicality effects first and foremost among the data it aims to explain.

Prototype theory has many adherents, and many well-motivated critics. For while we may often rely on statistically weighted features in categorization, particularly in rapid, perceptually based categorization, it seems that this cannot be the whole story. Imagine, for example, that you are presented with a raccoon. A perverse scientist then comes along and alters the creature, dying its fur so that it takes on the markings that are typical of a skunk, and even goes so far as to implant a sac of smelly liquid that the creature can use to spray smells when it is under stress. How would you categorize this creature? It now has all the typical features of a skunk, yet overwhelmingly, from elementary school on up, people say this is still a raccoon (Keil 1989). This finding has proven difficult for standard prototype theories to accommodate.

Furthermore, it seems increasingly clear that typicality ratings are not solely driven by statistical facts; crucially the causal status of features also matters. Imagine that two features are equally prevalent among members of a kind and have the same cue validity, but that one is understood as generally being the cause of the other. Suppose then that an instance of the kind has one feature but not the other. Since the relevant statistical facts are the same in both cases, prototype theory would seem to predict that typicality ratings of the individual would not be affected by *which* feature is lacking. However, individuals exhibiting the effect but not the cause are rated as less typical than those exhibiting the cause but not the effect (Ahn et al. 2000).

These results, along with many others, suggest that our ways of categorizing things, and reasoning about things in categories, involve a richly structured knowledge base that is responsive to causal-explanatory factors as well as statistical factors (e.g., Carey. 1985, 2009; Gelman 2003; Gopnik and Meltzoff 1997; Keil 1989). As we will use the term here, this is the outlook typical of the so-called theory-theory of concepts.[2] Since theory-theory, so construed, posits that our concepts or criteria for

[2] Sometimes 'theory-theory' is reserved for the specific view that our psi-concepts, including young children's psi-concepts, are *very like* scientific theories. We follow many in the field by using the term more inclusively.

categorization and generalization are sensitive to causal-explanatory structure, there is not too much to be said beyond that about the *general* features of our concepts. Rather, it may be empirically useful to go on to consider concepts within each broad domain, e.g., natural kind concepts, artifact concepts, social concepts, mental state concepts, mathematical concepts, and so on and so forth.

For example, a view known as *psychological essentialism* seems to provide a great deal of insight into how our natural kind concepts are structured, indeed from a very young age (e.g., Gelman 2003; Leslie 2013). However, this view is very likely not applicable to artifact concepts or to mental state concepts, and certainly not to mathematical concepts. This sort of domain sensitivity should not be seen as a failing of theory-theory, but as an upshot of the complex and myriad ways we have of categorizing things, and of generalizing on the basis of those categories.

Probably the view that fits best with the mass of empirical material on concepts is a hybrid of theory-theory with some elements drawn from prototype theory. There are terms like "red" or "dog" which we can apply rapidly and without reliance on theory, at least in some circumstances. It is natural to think that this goes by way of subpersonal processing of sensory and perceptual information with subsequent comparison with paradigmatic or prototypical sensory and perceptual profiles—how red things look, how the varieties of dogs look, etc. The characterization of such prototypical profiles—in particular whether and to what extent they use prevalence and cue validity—is a complex piece of empirical psychological theorizing, yet to be completed.

7.2 THE GENERIC ENCODING HYPOTHESIS

One feature that the foregoing psychological theories of concepts have in common is that they all make some reference to properties that are possessed by members—*plural*—of the target category; they all involve focus on forms of *generalization* concerning the category and its properties. According to the classical theory, the relevant generalizations are universal generalizations; the prototype view treats them as probabilistic generalizations; while on the theory-theory, they are causal-explanatory general beliefs. These observations suggest a possible alternative route to studying the nature of our classificatory and inferential heuristics: we should look to the empirical investigation of our earliest and most fundamental types of generalization.

Suppose, for example, that it was possible to identify and describe our most basic and persistent way of forming general judgments about kinds or categories, our most basic and persistent way of moving from information concerning individual members of a category to judgments concerning the category in general. It would be quite surprising if this basic and persistent manner of generalization was not centrally connected with the heuristics for categorization and inference concerning kinds or categories. Thus a natural and conservative empirical hypothesis would be that our conceptual heuristics in large part consist of such cognitively basic and persistent types of generalizations.

Recent interdisciplinary research suggests an intriguing possibility along these lines; namely that our basic way of generalizing information issues in *generic* generalizations, which are articulated in language via generic sentences such as "tigers have stripes," "lions have manes," and "mosquitoes carry malaria" (e.g., Gelman 2010; Hollander et al. 2002; Leslie 2007, 2008, 2012; Leslie and Gelman 2012; Leslie et al. 2011; Mannheim et al. 2011; Tardif et al. 2011).

Such generic sentences exhibit a puzzling truth-conditional profile, as a few familiar examples quickly illustrate. Consider, for example, "lions have manes"—this strikes most people as obviously true, yet only mature male lions have manes. There are perfectly normal lions (e.g., female lions) who lack manes, and yet the generic seems true. Further, there are *more* male lions than there are maned lions (since some males are immature or lack manes for genetic or environmental reasons), yet the generic "lions are male" is widely rejected, which suggests that contextual restriction is not the source of the phenomenon. Perhaps even more puzzling are generics such as "mosquitoes carry malaria," which are accepted despite the fact that only about 1% of mosquitoes carry the virus. Yet generics such as "books are paperbacks" are robustly rejected, even though over 80% of books are paperbacks (for more discussion of generics, see Carlson and Pelletier 1995; Cohen 1996; Leslie 2007, 2008; for empirical investigation of people's judgments of these sorts of generics and others, see Prasada et al. 2013).

Most importantly for our purposes here, generic generalizations are obviously not equivalent to universal generalizations, as is already confirmed by the truth of "lions have manes" and "mosquitoes carry malaria." Obviously, even high-prevalence generics such as "tigers are striped" and "dogs have four legs" tolerate exceptions in a way that their universal counterparts do not. "All tigers are striped" is falsified by a single stripe-free albino tiger; similarly for "all dogs have four legs." The corresponding generics are more robust, however. They are true in the face of such exceptions.

If generic generalizations constitute our most basic and persistent way of making general judgments about categories, this raises still another empirical challenge for the classical view. A proponent of the classical view would have to argue that the general information employed in our classificatory heuristics does not originate from our most basic and persistent way of forming general judgments. The information that we use to identify members of a category would not come by way of our basic and persistent means of forming general judgments about the category. This is not incoherent, but given the overwhelming absence of empirical evidence in favor of the classical view, it stands out as the positing of yet another defensive epicycle.

Likewise for what we have stigmatized as the quasi-supernaturalist picture of language use: if our classificatory heuristics commonly exploit generics, then our classifications are *not* invariably guided by a grasp of meanings understood as determinants of universally necessary and sufficient conditions, i.e. conditions which would fix the extension of the classificatory term in question in every possible world.

Why should we think that generic generalizations are more cognitively basic and persistent than universal ones? Some of the relevant data come from the study of language acquisition. As noted, generics have a very complex truth conditional profile; providing an account of when generic sentences are true or false is a quite demanding task (see e.g., Carlson and Pelletier 1995; Cohen 1996; Leslie 2008). In contrast, it is very easy to provide an account of when universally quantified statements are true ("all Ks are F" is true iff the set of Ks is a subset of the set of Fs). In light of this, one would expect that universals would be easier for young children to acquire and process than generics; however, this is precisely the opposite of what we find. Generics are produced and understood by preschool-aged children, and the data collected to date suggest that these young children have a remarkably adult-like understanding of generics. For example, preschoolers who know that only "boy" lions have manes will accept "lions have manes" but reject "lions are boys"—despite implicitly understanding that there are at least as many "boy" lions as there are maned lions (Brandone et al. 2012; see also Gelman and Raman 2003; Gelman et al. 2002; Graham et al. 2011; for a summary of the available evidence on generic acquisition, see Leslie 2012).

Preschoolers are generally competent with the quantifier "all" *when it is applied to a specific set of individuals* (e.g., Barner et al. 2009). For example, if preschoolers are shown six crayons and asked, "Are all *of these* crayons in the box?" they are usually able to answer the question correctly. Most of the work on quantifier acquisition has focused on such situations; however, it should be clear that these sorts of limited, nonprojectable pseudogeneralizations are not the sort that are involved in conceptual heuristics. The question, then, is how young children fare with open-ended, category-wide universals—not "all of these crayons," but "all crayons." Several studies indicate that they have considerable difficulty processing universal quantifiers in such kind-wide generalizations. Most intriguing, though, is that when preschoolers are confronted with such kind-wide universals, they do not simply provide random, incorrect answers; instead *they treat the universals as though they were generics*. That is, preschool children not only consistently evaluate generics just as adults do, they also evaluate kind-wide universals as generics (Hollander et al. 2002; Leslie and Gelman 2012; Tardif et al. 2011; for a detailed review, see Leslie 2012). In addition to English-speaking children, such findings have also been documented among Mandarin Chinese- and Quechua-speaking children; similar results have also been found with other quantifiers (Brandone et al. 2015; Hollander et al. 2002; Mannheim et al. 2011; Tardif et al. 2011).

Importantly, these findings are just what one would expect on the hypothesis (Leslie 2007, 2008, 2012) that generics, unlike universals (and "some"- and "most"-quantified statements), articulate cognitively basic generalizations. If the cognitive system has a basic, default way of forming general, open-ended judgments, then it may sometimes fall back on this means of generalizing when asked to process a more taxing and sophisticated generalization. This tendency will be most pronounced in young children, who would be expected to struggle with the more taxing generalizations. Not only do young children not struggle with generic generalizations, they substitute their understanding of the generic when asked to consider category-wide quantified generalizations.

If generics do in fact articulate cognitively basic, default generalizations, then one would expect that these effects might not be limited to young children. Adults might also be susceptible to the error of treating quantified statements as generics. Indeed, under a variety of circumstances, adults do show a robust tendency to accept universally quantified statements such as "All ducks lay eggs," despite knowing that male ducks do not lay eggs (where the tendency to accept the universal was *not* due to participants interpreting the universal as quantifying over only females, or over subkinds of ducks; Leslie et al. 2011; see also Meyer et al. 2011). This finding would be explicable if adults were not always evaluating the universal claim, but were instead, like preschoolers, sometimes substituting their evaluation of the corresponding generic.

Further confirming evidence for the persistence of the generic mode of generalization can be found in the study of adult reasoning errors. For example, Steven Sloman (1993, 1998) investigated adults' evaluations of arguments that involve the quantifier "all," finding that their evaluations did not conform to the logic of universal quantification. For example, his participants judged that arguments such as (A) are strictly stronger than arguments such as (B), despite judging that reptiles are indeed animals:

(A) All animals use norepinephrine as a neurotransmitter; therefore all mammals use norepinephrine as a neurotransmitter
(B) All animals use norepinephrine as a neurotransmitter; therefore all reptiles use norepinephrine as a neurotransmitter

This pattern of judgment is simply mistaken given the logic of the universal quantifier; however, if we replace the universals in the arguments with generics, then the judgments of the participants would be very reasonable. Since generics tolerate exceptions, the claim "Animals use norepinephrine as a neurotransmitter" can be true even if some animals are exceptions to the claim. If one also judges that reptiles may not generically use norepinephrine while mammals probably do, then argument (A) is indeed stronger than argument (B). Hence these results are as one might expect if adults have a tendency to evaluate universals as generics.

Note that adults also judge that universals such as "All ravens are black" are more likely to be true than universals such as "All young jungle ravens are black," despite understanding that the latter are a subset of the former (Jönsson and Hampton 2006). Again, this is incoherent if one is really dealing with universally quantified statements; however, if one were instead evaluating these universals as generics, this would be a reasonable judgment, since for all one knows young jungle ravens may be exceptions to the generic "ravens are black." These results are thus naturally read as lending support to the hypothesis that adults are treating these universals as generics. As a further piece of converging evidence from another experimental paradigm, it has been found that both preschoolers and adults *recall* previously presented quantified statements as generics (Leslie and Gelman 2012).

The hypothesis that generics, unlike quantified statements, articulate cognitively basic and persistent generalizations thus has a fair amount of empirical support at this time. As a final observation in favor of the hypothesis, we might note that quantified statements require a phonologically articulated element, namely the quantifier itself. That is, we say "*All* tigers are striped" or "*Most* tigers are striped"; however, in the case of the generic, there is no corresponding articulated element (e.g., "*Gen* tigers are striped"). This is not an isolated fact about English; rather it would appear that few, if any, natural languages have a dedicated, articulated generic operator (Carlson and Pelletier 1995; Dahl 1985).

Here is an explanation for this otherwise puzzling fact: if one wishes to interact efficiently with a system, and the system has a basic, default way of proceeding or performing a task, then one need only issue an explicit instruction to the system if one wishes it to *deviate* from this default way of proceeding. To convey the idea in more intuitive terms, if one is dealing with a child who, say, by default does not pick up her toys, one only needs to say something if one wishes the child to *deviate* from her default and actually pick up her toys. If one does not wish the child to pick up her toys on a given occasion, it would be a waste of breath to say, "Don't pick up your toys!" since this is what will happen even if one remains silent. Thus, quantifiers may be articulated in language because one needs to *tell* the cognitive system, as it were, to deviate from its default, generic mode of generalizing, and instead generalize in the universal manner or the existential manner and so on and so forth (for more details, see Leslie 2008, 2012). Generics, by virtue of expressing the basic or default mode of generalization, require no such phonological marking. They represent the unmarked case.

More can be said about the centrality of generic generalization (see Leslie, forthcoming). Suffice to say, the best available theoretical construal of the data supports the view that our first inferences to generalizations about kinds produce generics, and that the generic form remains the default mode of generalization, even in adult life. Second, as already noted, on both the theory-theory and prototype theories, the heuristics which guide the application of our terms are general in form; indeed they must be so if they are to be available for use from one occasion to another.

Those two considerations lead us to propose a new, empirically motivated and philosophically consequential, amendment to both the theory-theory and the prototype theory, namely the *generic encoding hypothesis*: the heuristics which typically guide our use of terms by exploiting prevalence, cue-validity,[3] and causal explanatory structure[4] are properly formulated in generic terms.[5]

7.3 PSI-CONCEPTS ARE NOT MEANINGS

We are now in a position to see why the cluster of generic heuristics a speaker associates with a term should not be taken as specifying the semantic meaning of that term in the speaker's language.

For one thing, the exact heuristics employed may vary from speaker to speaker, each of whom may speak the same language L *because of their convergence on a pattern of conventional use, a pattern corralled by deference to correction by distinguished others, i.e. parents, teachers, and those in the know*. They may then come to mean the same by a given term of L, even though they employ different heuristics. The same for a single speaker over her lifetime: without changing the meaning of a term *F* she may simply learn more about *F*s and thereby come to have better recognitional heuristics when it comes to classifying *F*s; even so, she may continue to use *F* with its L-meaning, i.e. continue her earlier pattern of conventional use of *F*.

Second, even when an inference from

X is *F* and X is a *G*

to

X is an *FG*

is L-valid, i.e. truth-preserving in virtue of the meanings of the terms of L, it may not be the case that the common heuristic for an *FG* among L-speakers is a union of their individual heuristics for *F* and *G*. This can be so, even though *FG* is a term whose extension is just the intersection of the extensions of *F* and *G*. For *FG*s may lack one or another of the classificatory heuristic features had by either *F*s or *G*s.

Here is a nice example adapted from Jerry Fodor and Ernie Lepore (1996). If some specific thing is both a pet and a fish then it is a pet fish. But the prototype of a pet fish, or as we would put it, the most salient generic pet fish, is a goldfish, which is not well modeled by the union of the features of generic fish and generic pets.

[3] Here we have in mind such heuristics as "Collared animals are dogs."
[4] "An animal's outer appearance is due to its 'interior' properties."
[5] It is, perhaps, rather telling that psychological experiments concerned with concepts almost invariably use the generic form to articulate conceptual knowledge.

There is a third way in which concepts, understood as clusters of heuristics associated with a term, may come apart from meanings. Two people may associate the very same heuristics with *F*, and yet mean different things by *F* in the sense that their dispositions to use *F* in various speech acts to contribute to communicative ends, i.e. to things being asserted, commanded, inquired after, etc., can differ. For their patterns of deference to the usages of parents, teachers and experts can significantly differ. Ms. Self-Sufficient may not be into that sort of thing at all, while Mr. Go-Along-to-Get-Along may be totally open to correction by others.

Suppose that both Ms. S's and Mr. G's heuristics for applying the term "dog" are exhausted by the following down-and-dirty criteria:

Is it an animal?
Does it have one of the characteristic looks, smells, coat textures, etc., of one of the familiar kinds of animals we call 'dogs'?
Is it the offspring of an animal with one of the characteristic looks, smells, coat textures, etc., of one of those familiar kinds of animals we call 'dogs'?

As things actually go in suburban environments, these three criteria may be good heuristics for collecting together observed instances of the kind dog, the kind we now know to be the species *Canis familiaris*. However, this is due to the contingent fact that the canines around us are almost all of them from that species. The dogs have a sister taxon *Cana lupus* which includes all and only the wolves. Wolves are not dogs, but some wolves look very like German shepherds, some look very like huskies, some look very like malamutes, and some look very like the new dog breed Tamaskans, which were deliberately bred from German shepherds, malamutes, and huskies to resemble wolves.

However, wolves avoid human communities and so typically don't roam in suburban neighborhoods, like those frequented by Ms. S and Mr. G. So, both Ms. S and Mr. G invariably recognize the dogs around them as dogs, and since they do not encounter wolves they do not in fact count any wolves as dogs, though both would be likely to do this, since they lack distinguishing criteria, such as that dogs bark while wolves howl. What makes Mr. G's term "dog" refer to the dogs rather than to a group which includes both the dogs and the wolves is his pattern of deference to distinguished use—in this case to that system of biological classification which tells us that the dogs form a species *Canis familiaris*, a species which excludes the wolves. However, Ms. S, who has no such disposition to defer, has nothing that would make her term "dog" have such an extension. Given just her criteria or ways of telling, her term "dog" has both dogs and doglike wolves in its extension. In this respect, though their heuristics are the same, Ms. S and Mr. G speak a different language in the strict sense of a language understood as a potentially infinite list of pairings of sentences and (extension-determining) meanings. For Ms. S's term "dog" arguably has both dogs and wolves in its extension.

7.4 PHILOSOPHICAL CONSEQUENCES

Though we "know the meanings of our terms" this is typically not *by way of* grasping meanings. Semantic meaning properly figures in the individuation of a given language L, in the characterization of analytic validity in L, and collaterally in the individuation of "concepts" semantically conceived, i.e. as pairings of (subsentential) terms with their meanings. It has at most a minor role in a general explanatory account of how we use terms. A very small percentage of simple terms are like "vixen," "copse," etc., where speakers *are* in a position to articulate the terms' semantic meaning.

As one of us argued long ago (Johnston 1988), although questions of syntax are of great interest, the pure theory of semantic meaning has very little to it. Semantic meanings are functional indexes, suited to play certain quasi-model theoretic roles in the specifications of truth conditions and of analytic validity. To think that it is grasp of semantic meaning that quite generally guides our use of terms is a level-confusion. The truths about the semantic meanings of our terms are epiphenomena generated by our conventionally constrained use of those terms. And this use is guided by our generic, prototypical, and partly causal-theoretical criteria for applying and withholding the terms.

So what?

Well, to begin on exploring the philosophical consequences of this view, recall verificationism about meaning: the view that the meaning of a sentence is to be given by its methods of verification and/or falsification, which in turn decompose into the heuristics which guide our application of the subsentential expressions which make up the sentence. In the light of the foregoing, verificationism appears to be another form of the level-confusion we have been emphasizing. Methods of verification and falsification—heuristics or criteria—are one thing, meanings another.

When we apply any predicate *F*, we are using criteria or ways of telling—i.e. methods of fairly direct verifying and falsifying—whether such and so is an F. If our respective ways of telling quite generally coincide in their deliverances, and if dispositions to defer also coincide, we may come to speak the same language, i.e. to be disposed to use the sentences of some language with the same speech act potential, namely to assert, command, inquire after, etc., the same things. And yet, thanks to the language's rules of semantic composition, there may be meaningful sentences of the language, such as unqualified universal generalizations, which have no method of direct verification, along with other sentences, such as negative existentials, which have no method of direct falsification. There may be meaningful sentences of the language about events in the past, all of whose traces have washed away. Likewise for sentences about events in "pocket universes" that are forever outside our light cone. We could have a well-confirmed theory that tells us there are such events, while leaving their specific character open. The unverifiable and unfalsifiable sentences representing the *specific* characters of such events nonetheless will have a meaning compositionally conferred upon them. As a result, some of the things that speakers can do with the language is ask questions which have no accessible answers. The corresponding possible answers thus have meanings that transcend what can be settled by speakers' methods of verification and/or falsification for the sentences that express them.

Recall that Michael Dummett held that a theory of meaning for a language was a theoretical representation of a practical ability by way of a statement of what, knowledge of which, would suffice to be able to use the language. He then famously argued that a theory of meaning should take the form of a statement of conditions of verification and not an assignment of potentially evidence-transcendent truth conditions to the sentences of the language. For, he reasoned, the latter would overdescribe the practical ability we do in fact have, precisely because our criteria of use underdetermine such truth-conditions.

In the light of the forgoing, that argument looks like the familiar mixing of levels, the level of meaning and the level of what guides concept application. One thing we can do with our language is ask questions that have epistemically inaccessible answers. This ability is conferred on us not by our actually being guided by evidence-transcendent heuristics (whatever they would be), but by being able to produce speech acts with evidence-transcendent truth or satisfaction conditions, thanks to the semantically compositional character of the language we speak.

Recall that Jerry Fodor was keen to attribute verificationism to cognitive psychologists offering empirical theories of concepts. As against Fodor's useful provocations, we have argued here that the proper theoretical placement of that psychological work shows precisely why verificationism is mistaken and why the relevant cognitive psychologists were *not* in fact verificationists. Fodor's wholesale rejection of psychological theorizing about concepts, so far as we can see, consisted of making just the right points about concepts in the philosophical sense and then supposing that psychologists were theorizing about that. Again, the levels have to be confused to get the argument going.

Despite the storytelling that presents prototype theory as an improvement on the classical theory, which in its turn *seems* like a meaning theory, the psi-concepts are not an empirical account of the phi-concepts, nor were they meant to be.

A related line of thought emerging from our reflections suggests that names could have meanings, over and above their denotations, even though Saul Kripke (1980) was entirely right that the meaning of a name is not given by our ways of identifying the bearer of the name. Setting aside Kripke's modal argument, the success of which turns on the question of whether our heuristics are properly understood as tied to how things go in actuality, Kripke's other arguments—that our heuristics differ, that they are often impoverished, and that they are ill-suited in the specification of universal necessary and sufficient conditions for being the bearer of the name—are from the present point of view best understood as perfectly correct antiverificationist points. As Kripke himself was careful to note, those arguments do not in themselves take us all the way to "Millianism" about names, i.e. the thesis that names have no meaning over and above their denotation. They simply rule out identifying the meaning of a name in our language with the heuristics we use to apply it. In our terms, Kripke highlighted the difference between psi-concepts and phi-concepts in the case of proper names.

7.5 THE NORMATIVITY OF MEANING

Another upshot of distinguishing psi-concepts and phi-concepts has to do with the so-called normativity of meaning, also brought into sharp relief by Kripke (1982). To be sure, semantic meaning is normative in that there are correct and incorrect uses of terms. But this is so only *relative* to some specific language. "To table" in American English means to remove from the main line of discussion. In British English, it means the opposite: to introduce into the main line of discussion.[6] (Apparently, the mutual failure to understand this led to a temporary rupture between Churchill and Roosevelt during an important campaign of the Second World War.) Clearly the norms for correct use have to be stated relative to a language; they are conditional norms, such as

> If you aim to use 'to table' as an expression of American English, e.g. in a conversation with speakers who only have American English, then you should use it to mean: to remove from the main line of discussion.
>
> If you aim to use 'to table' as an expression of British English, e.g. in a conversation with speakers who only have British English, you should use it to mean: to introduce into the main line of discussion.

There is accordingly nothing puzzling about the so-called normativity of meaning; it is just a case of the normative principle that tells one to use a means appropriate to one's end. The intuition that the normativity of meaning is somehow more than this, indeed somehow *categorical* in nature, can be traced, we suggest, to the relative paucity of homophones with differing meanings across different languages. Because of this contingent fact, the always correct antecedent reference to a language is almost always harmlessly left out. The "normative force" you feel is the internalization of conventional pressure towards local conformity of use; it is not the supernatural influence of a meaning rule on your psychological life. There is no general semantic conscience to guide us. ("Vixen," "copse," and the like notwithstanding.) What we sometimes feel to be guiding our use are our heuristics, not antecedent *graspings of meaning*.

As far as following a linguistic rule goes, the psychology of language use is the place to look. It implies that there is no guarantee of singularity or of determinacy

[6] Thanks to Christopher Peacocke for the example.

when it comes to the rule to which our use of a term conforms. More generally, there is no guarantee that there is a unique language in the abstract sense that we are related to by the actual language relation.

This is not "meaning skepticism" in any worrying sense, precisely because meaning is not the source of use. Meaning simply figures in the systematic *registration* of that use for formal semantic purposes, i.e. individuating a language L, specifying conditions of truth-in-L, and characterizing analytic validity-in-L.

7.6 THE GENERIC ENCODING HYPOTHESIS AND THE METHOD OF CASES

As late as 2010, Michael Dummett can be found defending the view that the proper method of philosophy is the analysis or articulation of the conditions of application of our concepts. Several of our own colleagues, especially among those working in ethics, treat their topic as nontrivial conceptual truth and their method as the codification of conceptually clearheaded reactions to cases.

The background thought may be developed as follows. As masters of concepts we have at least an implicit grasp of their application conditions; this tacit knowledge of when they apply and when they should be withheld can be manifested equally well in real and imaginary cases. This must be so, since the master of a concept is antecedently armed with a capacity to tell whether or not to apply the concept, however reality might turn out to be (perhaps within certain limits of normality). Here then is a method for articulating our tacit knowledge of the application conditions of our concepts.

In the best scenario, the method delivers a "conceptual analysis"; that is, an account of a special sort of universally necessary and sufficient condition or set of conditions for the application of the relevant concept, namely a universally necessary and sufficient condition or set of conditions that could be recognized as correct simply on the basis of a certain sort of ideal reflection on our tacit understanding of when to apply and when to withhold the concept in question.

Therefore, the relevant verdicts and the resultant analysis can be delivered from the armchair, i.e. without any significant empirical investigation; so it is sometimes said that the relevant analyses could be known a priori; roughly, in a condition approximating to blissful ignorance of the empirical facts.

There were a few promising victories for this kind of method, but they were skirmishes rather than major battles. The analysis of the concept of knowledge was, at least for a while, considered a paradigm of this kind of investigation, one which neatly exemplified how "the method of cases" could lead us to an analysis of a concept. "Intuitions"—that is judgments—as to whether the case at hand was, or was not, a case of knowledge were collected by visiting real and imaginary cases alike, and then those intuitions were brought into some sort of reflective equilibrium that bore on the question of the universally necessary and sufficient conditions for someone's knowing some arbitrary proposition. Imagined cases were naturally treated as on a par with real cases; for if we are interested in articulating our tacit understanding of the application conditions of our concept it would be odd to restrict our evidence base to the adventitious experiments of stepmotherly nature, when we could also avail ourselves of the full range of ingeniously designed thought experiments. Wouldn't that be like only considering the moves that have been made in actual chess games, rather than the full range of moves that *could have been* made? As in chess, so with our concepts: imagination is a reliable guide to what could happen. It thereby provides us with cases that are just as helpful as the actual cases so far as rendering explicit our implicit understanding of the application conditions of our concept; as it might be, the concept of a mate in four or the concept of knowledge.

The ideology behind the method of cases thus offers to explain how the imagination can have a probative status, how it could have a kind of evidential significance which mere fancy could not. The imagination's philosophically interesting function is to generate a wider than actual range of cases, across which our conceptual competence can express itself.

For a good while, this method looked attractive when it came to the concept of personal identity. The Anglophone philosophy of personal identity emerged as a going concern in the 1960s thanks to the work of such philosophers as David Wiggins, Bernard Williams, Sydney Shoemaker, John Perry, Derek Parfit, and others inspired by them. These philosophers worked explicitly within the idiom of analytic philosophy and supposed that the real task of the philosophy of personal identity was to illuminate our *concept* of personal survival by means of organizing our intuitions about survival or continued existence, intuitions gleaned from a wide range of real and imaginary cases.

The fact that the target was a *concept* made the method of cases look like a viable approach in the case of personal identity. We are highly competent with the concept of personal identity; we have applied it successfully in a wide range of cases throughout human history, and in the common run of cases we appear to have a mass of accumulated knowledge of who is, and *was*, whom. So we must have at least an implicit grasp of the application conditions of the concept of personal identity, and this tacit knowledge of the concept's application conditions can be manifested equally well in real and imaginary cases.

Thus in the case of the concept of personal identity the dominant method in analytic philosophy was then to collect intuitions about real and imaginary cases of personal survival and ceasing to be, and then bring those intuitions into some sort of reflective equilibrium that bore on the question of the necessary and sufficient conditions for an arbitrary person's survival. The result would be the filling in of the details of the relation *R* in an a priori (and necessary) biconditional of this form:

x, considered at t, is numerically the same person as y, considered at t*, if and only if x*R*y.

A specification of *R* would entail a specification of the "identity" or, better, "persistence" conditions of persons, i.e. a specification of what changes they could and could not survive. Thus arose the old analytical question: Is *R* a matter of x and y having the same body, or being the same organism, or having the same consciousness, or having the same mind (however that mind might be embodied), or having the same separable immaterial soul?

There are many worries that have been and can be raised against this whole approach to the question of personal identity,[7] but we are now in a position to present a dilemma that threatens the whole approach: *is the target concept a phi-concept or a psi-concept?*

If the concept of personal identity is a psi-concept, shouldn't we admit that it is the techniques of cognitive psychology, some of whose interesting results have been already described, which provide the best way to get at our concept, i.e. the cluster of heuristics which guide us in the use of our terms?

On the other hand, if it is a phi-concept or meaning, then absent a quasi-supernatural picture which depicts our use of terms as quite generally the upshot of grasp of meaning, why should we suppose we are in a position to articulate those meanings just from armchair reflection? We do not know how to use our terms by way of knowing a theory of meaning for our language. It is not just that, as Hilary Putnam once famously said, "Meanings are not in the head." Meanings are not anywhere at all; they are theoretical posits for the purpose of individuating languages, and defining language-relative truth conditions and validity. Meanings register usage; they do not generate it.

In response to this dilemma, many philosophers today would abandon the second horn. After all, there has been a widespread loss of hope that any true biconditional of the above form could be analytic, so that those who denied it would either be failing to understand their language or speaking another language.

But then, whence the conviction that it is philosophers who are best placed to articulate the relevant heuristics which have provided us with knowledge of who is, and *was*, whom in the common run of cases? Whence the conviction that these heuristics, taken individually or collectively, apply a priori? Couldn't they all be empirically falsified, so that the articulation of our psi-concept might amount to simply a folk-theory of personal persistence, one open to empirical refutation?

The best answer respects the good thought that we cannot always have been totally wrong about something, on pain of losing the something to have been wrong about. For how then would we have focused on a topic to be totally wrong about? (Setting aside the infrequently used method of bare perceptual demonstration of an item as "this" or "that" without having any controlling conception in play.)

However, the good thought is a holistic one, with minimal constraining power. The good thought leaves room for massive error.

[7] For some others, see Wilkes 1988 along with Johnston 1987 and 2010, 44–47.

More particularly, the generic encoding hypothesis allows for the possibility that we could be right enough about a topic without knowing any very interesting universal truths about it, but only useful generics involving it. This suggests that the urge to articulate universal necessary and sufficient conditions concerning the topic by way of articulating our criteria or heuristics has no general theoretical justification, and may just be misguided.

That this is not just an abstract possibility is suggested by the case of personal identity itself. It may be that our heuristics for tracking a persisting person involve the generic criterion to the effect that persons survive if their individual minds continue on, and the generic criterion to the effect that persons survive if their bodies are kept alive and functioning. But it may also be that when these two heuristics point in different directions we should shrug our shoulders, and are inclined to do so, unless operated on by some misleading presentational feature of the case. So many of the imagined cases in the philosophical literature, such as teletransportation and brain transplantation, may simply be cases in which our psi-concept idles thanks to the generic character of the heuristics involved. The same may hold for some actual cases, such as the persistent vegetative condition. The result is that there is simply no analysis of the concept of personal identity, as traditionally understood.

So also with other concepts of philosophical significance. For all Gettierology showed, "Knowledge is true justified belief" and "Justified true belief is knowledge" may both be true generics. There is no general reason, arising from the nature of grasping meaning, to suppose that there is an a priori universal statement of necessary and sufficient conditions somewhere in the offing. The generic encoding hypothesis makes the ambition of traditional philosophical analysis seem quaint for most concepts.[8]

We take these reflections to reinforce the idea that the proper philosophical method, here as elsewhere, is not to limit oneself to the impoverished realm of conceptual or a priori knowledge, knowledge somehow deriving from, or embedded in, our competence with the meaningful terms of our language. The proper method is to use *all* one knows and all one can find out, in the most ingenious ways one can. Philosophy is not, and certainly not exclusively, the analysis of concepts. Philosophy is integrative theoretical vision combined with argumentative ingenuity, deployed at a fairly abstract level. Who could fail to love that, at least when it is done well?

On this view, philosophy has *no* special "resource" in Michael Dummett's intended sense. But so far from marginalizing philosophy, this liberates it. The upshot is merely a clarification of what was always anyway our obligation as philosophers: we must learn a lot more science than the analysts of yore deemed relevant.

We need to get out of the armchair and again start looking into things.

[8] Similarly, "Generically speaking, you do not make a hirsute man bald by removing a hair" seems a good heuristic guiding our use of 'bald'. But . . . well, you know how that goes. A question: how much of how that goes turns on the confusion between psi-concepts and phi-concepts?

7.7 THE END OF MEANING

Once psi-concepts are separated off from semantic meaning, a question remains as to the interest of semantic meaning. Suppose that as a matter of fact, the actual examples of analytic validity are rather few and far between, and that conventionally constrained use is not what it was (if it was ever what it was!) in the sense of producing a stable conformity of use over significant periods of time. Then the notion of a population sharing a language in the semantically tight sense, though perfectly coherent, may find little de facto application.[9] We would then look back on the theory of meaning as a massive overapplication of the tools of model theory, without much in the way of results to show for it.

Less pessimistic but still important is the claim we have argued for here. Cognitive psychology undermines the metaphysics of meaning, the idea that use is *guided by* grasp of meaning. In doing so, it also lays to rest the analytic conception of philosophy with its characteristic target of analyses facilitated by the method of cases. This result stands even if there remains an analytic/synthetic distinction, along with some inferences that are valid in virtue of meaning.

REFERENCES

Ahn, W., Kim, N. S., Lassaline, M. E., and Dennis, M. J. (2000). Causal status as a determinant of feature centrality. *Cognitive Psychology 41*: 361–416.

Armstrong, S. L., Gleitman, L. R., and Gleitman, H. (1983). What some concepts might not be. *Cognition 13*: 263–308.

Barner, D., Chow, K., and Yang, S. (2009). Finding one's meaning: A test of the relation between quantifiers and integers in language development. *Cognitive Psychology 58*: 195–219.

Braddon-Mitchell, D., and Nola, R. (2009). *Conceptual Analysis and Philosophical Naturalism*. Cambridge, MA: MIT Press.

Brandone, A., Cimpian, A., Leslie, S. J., and Gelman, S. A. (2012). Do lions have manes? For children, generics are about kinds, not quantities. *Child Development 83* (*2*): 423–33.

Brandone, A., Gelman, S. A., and Hedglen, J. (2015). Young children's intuitions about the truth conditions and implications of novel generics. *Cognitive Science 29*: 711–38.

Carey, S. (1985). *Conceptual Change in Childhood*. Cambridge, MA: MIT Press.

Carey, S. (2009). *The Origin of Concepts*. New York: Oxford University Press.

Carlson, G. N., and Pelletier, F. J. (1995). *The Generic Book*. Chicago: Chicago University Press.

Cimpian, A., Brandone, A. C., and Gelman, S. A. (2010). Generic statements require little evidence for acceptance but have powerful implications. *Cognitive Science 34* (*8*): 1452–82.

Cimpian, A., Gelman, S. A., and Brandone, A. C. (2010). Theory-based considerations influence the interpretation of generic sentences. *Language and Cognitive Processes 25* (*2*): 261–76.

Cohen, A. (1996). *Think Generic: The Meaning and Use of Generic Sentences*. PhD dissertation, Carnegie Mellon University.

[9] This *may* the view that Donald Davidson (1986) expresses in his important but elusive essay "A Nice Derangement of Epitaphs" by provocatively suggesting that there are no languages. Whether the contemplation of malapropisms is the most direct route to this result is another question.

Dahl, O. (1985). *Tense and Aspect Systems.* Oxford: Blackwell.

Davidson, D. (1986). A nice derangement of epitaphs. In E. Lepore, ed., *Truth and Interpretation: Perspectives on the Philosophy of Donald Davidson.* Oxford: Blackwell, 433–46.

Davies, M. (1981). *Meaning, Quantification, Necessity: Themes in Philosophical Logic.* London: Routledge and Kegan Paul.

Dummett, M. (2010). *The Nature and Future of Philosophy.* New York: Cambridge University Press.

Fodor, J. A. (1998). *Concepts: Where Cognitive Science Went Wrong.* Oxford: Oxford University Press.

Fodor, J. A., Garrett, M. F., Walker, E., and Parkes, C. (1980). Against definitions. *Cognition 8*: 263–367.

Fodor, J. A., and Lepore, E. (1996). The pet fish and the red herring: Why concepts aren't prototypes. *Cognition 58* (2): 243–76.

Fodor, J. A., and Lepore, E. (2002). *The Compositionality Papers.* New York: Oxford University Press.

Gelman, S. A. (2003). *The Essential Child: Origins of Essentialism in Everyday Thought.* New York: Oxford University Press.

Gelman, S. A. (2010). Generics as a window onto young children's concepts. In F. J. Pelletier, ed., *Kinds, Things, and Stuff: The Cognitive Side of Generics and Mass Terms.* New York: Oxford University Press, 100–123.

Gelman, S. A., and Raman, L. (2003). Preschool children use linguistic form class and pragmatic cues to interpret generics. *Child Development 74*: 308–25.

Gelman, S. A., Star, J., and Flukes, J. (2002). Children's use of generics in inductive inferences. *Journal of Cognition and Development 3*: 179–99.

Gopnik, A., and Meltzoff, A. N. (1997). *Words, Thoughts, and Theories.* Cambridge, MA: MIT Press.

Graham, S. A., Nayer, S. L., and Gelman, S. A. (2011). Two-year-olds use the generic/nongeneric distinction to guide their inferences about novel kinds. *Child Development 82*: 493–507.

Hollander, M. A., Gelman, S. A., and Star, J. (2002). Children's interpretation of generic noun phrases. *Developmental Psychology 38*: 883–94.

Johnston, M. (1987). Human beings. *Journal of Philosophy 84*: 59–83.

Johnston, M. (1988). The end of the theory of meaning. *Mind and Language 3* (1): 28–42.

Johnston, M. (2010). *Surviving Death.* Princeton, NJ: Princeton University Press.

Jönsson, M. L., and Hampton, J. A. (2006). The inverse conjunction fallacy. *Journal of Memory and Language 55*: 317–34.

Keil, F. (1989). *Concepts, Kinds, and Cognitive Development.* Cambridge, MA: MIT Press.

Khemlani, S., Leslie, S. J., and Glucksberg, S. (2009). Generics, prevalence, and default inferences. *Proceedings of the 31st Annual Cognitive Science Society.* Amsterdam: Cognitive Science Society.

Kripke, S. (1980). *Naming and Necessity.* Cambridge, MA: Harvard University Press.

Kripke, S. (1982). *Wittgenstein on Rules and Private Language.* Cambridge, MA: Harvard University Press.

Laurence, S., and Margolis, E. (2003). Concepts and conceptual analysis. *Philosophy and Phenomenological Research 67* (2): 253–82.

Leslie, A. M., and Keeble, S. (1987). Do six-month-old infants perceive causality? *Cognition 25*: 265–88.

Leslie, S. J. (2007). Generics and the structure of the mind. *Philosophical Perspectives 21*: 375–403.

Leslie, S. J. (2008). Generics: Cognition and acquisition. *Philosophical Review 117* (*1*): 1–47.

Leslie, S. J. (2012). Generics articulate default generalizations. *Recherches Linguistiques de Vincennes 41*: 25–45.

Leslie, S. J. (2013). Essence and natural kinds: When science meets preschooler intuition. In T. Gendler and J. Hawthorne, eds., *Oxford Studies in Epistemology*, vol. 4. New York: Oxford University Press, 08–65.

Leslie, S. J. (in press). The original sin of cognition: Fear, prejudice, and generalization. *Journal of Philosophy*.

Leslie, S. J. (forthcoming). New York: Oxford University Press.

Leslie, S. J., and Gelman, S. A. (2012). Quantified statements are recalled as generics: Evidence from preschool children and adults. *Cognitive Psychology 64*: 186–214.

Leslie, S. J., Khemlani, S., and Glucksberg, S. (2011). All ducks lay eggs: The generic overgeneralization effect. *Journal of Memory and Language 65*: 15–31.

Lewis, D. (1969). *Convention*. Cambridge, MA: Harvard University Press.

Lewis, D. (1975). Languages and language. In K. Gunderson, ed., *Language, Mind and Knowledge*. Minneapolis: University of Minnesota Press.

Machery, E. (2009). *Doing without Concepts*. New York: Oxford University Press.

Mannheim, B., Gelman, S. A., Escalante, C., Huayhua, M., and Puma, R. (2011). A developmental analysis of generic nouns in southern Peruvian Quechua. *Language Learning and Development 7* (1): 1–23.

Margolis, E., and Laurence, S. (1999). *Concepts: Core Readings*. Cambridge, MA: MIT Press.

Meyer, M., Gelman, S. A., and Stilwell, S. M. (2011). Generics are a cognitive default: Evidence from sentence processing. In *Proceedings of the 33rd Annual Conference of the Cognitive Science Society*. Boston: Cognitive Science Society.

Murphy, G. (2002). *The Big Book of Concepts*. Cambridge, MA: MIT Press.

Peacocke, C. (1975). Finiteness and the actual language relation. *Proceedings of the Aristotelian Society 75*: 147–65.

Pelletier, F. J., and Asher, N. (1997). Generics and defaults. In J. van Benthem and A. ter Meulen, eds., *Handbook of Logic and Language*. Cambridge, MA: MIT Press, 1125–79.

Prasada, S., Khemlani, S., Leslie, S. J., and Glucksberg, S. (2013). Conceptual distinctions amongst generics. *Cognition 125* (3): 405–22.

Ramsey, W. (1992). Prototypes and conceptual analysis. *Topoi 1*: 59–70.

Rosch, E. (1973). Natural categories. *Cognitive Psychology 4*: 328–50.

Rosch, E. (1978). Principles of categorization. In E. Rosch and B. B. Lloyd, eds., *Cognition and Categorization*. Hillsdale, NJ: Lawrence Erlbaum Associates, 27–48.

Rosch, E., and Mervis, C. B. (1975). Family resemblances: Studies in the internal structure of categories. *Cognitive Psychology 7* (*4*): 573–605.

Schiffer, S. (1972). Meaning. Oxford: Clarendon Press.

Schroeter, L. (2004). The limits of conceptual analysis. *Pacific Philosophical Quarterly 85*: 425–53.

Sloman, S. A. (1993). Feature-based induction. *Cognitive Psychology 25*: 231–80.

Sloman, S. A. (1998). Categorical inference is not a tree: The myth of inheritance hierarchies. *Cognitive Psychology 35*: 1–33.

Smith, E. E., and Medin, D. L. (1981). *Concepts and Categories*. Cambridge, MA: Harvard University Press.

Tardif, T., Gelman, S. A., Fu, X., and Zhu, L. (2011). Acquisition of generic noun phrases in Chinese: Learning about lions without an "-s." *Journal of Child Language 30*: 1–32.

Taylor, B. (1982). On the need for a meaning-theory in a theory of meaning. *Mind 91*: 183–200.

Wilkes, K. (1988). *Real People*. New York: Oxford University Press.

8 Natural Language and Its Ontology

Friederike Moltmann

8.1 INTRODUCTION

Natural language, it appears, reflects in part our conception of the world. Natural language displays a great range of types of referential noun phrases that seem to stand for objects of various ontological categories and types, and it also involves constructions, categories, and expressions that appear to convey ontological or metaphysical notions, for example ontological categories of various sorts, plurality, quantity, identity, causation, parthood, truth, and existence. But it also appears that natural language reflects ontological categories, structures, and notions that not everyone may be willing to accept, certainly not every philosopher, but often not even an ordinary person when thinking about what there is and the general nature of things.

Most notably, natural language appears to involve a wealth of terms that display a rich ontology of artifactual, derivative, minor, abstract, and even nonexistent entities that philosophers generally tend to reject and even an ordinary person when thinking about what there really is. They include terms for qualities or properties (*wisdom, the property of being wise*), tropes (*Socrates's wisdom, John's tallness*), nonworldly facts (*the fact that John won the race or Bill did*), variable objects, as I call them (Moltmann 2013b, forthcoming) (*the increasing number of students, the book John needs to write*), and intentional objects (that is, merely conceived or nonexistent objects) (*the building mentioned in the guide that does not exist*). Whether abstract objects such as qualities or properties really exist is disputed, as is the existence of tropes and facts, and even more so, of course, the existence of variable and intentional objects. But while a philosopher may reject the disputed entities, she is likely to use natural language terms for them when engaging in ordinary conversation.

Also metaphysical concepts may be reflected quite differently in natural language predicates than philosophers would expect. For example, while it is nowadays a common view that existence is a univocal concept applying to everything there is, the predicate *exist* is actually restricted to material objects and abstract objects of certain sorts and is inapplicable to events (*The building described in the guide exists* and *the smallest prime number exist* are acceptable, but not so for *the inauguration existed six months ago*). Again, this holds not just for ordinary speakers (nonphilosophers), but also philosophers, who will not be able to suspend the restrictions when using the predicate *exist* in ordinary discourse.

There are different reactions philosophers take in view of such discrepancies between the ontology implicit in language and the reflective ontology that a philosopher or nonphilosopher is willing to accept explicitly. For some philosophers, it means a complete rejection of language as a guide to metaphysics, to what there really is and the real nature of things. Many contemporary philosophers take that position, aiming to focus just on the metaphysics of the real, not caring about what is reflected in natural language or even other common-sense judgments. Other philosophers, in the tradition of ordinary language philosophy, focused entirely on natural language (Strawson, Austin). Yet other philosophers took language seriously for some purposes, but not others, for example Frege. Frege was guided by language for his view of numbers as objects, but not for his view of truth values having that status. Not all philosophers that appealed to natural language, of course, did so having in mind while recognizing a discrepancy between the metaphysics displayed by natural language and the metaphysics of what there really is. Aristotle, medieval philosophers, Frege, Twardowski, and even Strawson, it seems, did not.

In any case, throughout history philosophers made appeal to language for the purpose of an ontological view. But they generally did so without an explicit methodology and in a nonsystematic way, citing one or two sentences without ensuring that a real linguistic generalization was at hand. Arguments for numbers and propositions from putative referential terms are examples. Thus, Frege (1884) took it to be evident that a sentence like *the number of planets is eight* is an identity statement in which the subject and the postcopula term could stand only for numbers as abstract objects. However, then the sentence *the number of planets is the number eight* should be equally good, which it is not (Moltmann 2013b). Similarly, following Frege (1918), many philosophers consider *that*-clauses to be proposition-referring terms and attitude verbs predicates that take propositions as arguments. But then a sentence like *John claimed that hen won* should permit the inference to *John claimed the proposition that he won*, which it does not, at least for most attitude verbs.[1]

The development of theoretical linguistics in the twentieth century, both natural language semantics and syntax, forces a revision of this way of appealing to natural language for the purpose of a philosophical argument. It is no longer convincing or appropriate to make arbitrary appeal to some linguistic examples or others at a time when establishing linguistic generalizations, including philosophically relevant ones, has become the domain of a highly developed theoretical discipline. Moreover,

[1] There are a few verbs with which such substitution is possible, such as *assert, believe*, and *prove*, but they are the exception rather than the general case. See, for example, Moltmann 2013b, chap. 4.

it is no longer only a matter of a philosopher's choice of being more or less interested in the ontology reflected in language. Rather the ontology implicit in natural language emerges as a domain of study in itself, as the subject matter of *natural language ontology*, as a branch both of linguistics and of metaphysics.[2]

This brings with it a range of issues and challenges. One issue is how natural language ontology situates itself within metaphysics and how it relates to pursuits in metaphysics that are not focused on natural language. One important challenge is to make explicit the methodology that natural language ontology should pursue, which in part is the methodology that philosophers throughout history have relied on when making appeal to natural language. This involves the question what sorts of linguistic data can and cannot be considered reflective of the ontology of natural language, and what part of natural language or its use should matter. Another challenge is the characterization of the ontology of natural language, given that it may significantly differ from the reflective ontology of an ordinary speaker (or philosopher).

This chapter aims to contribute to the issues and challenges of natural language ontology by clarifying its subject matter in relation to other branches of metaphysics, by making the criteria explicit that distinguish linguistic data that may reflect the ontology of natural language from those that may not, by laying out some core cases that show a discrepancy between the reflective or philosophical ontology and the ontology implicit in language, and by giving a general characterization of the ontology of natural language and the domain of real or conceived objects it includes.

8.2 THE ONTOLOGY OF LANGUAGE AS THE SUBJECT MATTER OF NATURAL LANGUAGE ONTOLOGY

Throughout the history of philosophy, there has been a practice of making appeal to linguistic examples or generalizations in support of a particular ontological notion or view. In this sense, natural language ontology has been practiced to a greater or lesser extent throughout the history of philosophy. Aristotle and medieval metaphysicians such as Aquinas and Ockham made appeal to natural language in support of a metaphysical notion or view. In early analytic philosophy, we find explicit appeal to natural language in Frege as well as Twardowski ([1911] 1999).

[2] I choose the term "natural language ontology" rather than "natural language metaphysics." The latter would be more accurate since metaphysics is generally understood as the more general term whose subject matter is not just what there is, but also the nature of things. However, "ontology" is a term more suited to talk about the subject matter of the discipline, the ontology of natural language. Moreover, there are irrelevant connotations associated with "metaphysics" (the surreal, the metaphysical). In addition, "ontology" is a term more customary in linguistics or other "applied" domains ("applied ontology"). "Ontology" is then understood to be as general as "metaphysics," having as its subject matter not just what there is but also the nature of things.

Appeal to natural language then plays an important role in the philosophy of Austin, Strawson, and Vendler and various contemporary philosophers.

Along with the development of theoretical linguistics, it is becoming clear that the ontology of natural language is also a subject matter of study in itself, the subject matter of natural language ontology, and that natural language ontology needs to be recognized as a discipline in itself, as part of both natural language semantics and metaphysics. But then how does natural language ontology situate itself within metaphysics in general?

Natural language ontology would be part of descriptive metaphysics in Strawson's (1959) sense. Descriptive metaphysics has as its aim to uncover our shared conceptual scheme, as Strawson characterizes it, or better, since metaphysics is not about concepts but objects, the ontological categories, structures, and notions represented by our shared conceptual scheme. Like descriptive metaphysics in general, natural language ontology would be characterized in terms of the data with which it concerns itself. Those data, however, are not so much common-sense judgments, but linguistic intuitions, judgments about the acceptability or grammaticality of natural language sentences and constructions.

Should natural language metaphysics be based exclusively on linguistically reflected intuitions? The answer must be negative. Natural language may underspecify the ontology that is involved in its semantics. Theories within natural language ontology cannot and should not be developed solely being based on linguistic data, but must be pursued in conjunction with purely philosophical considerations and arguments. To give examples from my own work, the theory of parts and wholes in Moltmann (1997) is motivated also by language-independent intuitions about form and integrity and the theory of attitudinal objects in Moltmann (2014a) by language-independent intuitions about artifacts.

Unlike other branches of metaphysics, however, natural language ontology should give priority to linguistically reflected intuitions, rather than common-sense judgments. Common-sense judgments may represent a speaker's reflective ontology, not the ontology implicit in language. Other branches of descriptive metaphysics may be based on common-sense judgments that are not reflected in language and may even contradict intuitions reflected in language, say applied ontology and what one may call "folk metaphysics."

Strawson (1959) contrasts descriptive metaphysics with what he calls "revisionary metaphysics." The aim of revisionary metaphysics, for Strawson, is to conceive of a better ontology than the one we ordinarily accept. Strawson does not specify further how "better" is supposed to be understood, say, whether for the purpose of getting a better understanding of the nature of reality or for the purpose of better developing particular scientific theories.

A notion similar to that of descriptive metaphysics has recently been proposed by Fine (2017a), namely *naive metaphysics* or the *metaphysics of appearances* (see also Fine 2001). Naive metaphysics for Fine concerns itself with how things appear, without trying to address the question of whether they are real. Naive metaphysics deals with our ordinary judgments and has as its subject matter the things that we ordinarily take there to be as well as their nature. Fine contrasts naive metaphysics with what he calls "foundational metaphysics." Foundational metaphysics has as its subject matter what there really is. Foundational metaphysics, as Fine emphasizes, presupposes naive metaphysics. Foundational metaphysics has to make use of the concepts and statements that naive metaphysics aims to clarify and it has as one of its tasks to explain them, if possible, in more fundamental terms, be they materialist or mentalist, or whatever. Naive metaphysics as the metaphysics of appearances should not be guided by considerations of foundational metaphysics, but rather, as Fine argues, foundational metaphysics must take naive metaphysics as its starting point in order to do its foundational work. Again, natural language metaphysics would be part of naive metaphysics and as such, given Fine's perspective, would have an important role to play even in the interest of foundational metaphysics.

For Fine, the subject matter of naive metaphysics is the things that appear to be and how they appear to be in most general terms. No considerations of what is real or fundamental should come into play when pursuing naive metaphysics. Such a separation of naive and foundational metaphysics may not be entirely unproblematic. Considerations of truth and thus reality play an important role for naive metaphysics as well. Common-sense judgments are directed toward reality, and only those held to be true should be taken into account by naive metaphysics. Also, the objects among the "appearances" that are the subject matter of naive metaphysics, even if highly derivative, may be grounded in reality and have to be understood in the way they are so grounded (section 8.5.2). To figure out what sorts of object a referential noun phrase stands for requires considerations as to what grounds the truth or falsehood of sentences in which that noun phrase occurs, and thus conceiving of the object in relation to what there really is. Certain considerations of foundationalist metaphysics therefore may have to play a role even for pursuing naive metaphysics. Thus, the two metaphysical projects cannot be strictly separated, with one building upon the other.

If natural language metaphysics is part of naive metaphysics in Fine's sense, the choice of the predicate "naive" is misleading. Natural language ontology does not concern itself with what the ordinary person naively takes there to be (if one can even speak of a generic ordinary person, the generic nonphilosopher).[3] Rather, it deals with the ontological categories, notions, and structures that only a deep and systematic analysis of natural languages may uncover.[4] It is the ontology implicit in language, not the ontology displayed by "naive" ontological reflections of

[3] Note, though, that this is not the notion of "naive" that Fine has in mind. Fine means simply "uncritical."

[4] Fine (personal communication) also uses the term "shallow metaphysics" instead of "naive metaphysics," perhaps a better choice when it comes to natural language ontology. "Shallow metaphysics" is the term I use, following Fine (personal communication), in Moltmann (2014b) for the branch of metaphysics that natural language ontology belongs to.

nonphilosophers. For that reason, the term "descriptive metaphysics" is less misleading and is also to be preferred since it is a more established term.

Natural language ontology thus is the branch of descriptive metaphysics that gives priority to linguistic data. There are also areas where natural language may play no role and descriptive metaphysics would cover metaphysical analyses that are based on common-sense intuitions that are not as such reflected in language. Fine's (2017b) paper on form, for example, falls explicitly within descriptive metaphysics, but does not take into consideration any linguistic data.

Linguistically reflected intuitions need not coincide with common-sense judgments about ontological issues. There may be discrepancies between a metaphysical notion reflected in language and the one a philosopher, or a nonphilosopher upon reflection, may be willing to accept. An example mentioned already is existence. Existence is commonly held to be a notion that trivially applies to everything there is, or at least every actual thing, a view held by a great number of contemporary philosophers and defended particularly by Inwagen (2014). However, in natural language the predicate *exist* is subject to strict conditions on the type of entity to which it can apply: it applies to material and abstract objects, but not to events (Hacker 1982; Cresswell 1986; Fine 2006; Moltmann 2013d):

(1) a. The house still exists.
    b. The largest prime number does not exist.
(2) a.??? The rain still exists.
    b.??? The protest existed yesterday.

Events do not exist, but take place, happen, occur, or last. Existence in the ontology of natural language thus divides into different modes of being for different sorts of entities. Not only philosophers, but also an ordinary person, may have a reflective notion of existence that is not the one conveyed by the verb *exist*, which conveys the particular mode of being of enduring objects.

Later I will present a range of particularly striking cases of objects that act as semantic values of referential terms and that philosophers but also nonphilosophers upon reflection are not likely to accept. These are entities, though, that speakers implicitly accept—at least when using the language.

Characterizing the ontology of natural language thus requires a distinction between *implicit acceptance* of an object or ontological notion and *explicit acceptance* and perhaps even degrees of explicit acceptance. Implicit acceptance defines the ontology implicit in language; explicit acceptance defines the reflective ontology of speakers.[5] The ontology of natural language thus needs to be distinguished from both the reflective ontology of speakers and the ontology of what there really is.

[5] The implicit-explicit distinction also plays a role in ethics. In ethics, though, what is explicit is given priority over what is implicit, which includes prejudice and bias (Brownstein 2015). By

It is not quite correct to speak of *the* reflective ontology, actually, since there may be various reflective ontologies of speakers, at least in certain areas. The data for descriptive metaphysics to take into account when directed toward a reflective ontology are much less clear and stable than the data relevant for the ontology of natural language, as seems the case, for example, for the ontological categories of events and tropes (section 8.3). Natural language ontology raises its own issues, though. One issue is to what extent the ontology may be specific to a particular language. The latter touches upon the Safir-Whorf hypothesis and it raises the question whether ontological differences associated with different languages may be compatible with a universal ontological core. These questions be important themes for the pursuit of when pursuing natural language ontology.

Another issue is to what extent the ontology of a natural language may be driven by language itself. For example, certain objects in the ontology may be generated by the semantics of complex terms in the language (see section 8.6). This means that implicit acceptance of that part of the ontology will be tied to the use of language itself, whereas other parts may belong to a less language-driven implicit cognitive ontology.

A final issue that the ontology of natural language presents is that there are different ways for concepts or entities to be involved in the semantics of natural language and they may reflect different degrees of implicit acceptance or reflection. This is the case, for example, for the notion of existence. The noun *existence* obviously can be used for the reflective notion, as I have used it in this chapter. However, the verb *exist* conveys a restricted notion, regardless of the language user's reflective ontology. A more general distinction discussed in section 8.6 is that between the core of language, roughly consisting of expressions whose use does not involve ontological reflection, and the periphery of language, which consists of expressions that imply a certain degree of ontological reflection (see section 8.6).

To summarize, various levels of judgments associated with different degrees of reflection need to be distinguished for pursuing metaphysics: linguistically manifest intuitions (of possibly different linguistic levels), common-sense judgments that reflect a shared conceptual scheme (or perhaps a partly shared conceptual scheme), and judgments that belong to a particular philosophical view.

8.3 HOW IS ONTOLOGY REFLECTED IN NATURAL LANGUAGE?

There has hardly been an explicit effort of clarifying the methodology of natural language ontology, as a practice philosophers have pursued throughout the history

contrast, in the case of ontology it is rather the converse. When pursuing natural language ontology the reflective, explicitly accepted ontology may present prejudice, rather than the ontology implicit in language.

of philosophy as well as an emerging discipline that is part of both theoretical linguistics and metaphysics.

The philosophical practice of natural language ontology, as a matter of fact, follows rather strictly certain implicit assumptions as to what data could support an ontological view and what data couldn't. Not just philosophers pursuing natural language ontology for the purpose of particular philosophical arguments make use of certain types of linguistic data but not others. This also holds for linguists and philosophers in pursuit of the study of the ontology of natural language by itself. That is, the actual practice of natural language ontology follows an implicit methodology. Making those assumptions explicit is very important for an appropriate understanding of the ontology of natural language and the project of natural language ontology as such.

Natural language ontology obviously should not take into account statements that only a particular philosopher would accept—statements conveying a particular philosopher's ontological view, say. Natural language ontology likewise should not take into account statements conveying what may be the ontological view of a nonphilosopher. Statements of this sort may articulate what one may call the *reflective ontology* of an ordinary person, but they are not indicative of the ontology implicit in natural language.

What sorts of criteria distinguish statements that reflect the ontology implicit in natural language from those that do not? One paramteter that plays a role in distinguishing relevant and irrelevant data for natural language ontology is the difference between ontological assertions and presuppositions. Sentences that themselves make ontological assertions can hardly be taken as indicative of the ontology implicit in natural language. Philosophers never make use of such sentences when drawing on natural language for the purpose of a metaphysical argument. Thus, no practitioner of natural language ontology would appeal to sentences like (3) in order to argue for events being part of the ontology of natural language:

(3) There are events.

Also, philosophers who would *not* want to endorse events as an ontological category would generally be unimpressed by statements like (3) (just as atheists would not be impressed by sentences such as *There is God*).

Similarly, Platonists that seek support from natural language for properties being objects would never appeal to statements of the sort in (4):

(4) There are properties/qualities/virtues.

Similarly, nominalists generally would be unimpressed by the availability of statements like (4).

One type of statement that *is* used to argue for natural language supporting an ontological category is a statement in which a particular referential noun phrase occurs that stands for an entity of the relevant ontological category. Thus, statements like (5a, b, c) support events as part of the ontology of natural language:

(5) a. John's walk was slow.
b. The rain caused the roof to fall down.
c. The war took place a decade ago.

Why should the semantic values of nouns such as *walk*, *rain*, and *war* stand for events? That is because the predicates that are applicable to what such nouns stand for express properties characteristic of events, such as properties of duration, causation, and perdurance (*take place*), properties that, taken together, are applicable only to events.

Similarly, statements such as (6) are suited to motivate properties or qualities being part of the ontology of natural language:

(6) a. Socrates has wisdom.
b. Socrates and Plato share great wisdom.
c. Wisdom is everywhere in this book.

Here a referential noun phrase (*wisdom*) stands for something that can be possessed, shared by individuals, and be at different locations at once, properties characteristic of qualities.

Being the semantic value of a *referential noun phrase*, in general, is considered the primary criterion for an object to be part of the ontology of natural language. Referential noun phrases, that is, names and referential (nonpredicative) occurrences of definite noun phrases, presuppose the existence of their semantic value, an object. Sentences such as (5a–c) and (6a–c) presuppose rather than assert the existence of events and qualities. Thus, here it is presuppositions that are indicative of the ontology implicit in language.

Also, quantification over individual members of an ontological category is generally taken to be a reflection of that ontological category being part of the ontology of language, for example (7), for the category of events:[6]

(7) Heavy rain caused the roof to fall down.

Davidson (1967), the most influential advocate of events as part of the ontology of natural language, did not so much appeal to statements such as (7), though. Rather

[6] This is an addition to Moltmann 2017, where only presuppositions are considered indications of the ontology implicit in natural language.

he argued that verbs take events as implicit arguments and adverbial modifiers are predicates of those events. That was to account for valid inference such as from (8a) to (8b):

(8) a. John walked slowly.
    b. John walked.

Thus, for Davidson, the verb *walk* describes a two-place relation between events of walking and agents and *slowly* acts as a predicate of an events, as in the logical form of (8a) in (9):

(9) $e(walk(e, John & slowly(e))

Of course, *slowly* as an adverbial event predicate has the very same meaning as the adjectival event predicate *slow* in this sentence:

(10) John's walk was slow.

Why did Davidson's data qualify as reflecting the ontological category of events? For Davidson, adverbials act as predicates, and they act as predicates of the sort of entities described by verbs, which can only be events, the sorts of things the corresponding deverbal nominalization stands for.

The very same arguments that had motivated events being part of the ontology of language motivate tropes, that is, particularized properties, being *to be* part of it.[7] Tropes have played an important role since Aristotle, for whom they made up an ontological category besides substances, secondary substances, and qualities. Tropes have traditionally been considered the semantic values of referential NPs formed with adjective nominalizations such as *Socrates's wisdom, John's happiness*, or *the redness of the apple* (Strawson 1959; Woltersdorff 1970; Moltmann 2004; 2013b, chap. 2). But tropes also play a role as implicit arguments of adjectives. Modifiers of adjectives, at least to an extent, also occur as predicates of the corresponding trope-referring term obtained from the adjective (Moltmann 2009, 2013b):

(11) a. Socrates is extremely wise.
     b. Socrates's wisdom is extreme.

This means that *wise* describes a relation between wisdom tropes (manifestations of wisdom) and agents, so that (11b) will have the logical form in (12):

(12) $t(wise(t, Socrates) & extreme(t))

[7] For the notion of a trope see Williams 1953; Campbell 1990; and Woltersdorff 1970.

Natural language gives equal support for tropes and for events. In a sense, though, the reflective ontology of speakers does not support both categories equally well. While events are an established category in contemporary linguistic semantics (and syntax) and in philosophy, tropes are much less so. This was different, however, in earlier periods in the history of philosophy, as already mentioned. Tropes during those periods played a much more important role in metaphysics and presumably the reflective ontology of speakers. Thus, for events and tropes, and perhaps in general, natural language gives a more stable ground of judgments indicative of ontology than the more explicit assumptions underlying a reflective ontology.

8.4 REIFYING TERMS AND THE ONTOLOGICAL CORE-PERIPHERY DISTINCTION

While deverbal and deadjectival nominalizations generally are considered good support for events and tropes being part of the ontology of natural language, there is a type of referential noun phrase in English that is generally not considered a reflection of the ontology implicit in language. These are what I call "reifying terms" (Moltmann 2013b, chap. 6). An example of a reifying term is *the property of wisdom*:

(13) Socrates has the property of wisdom.

Whereas properties may seem unproblematic to some, the very same construction also permits reference to numbers, degrees, truth values, and propositions:

(14) a. the number eight
b. the truth value true
c. the degree of John's happiness
d. the proposition that S

Philosophers hardly ever appeal to constructions of this sort when arguing for ontological categories. Thus, Frege did not appeal to terms like (14a), but rather to terms like *the number of planets* when arguing for numbers being objects, and he certainly did not motivate truth values by appealing to terms like (14b). Semanticists that posit degrees as objects in the semantics of natural language generally do not appeal to noun phrases like (14c) but rather to constructions with positive or comparative adjectives (Cresswell 1977). Finally, the linguistic motivation for propositions generally comes from the apparent referential status of simple *that*-clauses, not noun phrases like (14d).

Reifying terms, as in (14), introduce new entities on the basis of a sortal and, generally, a nonreferential expression or use of an expression (a noun or adjective, for example), leading to a reified concept or propositional meaning (Moltmann 2013b, chap. 6). Thus, *the number eight* introduces a number as an object on the basis of a number adjective or quantifier *eight*, and *the truth value true* introduces a truth value as an object on the basis of the truth predicate *true*.[8] Reifying terms are part of English. But they can only be taken as evidence for certain types of objects playing a role in a special discourse, not as a reflection of objects being part of the ontology of ordinary language.

In addition to reifying terms, natural language also permits extensions with terms used for special, philosophical or scientific, discourse. Those terms should have the same semantics as other referential noun phrases, namely that of standing for an object. But the objects they stand for need not belong to the ontology of natural language, but only to an extension of it (and like all referential noun phrases, the terms in the periphery may stand for merely conceived objects, not actual ones; see section 8.7).

Two sorts of linguistic data thus need to be distinguished: linguistic data that in some sense belong to the *core* of language and data that belong to a part that conveys an ordinary speaker's or philosopher's ontological reflections, its *periphery*. The periphery in that sense includes reifying terms as well as (technical) terms that may have been introduced by extending the language. The periphery involves entities that some philosophers or nonphilosophers may accept, but which would not be part of the ontology of natural language, the ontology *any* speaker implicitly accepts when using the language. The latter is what is reflected in the core of language. Thus a condition of the following sort obtains:

(15) The ontology of natural language is reflected in the core of language, not its periphery.

Certainly, the periphery of language also has a semantics, and it also reflects an ontology with its referential terms standing for entities of some sort. Using Fine's (2017a) notion, this ontology would be part of the subject matter of the metaphysics of appearances as well. This holds even for terms in the periphery that belong to foundational metaphysics.

The terms "core" and "periphery" recall a distinction of the same name that Chomsky (1986) made for the syntactic structure of languages. The Chomskyan distinction is in a way the analogue for syntax of the present ontology-oriented distinction. For Chomsky (1986), roughly, the "core" of a language consists in what is determined by (innate) universal grammar, that is, universal principles together

[8] Predicate-initial sortals also have a reifying function:

(i) a. Wisdom is a property few have.

b. Eight is a number that is divisible by two.

See Moltmann 2013b, chap. 6, for discussion.

with the way the parameters of universal grammar are set for that language. By contrast, what Chomsky calls the "periphery," roughly, consists in idiosyncratic rules unique to that language and added on in the historical development of the language. Chomsky's distinction does not in any way coincide with the present one since, clearly, the formation of new philosophical terms may take place entirely within what Chomsky would call the "core" of language. This certainly is the case with reifying terms, which are formed productively across European languages.

Universality is not what matters for what terms are in the periphery in the present sense, only ontological reflection does. Entities that are semantic values of terms in the periphery may be the outcome of applying universal conditions for generating objects.[9] What matters is that they are the semantic values of terms whose use involves ontological reflection.

Not all noun phrases that are reifying terms belong to the periphery in the present sense. Thus explicit fact-referring terms, which stand for nonworldly facts (*the fact that John won the race or Bill did*), and explicit property-referring terms (*the property of wisdom*) have been used by philosophers practicing natural language ontology , for example, Vendler (1967) when arguing for a fundamental distinction between facts and events, and myself in Moltmann (2004) when arguing for a distinction between qualities and properties.

8.5 THE ONTOLOGY OF NATURAL LANGUAGE REFLECTED IN ITS REFERENTIAL TERMS

8.5.1 Referential Terms and the Notion of an Object

It is one of the most striking features of natural language that it contains a wealth of referential terms displaying a great range of abstract, "derivative," and "minor" entities, many of which would not be part of the reflective ontology of an ordinary speaker or philosopher. In the following subsections, I will present a number of cases of a particularly striking discrepancy between the ontology displayed by certain referential noun phrases in natural language and what is likely the reflective ontology of ordinary speakers or philosophers.

First a few clarifying remarks are in order concerning the notion of a referential noun phrase. The notion of a referential noun phrase (or referential term) is used by philosophers and linguists alike as a criterion for the kinds of objects that are part of the ontology of natural language.[10] The notion of a referential

[9] Such conditions, in particular for the semantics of reifying terms, may take the form of abstraction principles (Wright 1983; Hale 1987) and conditions of generating pleonastic entities (Schiffer 1996, 2003). See Moltmann 2013, chap. 6, for an approach to reifying terms of that sort.

[10] For Frege referential terms serve as a criterion for objecthood itself. According to that criterion, an object is what can be the semantic value of referential term (Wright 1983; Hale 1987). However, objects may also play other semantic roles in the ontology of natural language without

noun phrase or term (when applied to natural language rather than a formal language) should be understood as a syntactic role of occurrences of expressions in sentences, rather than as a syntactic category. Definite noun phrases as such can also occur predicatively and as complements of intensional transitive verbs (*need, look for*), and then they do not stand for objects. While there is no agreed-upon set of criteria for identifying an occurrence of a noun phrase in a sentence as a referential term, there are two sorts of criteria for referential terms that philosophers and linguists generally make use of. This is the ability of supporting anaphora and the ability of being replaceable by a quantifier (under suitable circumstances).[11] In addition to the anaphora and quantifier criterion, a criterion for referential terms that is implicitly used in the philosophical and linguistic literature is the uniformity of the meaning of (extensional) predicates. If an extensional predicate with a range of referential noun phrases can yield a true sentence with another noun phrase *X*, then *X* should also be a referential noun phrase. This criterion highlights the connection between ontology and compositional semantics as well as the connection between truth and ontology (section 8.3), and it makes clear that the question of what objects are involved in the semantics of sentences can be pursued only together with the question under what circumstances a sentence is true.

8.5.2 Referential Terms and the Discrepancy between Implicit and Reflective Ontology

The referential terms of natural language do not display an ontology of what is ordinarily understood as "real" objects, but rather they present a great range of cases of a discrepancy between the ontology of natural language and the reflective ontology of speakers, philosophers as well as nonphilosophers.

acting as semantic values of referential terms, for example as implicit arguments and as parameters of evaluation.

[11] The anaphora and quantifier criteria need to be applied with caution, though. Not all quantifiers and pronouns that are able to replace an occurrence of an expression in a sentence are indicative of the expression acting as a referential term. There is a class of special quantifiers and pronouns that characteristically are able to replace nonreferential occurrences of expressions, which in English consists of quantifiers like *something, everything, nothing, several things*, and *that.* They can replace predicative complements, for example:

(i) a. Socrates is wise.

b. Socrates is something admirable.

Such special quantifiers and pronouns arguably are nominalizing expressions introducing a "new" domain of entities into the semantic structure of sentences, entities that would be referents of corresponding nominalizations (*wisdom* in (ib)) (Moltmann 2003, 2013b, chap. 3).

8.5.2.1 Definite NPs and the Notion of a Construction-Driven Variable Object

In Moltmann (2013b, forthcoming), I discussed and analyzed a construction with definite NPs that also shows a discrepancy between the ontology implicit in natural language and ordinary speakers' reflective ontology. The construction takes as its semantic value what I call *variable objects*, following Fine's (1999) notion of a *variable embodiment* (Moltmann 2013b, forthcoming). It comes in different varieties:

(15) a. The president of the United States is elected every four years.
b. The water in the container has increased.
c. The height of the water level has increased.
d. John's happiness has not changed.

The variable objects may be variable individuals as in (15a), variable quantities as in (15b), variable degrees (or quantitative tropes) as in (15c), or variable tropes as in (15d). This means, following Fine (1999), that (15a) and (15b) stand for objects that have possibly different individuals/quantities as manifestations at different times. This account can be carried over to (15c) and (15d). If heights are degrees (or quantitative tropes), then (15c) stands for a variable object whose manifestations are degrees (or quantitative tropes), and (15d) for one whose manifestations are tropes (Moltmann 2013b, forthcoming). Variable objects are objects that are associated with a function mapping a circumstance (time and world or situation) to an entity that is the manifestation of the object at the circumstance.

With the use of an intensional verb inside the relative clause, the same construction permits reference to variable objects that may lack actual manifestations but have manifestations only in counterfactual circumstances, as in (16) (Moltmann 2013b, forthcoming):

(16) The book John needs to write must be two hundred pages long.

The book John needs to write is a variable object that has as its manifestation a book John has written in any circumstance satisfying the need in question (Moltmann 2013b, forthcoming).

The motivations for positing variable objects as semantic values of the noun phrases in (15) and (16) are the same as for positing any other objects as semantic values. Variable objects noun phrases permit replacement by quantifiers, support anaphora, and may provide arguments of the very same predicates as noun phrases standing for ordinary objects ((16), for example, can be continued by *It cannot be any shorter or longer*).

For Fine (1999), variable embodiments include organisms and artifacts, as entities that permit a replacement of parts and thus have different material manifestations in different circumstances. The notion of a variable object would then be an extension of notions that are well accepted in the ontology speakers explicitly accept. As semantic values of the definite noun phrases in (15b–d), though, they would be closely tied to the content of the construction, and as such belong to a language-driven creative part of the ontology implicit in natural language.

Speakers certainly are not likely to accept variable objects in their reflective ontology (apart from perhaps presidential roles). But the construction in (15) – (16) is entirely productive and speakers make use of it whether or not they accept or would accept variable objects when thinking about what there is.

What is also important about variable objects is that they are not just conceptual creations, but rather are based on manifestations in actual or perhaps possible circumstances, which in turn determine their properties. This means that variable objects could not just be items in some conceptual-lexical structure.

8.5.2.2 Kind Reference with Bare Nominals

Another construction in English that appears to generate derivative objects in a fully productive way are determinerless or bare mass or plural nouns or modifier-noun combinations. In the contexts in (17a, b), the bare nouns are generally taken to stand for kinds (17a) (Carlson 1977) or qualities (17b) (Moltmann 2004, 2013b):

(17) a. Giraffes are not extinct.

b. Wisdom is better than cleverness.

Giraffes in (17a), on that view, stands for a kind of individual, a kind whose instances are semantic values of corresponding definite or quantificational noun phrases, of the sort *that giraffe* or *some giraffe*. *Wisdom* in (17b), on that view, stands for a kind whose instances are tropes, semantic values of corresponding definite or quantificational noun phrases, of the sort *Socrates's wisdom* or *some wisdom*. The referential status of the bare nouns in (17a, b) is supported by the usual criteria (e.g. anaphora support, as when (17a) is continued by *they are widespread in Africa*).

Bare nouns as kind terms (*giraffes*) with their associated terms for instances (*that giraffe*) seem to support the Aristotelian view of two sorts of universals with two sorts of particulars: secondary substances with primary substances as instances and qualities with tropes as instances (Moltmann 2004). Moreover, kinds seem to share other characteristics that Aristotle attributes to them. A universal for Aristotle exists only if instantiated. This seems to be reflected in the behavior of existence predicates:

(18) Wisdom exists.

Exist when predicated of a kind can only state the existence of an instance, not the existence of a possibly uninstantiated kind. Moreover, kinds inherit nonepisodic properties from their instances: men have legs because individual men have legs etc. (Moltmann 2004).[12]

However, kinds as semantic values of bare nominals differ significantly from any notion of a kind relevant in philosophical or scientific contexts. Any adjective-noun combination can make up a kind term, regardless of its content. Nominals like *tired giraffes* can be used semantically in the same way as kind terms like *giraffes*. Similarly, any combination modifier-adjective nominalization can serve as a term standing for a quality, for example *acquired wisdom*. The kinds in question thus do not match natural kinds or even kinds of artifacts, and the qualities do not match natural properties but non-natural ones.

In addition to kind terms or quality terms in this sense, natural language displays explicit property-referring terms:

(19) the property of being wise

Explicit property-referring terms can be formed from any complex predicate (in the form of a gerund), regardless of its content (*the property of being wise and tall, the property of being wise or not wise, the property of being extremely wise* etc.).

Natural language appears to display an ontological distinction between kinds as semantic values of kind terms and properties as semantic values of such property-referring terms (Moltmann 2004, 2013b). Kind terms differ from property-denoting terms with respect to the way predicates are understood. Thus, *exist* with kind terms can claim only the existence of an instance as in (18), but with property-denoting terms it claims the existence of a possibly uninstantiated kind (*The property of wisdom exists*). *Look for* when applied to a kind term requires just an instance to satisfy the search, as in (20a); but when applied to a property-referring term, it requires the property itself to play that role, as in (20b):

(20) a. John is looking for wisdom.
b. John is looking for the property of wisdom.

The two readings are also reflected in the way *find* is understood in (21a) and in (21b):

(21) a. John found wisdom.
b. John found the property of being wise.

[12] The view of bare plurals and mass nouns standing for kinds as objects goes back to Carlson (1977). This view was taken over to quality terms as terms for kinds of tropes in Moltmann (2004). Moltmann (2013b), by contrast, explores an account of kinds as semantic values of kind terms in terms of plural reference (Oliver and Smiley 2013).

Both kind terms and property-referring terms stand for things that depend strictly on the descriptive content of the terms, not on what kinds or properties are real or natural. The quality-property distinction again is a distinction that is generally not part of the reflective ontology of philosophers or nonphilosophers. A related distinction, not involving a reifying term as in (19), is the distinction between qualities and conditions, semantic values of gerunds like *being wise* (Levinson 1978), again a distinction that is not generally part of a speaker's reflective ontology.

8.5.2.3 Reference to Tropes

Also tropes (particularized properties) display a significant discrepancy between the ontology many contemporary philosophers accept and the ontology reflected in natural language. Tropes, more recently, have come to play a central role within foundational metaphysics. Since Williams's (1953) influential article, a number of philosophers have pursued a trope nominalist one-category ontology, proposing to conceive of properties as classes of similar tropes and individuals as bundles of colocated tropes (Campbell 1990; Bacon 1995; Simons 1994).

Natural language displays a wealth of trope-referring terms, namely noun phrases with adjective nominalizations such as *Socrates's wisdom* or *the redness of the apple* (Woltersdorff 1970; Moltmann 2004, 2009, 2013b). However, tropes as part of the ontology of natural language differ significantly from tropes as discussed in contemporary foundational metaphysics.

First of all, tropes in foundational metaphysics are generally taken to be manifestations of natural (or sparse) properties. By contrast, adjectives hardly ever express natural or fully specific properties and thus adjective nominalization could not stand for instances of natural properties. Instead they generally stand for complex tropes instantiating the nonnatural property the adjective expresses. Thus, *John's happiness* refers to a complex trope composed of the very specific things that together constitute John's happiness. In that respect, John's happiness differs from the state or condition that is the referent of *John's being happy* and the nonworldly fact that is the referent of *the fact that John is happy*. The latter are not grounded in specific features of reality, but rather are just constituted by the property of being happy and John, as the holding of that property of John (at a time) (Moltmann 2013c). For example, one can describe John's happiness (in detail), but not John's being happy or the fact that John is happy.

Natural language moreover displays a difference between a trope that is the referent of *John's height* and a trope that is the referent of *John's tallness*. The former can exceed Bill's height, but not really the latter, for example. John's tallness is a considerably more complex trope than John's height; it is something like John's height qua exceeding the contextual standard suitably construed (Moltmann 2009). Natural language furthermore displays a difference between the referent of *John's strength* and the referent of *John's weakness*, and that even in a case where John is in a way both strong and weak (Moltmann 2009). In such a case, *John's strength* and *John's weakness* could not possibly refer to the same physical condition, the same simple trope that has John as bearer. John's strength may exceed Mary's strength, but then that could not be true of John's weakness. This means that John's strength and John's weakness are both complex entities involving an inherently ordering with respect to the same sorts of physical conditions, but in different directions.

The ontology of tropes or trope-related entities that natural language reflects is thus considerably richer than the foundationalist trope ontologies that philosophers more recently have pursued. Still, trope terms in natural language stand for entities grounded in fully specific tropes, unlike terms for states, conditions, or nonworldly facts, which stand for entities strictly constituted by the descriptive content of the noun phrases making reference to them.[13]

8.5.2.4 Difficulties for Reference to (Real) Objects with Referential Noun Phrases

The discrepancy between the reflective ontology of speakers and the ontology reflected in language may also extend to nouns for artifacts, cities, organisms, and the like . They are the main target of Chomsky's (1998, 2013) critique of the traditional notion of reference as a relation to mind-independent, real objects.

Chomsky (1998, 2013) points out a range of examples of referential noun phrases which, allowing inconsistent property ascriptions, could not stand for real objects. For example, what we refer to as a "door" could be painted, replaced, and walked through, properties that could not be attributed jointly to material objects as standardly understood. Another example is a home (Chomsky 1998). What we refer to as a "home" may have peculiar combinations of properties: one can own or

[13] Another potential example of a discrepancy between the reflective ontology of speakers and the ontology implicit in natural language is noun phrases of the sort *the average American*. *The average American* is a noun phrase that appears to satisfy standard criteria for referential terms, allowing for a range of predicates that apply to clearly referential terms (*the average American likes hamburgers*) and supporting anaphora (*the average American likes hamburgers; he also likes French fries*). But it is not likely to stand for an object speakers would accept explicitly. The *ideal student* is a similar case. It exhibits criteria of referentiality and thus should stand for an object in the ontology of natural language. The object it stands for may have properties that are just based on counterfactual assumptions about students meeting certain standards. Noun phrases with modifiers such as *average, typical, ideal*, or *perfect* may be viewed as standing for highly derivative entities (whose properties, though, are grounded in facts or assumptions about particular entities). Such entities would be construction-driven, generated by the semantics of noun modification with certain sorts of modifiers. However, alternative analyses have been proposed of *the average American* not making use of derivative objects (Kennedy and Stanley 2009). Chomsky (1998), moreover, takes such noun phrases to be example of the traditional notion of reference as a relation to entities in the real world not applying to natural language.

sell a home, but not, for example, paint a home, in contrast to a house. Thus a home cannot be an object as standardly understood. Other examples Chomsky (1998, 2013) gives are cities, which can be destroyed and rebuilt at a different location, artifacts, which can undergo complete replacement of their parts, and persons with the relative independence of their bodies. Chomsky's point even extends to simple nouns like *water*.

Chomsky's (1998, 2013) overall conclusion is that the semantics of natural language terms does not involve the reference relation as traditionally conceived, as a relation to real objects. The semantics of referential noun phrases rather involve lexical/conceptual structures deployed by speakers in particular contexts to refer to particular aspects of reality. Instead of a semantics with the (traditional) notion of external reference as its central notion, referential terms should have an internalist semantics, involving another level of syntactic representation, that of lexical-conceptual structure.

Chomsky's critique targets the traditional notion of reference as a relation to the external world as well as the notion of a mind-independent object that goes along with it. Chomsky does not consider the reference relation in the sense in which I am introducing it in this chapter, as a relation that would permit mind-dependent and merely conceived objects as relata (entities in the ontology of appearances). This, on the present view, is the semantically relevant reference relation and is the reference relation that is reflected in language itself, namely in the intentional verb *refer to*. *Refer to* clearly can take as arguments mind-dependent objects and even intentional (nonexistent) objects (see section 8.7 and Moltmann 2016).

Chomsky stays within the philosophical tradition, not drawing a distinction between the ontology of the real and the ontology of appearances in Fine's sense. His critique is based on a single ontology of the real, that is, of a mind-independent reality. (Note that Chomsky's level of lexical conceptual structure consists in representations, not things represented, unlike the (intentional or real) entities in the ontology of appearances).

This then invites an exploration of ways of conceiving of objects, as part of the ontology of appearances, that would be suited as the semantic values of the noun phrases Chomsky discusses. Chomsky (personal communication) would deny that this was possible, as no object could be bearer of inconsistent properties. Yet there have been various attempts of making sense of apparent polysemies of the noun phrases in question, in particular within applied ontology, posting objects with different facets (serving as bearers of different predicates) or complex ontological categories (Arapinis and Vieu 2015).

In any case, the semantic behavior of certain referential noun phrases with different predicates challenges what appears to be the reflective ontology of speakers, and they impose an important task for natural language ontology as well as perhaps truth-conditional semantics.

8.6 THE ONTOLOGY OF NATURAL LANGUAGE

We have seen with a range of cases that the ontology displayed by natural language may differ from the sort of ontology that a philosopher may be willing to accept, and of course from the ontology of what there really is. But it may also differ from the ontology that an ordinary speaker of the language may be willing to accept when reflecting upon what there is. Natural language ontology concerns itself with the ontological categories, notions, and structures involved in "ordinary" statements of a nonphilosophical discourse, not those that are part of the assertive content of philosophical or quasi-philosophical reflections.

How then should the ontology of natural language, as an ontology distinct from the reflective ontology of both philosophers and nonphilosophers, be characterized? Certainly, such a characterization will involve the object-related attitude of implicit acceptance. A first proposal then would be this:

(22) The ontology of a natural language is the ontology that speakers of the language implicitly accept.

Given that characterization, the ontology implicit in natural language should be an ontology that is part of our cognitive system, as an ontology that goes along with speakers' implicit beliefs, while using or not using the language.

However, this is too strong a condition. We have seen that the ontology of natural language may in part be driven by language itself, by the semantics of particular constructions that make certain types of derivative objects as semantic values available.

In addition to variable objects and the quasi-kinds that act as semantic values of certain definite NPs and of bare nominals in English, the ontology of natural language may be language-driven in other respects. The mass-count distinction is a case in point. The mass-count distinction at first appears to reflect an ontological distinction. But as an ontological distinction, it would be a highly language-dependent distinction. *The rice, the rice grains*, and *heap of rice* are likely to stand for the very same thing not just for a philosopher, but for a nonphilosopher as well. However, the semantic values of the three terms bear different properties and thus appear to stand for different entities. The rice grains can be counted or be indistinguishable, but not the rice. The heap of rice may be small, but not the rice (and neither may the rice grains in the same sense).

Another example of a language-driven part of the ontology of natural language are discourse referents, semantic values of unbound anaphoric pronouns (or donkey pronouns), on a quasi-ontological conception. Discourse referents are presented as objects individuated by the flow of discourse in Karttunen (1976), Landman (1986), and Edelberg (1986) (though they do not have an ontological status as objects in most developments of dynamic semantics).

Yet another example are nonworldly facts described by noun phrases of the form *the fact that it is raining or it is snowing*—their canonical fact descriptions (Moltmann 2013b). Nonworldly facts are entities whose nature is fully displayed by the descriptive content of their canonical description.

The entities driven by constructions of the language or the discourse are not likely entities speakers implicitly accept as such, but rather only when using the language. More precisely, they are part of an ontology a speaker accepts when using the core of language and engaging in an ordinary use of language. This motivates the weaker characterization of the ontology of natural language in (23):

(23) The ontology of a natural language is the ontology a speaker implicitly accepts when using (in the ordinary way) the core of the language.

"Ontology" in (23) should not be understood as just consisting of a particular inventory of objects, but also includes methods or operations for introducing objects along with certain linguistic constructions. This would account for the creativeness of the ontology implicit in natural language, in particular, of its language-driven or construction-driven part. Acceptance of an ontological operation for generating objects then means potential acceptance of the objects that can be generated by that operation.

The characterization of the ontology of natural language in (23) makes reference to the ordinary use of the core of language. The notion of an ordinary use of language is not new; it had played already a central role in ordinary language philosophy with its focus on nonphilosophical, nontechnical uses of philosophically relevant expressions.

8.7 INTENTIONAL ("NONEXISTENT") OBJECTS AND THE ONTOLOGY OF NATURAL LANGUAGE

The objects that a speaker implicitly accepts when using natural language need not be real objects, but may also include merely conceived objects, objects the speaker mistakenly takes there to be. The relation of acceptance is an intentional relation, not requiring the existence of its object arguments. In fact, the domain of objects reflected in the core of language as well as its possible extensions cannot be characterized as a domain of "real" objects or objects grounded in reality in some way, but may include intentional (i.e. nonexistent) objects. The notion of a real object and the difference between a real and an intentional object should in fact not play a role at all for the ontology of natural language. Natural language does not semantically differentiate between terms standing for actual objects and terms standing for merely conceived ones, as long as the entities the terms purport to stand for are accepted as such by the language user.

Not only the object-related related notion of acceptance is an intentional notion, but also reference. Along with verbs like *accept, describe, and mention*, the object language predicate *refer to* is an intentional verb, in the sense of a verb that permits as arguments objects the speaker herself allows to be intentional objects (Moltmann 2016).[14] Making use of that notion of reference (as a notion that is part of the philosophy of language implicit in natural language) permits the semantic values of referential noun phrases to include intentional objects.

Note that intentional objects need to be distinguished from concepts or conceptions of objects. The latter are representations; the former are things meant to be represented. Intentional objects share properties with real objects. By contrast, representations have representation-specific properties, not the properties of what they represent.

Objects of reference that turn out to be nonexistent need to be distinguished from objects the speaker takes to be merely conceived objects. Natural language does allow referential noun phrases to stand for intentional objects which the speaker herself takes to be nonexistent. That is, natural language reflects a Meinongian view of there being objects that fail to exist (Parsons 1980; Salmon 1987, 1998; Fine 1982a; Priest 2005; Moltmann 2013a, 2016). To show this requires of course the right sorts of data. These could not be statements of the sort below, an assertion of the Meinongian view:

(24) There are things that do not exist.

Support for Meinongianism being reflected in language cannot take the form of assertions of the Meinongian view itself, but must take the form of presuppositions or quantification over particular sorts of intentional objects, and that in the form of statements from the core, not the periphery of language. The following sentences satisfy those requirements:

(25) a. The building described in the book does not exist.
b. A building described in the book does not exist.

(26) a. John thought about the building described in the book.
b. John thought about a house he wants to build.

(27) a. The person Bill referred to with *Annie* does not exist.
b. At least one person John referred to in his speech does not exist.

Complex definites with intentional verbs such as *describe, think about*, and *refer to* in (25a), (26a) and (27a) respectively presuppose intentional objects as semantic

[14] Their compositional semantics crucially involves the intentional verbs taking scope over the head noun (*book, building, house*) at the relevant linguistic structure; see the analysis in Moltmann 2006.

values resulting from their compositional semantics (Moltmann 2016).[15] Indefinite or quantified noun phrases as in (25b), (26b), and (27b) quantify over particular intentional objects, again required by the compositional semantics of those noun phrases. The noun phrases in (25, 26, 27) certainly are part of the core of language.

Reference to what the speaker herself takes to be merely conceived objects does not come for free, but requires particular lexical and syntactic conditions. While most predicates are existence-entailing, requiring as arguments entities that the speaker takes to be actual objects, there are certain predicates, especially predicates of existence and intentional predicates (*describe, think about, mention, refer to, imagine*), which permit as arguments entities the speaker takes to be intentional (nonexistent) objects (Moltmann 2016). Entities the speaker considers merely conceived objects would be part of the creative ontology of natural language on a view on which intentional objects are constituted by relations of coordination among acts of pretended or unsuccessful reference, reference acts that play a role explicitly or implicitly in the linguistic context (McGinn 2000; Moltmann 2016). [16]

A distinction thus needs to be made between the semantic values of noun phrases that speakers take to be objects and those that speakers allow to be merely conceived objects. The ontology of natural language is an ontology of real, derivative, mind-dependent, and intentional objects, including objects a speaker herself considers intentional objects. Intentional objects are of course entities that many philosophers are unwilling to accept (Inwagen 2001). However, they are clearly reflected in natural language and are indispensable for the compositional semantics of terms as in (25) and (26).

Besides being semantic values of referential terms, entities may play other roles in the semantic structure of sentences, for example as implicit arguments of predicates and as parameters of evaluation (times on the standard semantics of tense and possible worlds on the standard semantics of modals and conditionals). What sorts of semantic roles entities play in the semantic structure of sentences depends to an extent on particular semantic theories about relevant constructions or expressions, and it depends very much on the way the contribution of occurrences of expressions to the composition of the meaning of the sentence is conceived.

8.8 CONCLUSION

Natural language reflects its own ontology, an ontology that may diverge in different ways from the reflective ontology of philosophers and nonphilosophers. There are various distinctive features which set the ontology of natural language apart from the reflective ontology (or ontologies) of speakers. In particular, the ontology of

[15] Here "intentional verb" is meant to contrast with "extensional verb" as well as "intensional verb"; see Moltmann 2016.

[16] This would also account for the fact that in the semantics of natural language intentional objects can act only as semantic values of noun phrases, not as implicit arguments or as parameters of evaluation.

natural language is a rich, in part construction-driven, ontology, which includes various sorts of derivative objects, as well as intentional objects and objects recognized as intentional by the speaker.

The ontology of natural language, the ontology of appearances reflected in language, is a cognitive ontology and should play a role within the human cognitive faculty that is as important as that of other parts of language (syntax and phonology). The same expectations may then be set in regard to the ontology of natural language as for syntax: that a deeper analysis of a particular language or different languages with respect to the ontology they reflect should reveal highly systematic and universal features, shared ontological categories, operations, or notions.[17]

Not only metaphysics involves a distinction between what is implicit in natural language and what is explicitly present in philosophical or naive reflection. The same distinction applies to other branches of philosophy, such as philosophy of mind, philosophy of language, and epistemology. Thus, there is a philosophy of language to an extent reflected in language itself, as we have seen with the verb *refer*. For the philosophy of mind, what is implicit in language is particularly important, since its subject matter is, at least in part, the folk psychology of propositional attitudes and other notions, on which the syntax and semantics of verbs of propositional attitudes, perception, and emotion bear a lot.

ACKNOWLEDGMENTS

The chapter has benefited greatly from conversations with Chris Collins and an extensive email exchange with Noam Chomsky, as well as from audiences of talks at the University of Milan, Yale University, the Institute for History and Philosophy of Sciences and Technology in Paris, Princeton University, the 2016 Formal Ontology in Information Systems conference in Annecy, and the 2017 "Semantics and Philosophy in Europe" colloquium in Padua. Thanks also to Boban Arsenijevic, Mark Blechner, Susanne Bosche, John Collins, Jean-Claude Dumoncel, Kit Fine, Roegnvaldur Ingthorsson, Eve Kitsik, Brian McLaughlin, Spiros Moschonas, Benjamin Nelson, Miika Oksanen, Paolo Petricca, Maîk Suehr, and Tristan Tondino for comments on a previous version of this chapter.

REFERENCES

Arapinis, A., and Vieu, L. (2015). A plea for complex categories in ontologies. *Applied Ontology* 10 (3–4): 285–96.

Bacon, J. (1995). *Universals and Property Instances: The Alphabet of Being.* Oxford: Blackwell.

Brownstein, M. (2015). Implicit bias. In E. N. Zalta, ed., *The Stanford Encyclopedia of Philosophy*. Spring 2017 ed.

[17] Chomsky (2013) hints at such a perspective, though, of course, for the level of lexical-conceptual structure.

Campbell, K. (1990). *Abstract Particulars.* Oxford: Blackwell.
Carlson, G. (1977). A unified analysis of the English bare plural. *Linguistics and Philosophy* 1 (3): 413–57.
Chomsky, N. (1986). *Knowledge of Language: Its Nature, Origin, and Use.* Westport, CT: Praeger.
Chomsky, N. (1998). *New Horizons in the Study of Language and Mind.* New York: Cambridge University Press.
Chomsky, N. (2013). Notes on denotation and denoting. In I. Caponigro and C. Cecchetto, eds., *From Grammar to Meaning: The Spontaneous Logicality of Language.* New York: Cambridge University Press, 38–46.
Cresswell, M. J. (1977). The semantics of degree. In B. Partee, ed., *Montague Grammar.* New York: Academic Press, 261–92.
Cresswell, M. J. (1986). Why object exists, but events occur. *Studia Logica* 45: 371–75.
Davidson, D. (1967). The logical form of action sentences. In N. Rescher, ed., *The Logic of Decision and Action.* Pittsburgh: University of Pittsburgh Press, 81–95.
Edelberg, W. (1986). A new puzzle about intentional identity. *Journal of Philosophical Logic* 15: 1–25.
Fine, K. (1982). The problem of non-existents. I. Internalism. *Topoi* 1: 97–140.
Fine, K. (1999). Things and their parts. *Midwest Studies of Philosophy* 23: 61–74.
Fine, K. (2001). The question of realism. *Philosophers Imprint* 1 (1): 1–30.
Fine, K. (2006). In defense of three-dimensionalism. *Journal of Philosophy* 103 (12): 699–714.
Fine, K. (2017a). Naïve metaphysics. *Philosophical Issues* 27 (1): 98–113.
Fine, K. (2017b). Form. Journal of Philosophy 114 (10):509-535 Ms.
Frege, G. (1884): *Die Grundlagen der Arithmetik.* Breslau: W. Koebner. Translated as *The Foundations of Arithmetic: A Logico-Mathematical Enquiry into the Concept of Number,* by J. L. Austin. Oxford: Blackwell, 2nd ed., 1974.
Frege, G. (1918). Der Gedanke: Eine logische Untersuchung. *Beiträge zur Philosophie des deutschen Idealismus, (1918–1919)* 1: 58–77.
Hacker, P. M. S. (1982). Events, ontology, and grammar. *Philosophy* 57: 477–86.
Hale, B. (1987). *Abstract Objects.* Oxford: Blackwell.
Karttunen, L. (1976). Discourse referents. In J. D. McCawley, ed., *Syntax and Semantics,* vol. 7. New York: Academic Press, 363–86.
Kennedy, C., and Stanley, J. (2009). On average. *Mind* 471: 583–646.
Landman, F. (1986). Pegs and alecs. In *Towards a Theory of Information: The Status of Partial Objects in Semantics.* Dordrecht: Foris, 97–155.
Levinson, J. (1978). Properties and related entities. *Philosophy and Phenomenological Research* 39 (1): 1–22.
McGinn, C. (2000). *Logical Properties.* New York: Oxford University Press.
Moltmann, F. (1997). *Parts and Wholes in Semantics.* New York: Oxford University Press.
Moltmann, F. (2005). Part structures in situations: The semantics of *individual* and *whole. Linguistics and Philosophy* 28 (5): 599–641.
Moltmann, F. (2009). Degree structure as trope structure: A trope-based analysis of comparative and positive adjectives. *Linguistics and Philosophy* 32: 51–94.
Moltmann, F. (2013a). The semantics of existence. *Linguistics and Philosophy* 36 (1): 31–63.
Moltmann, F. (2013b). *Abstract Objects and the Semantics of Natural Language.* New York: Oxford University Press.
Moltmann, F. (2013c). On the distinction between abstract states, concrete states, and tropes. In C. Beyssade, A. Mari, and F. Del Prete, eds., *Genericity.* New York: Oxford University Press, 292–311.

Moltmann, F. (2014a). Propositions, attitudinal objects, and the distinction between actions and products. *Canadian Journal of Philosophy*, supplementary volume on propositions, edited by G. Rattan and D. Hunter, 43 (5–6): 679–701.

Moltmann, F. (2014b). Parts, wholes, abstracts, tropes and ontology. Interview by Richard Marshall. *3:AM*, February 14.

Moltmann, F. (2016). Quantification with intentional and with intensional verbs. In A. Torza, ed., *Quantifiers, Quantifiers, Quantifiers*. Dordrecht: Synthese Library, 141–68.

Moltmann, F. (2017). Natural language ontology. *Oxford Encyclopedia of Linguistics* Online.

Moltmann, F. (forthcoming). Variable objects and truthmaking. In M. Dumitru, ed., *The Philosophy of Kit Fine*. New York: Oxford University Press.

Oliver, A., and Smiley, T. (2013). *Plural Logic*. New York: Oxford University Press.

Parsons, T. (1980). *Nonexistent Objects*. New Haven: Yale University Press.

Priest, G. (2005). *Towards Nonbeing*. New York: Oxford University Press.

Pustejovsky, J. (1995). *The Generative Lexicon*. Cambridge, MA: MIT Press.

Schiffer, S. (1996). Language-created and language-independent entities. *Philosophical Topics* 24 (1): 149–67.

Simons, P. (1994). Particulars in particular clothing: Three trope theories of substance. *Philosophy and Phenomenological Research* 54 (3): 553–75.

Strawson, P. (1959). *Individuals: An Essay in Descriptive Metaphysics*. London: Methuen.

Twardowski, K. ([1911] 1999). Actions and products: Some remarks on the borderline of psychology, grammar, and logic. In J. Brandl and J. Wolenski, eds., *Kazimierz Twardowski: On Actions, Products, and Other Topics in the Philosophy*. Atlanta: Rodopi, 103–32.

Van Inwagen, P. (2001). Existence, ontological commitment, and fictional entities. In M. Loux and D. Zimmerman, eds., *The Oxford Handbook of Metaphysics*. New York: Oxford University Press, 131–57.

Van Inwagen, P. (2014). Modes of being and quantification. *Disputatio* 6 (38): 1–24.

Vendler, Z. (1967). *Linguistics in Philosophy*. Ithaca, NY: Cornell University Press.

Williams, D. C. (1953). On the elements of being. *Review of Metaphysics* 7: 3–18.

Woltersdorff, N. (1970). *On Universals*. Chicago: University of Chicago Press.

Wright, C. (1983). *Frege's Conception of Numbers as Objects*. Aberdeen: Aberdeen University Press.

PART V
MODALITY AND THE ONTOLOGY OF BODILY FEELINGS

9 Modal Prospection

John McCoy, L. A. Paul, and Tomer Ullman

9.1 THE METAPHYSICS OF MODALITY MEETS COGNITIVE SCIENCE

Recent work in metaphysics has been enriched and inspired by work in foundational physics and the philosophy of physics. We think there is even more potential for productive collaboration between cognitive science and metaphysics.[1] To this end, our chapter develops new connections between metaphysics and cognitive science, drawing together intuitive modeling and prediction in cognitive science with the metaphysics and epistemology of counterfactual reasoning.

Our focus is on how we interpret, represent, and understand possibilities. We are particularly interested in reasoning about certain sorts of self-involving possibilities, especially far-fetched possibilities for oneself or for individuals we are close to. However, our work connects to a wide variety of more general topics, such as those involving modal reasoning, simulation theory, the semantics of counterfactuals, modes of presentation, conceivability and possibility, decision-making, and debates about the nature of perspectival, or *de se*, thought and content.

This first section of our chapter develops a theoretical and empirical context for assessing certain kinds of possibilities that brings together metaphysics, epistemology, and computational cognitive science, and discusses ways to connect this richer perspective to other topics in philosophy, such as the concept of transformative experience, the semantics of counterfactuals, moral learning, and simulation theories. We start by drawing connections between counterfactual reasoning about physical possibilities and recent empirical work on intuitive physics judgments. We then extend this idea to reasoning about other people and to counterfactual reasoning about self-involving possibilities, and explore the parallels between intuitive self-involving judgments and intuitive physics judgments. We take the view that, when a person reasons about her self-involving possibilities, especially far-fetched possibilities, this reasoning may be supported by an underlying "self simulator," a kind of mental engine with an approximate understanding of who she is,

[1] See Goldman 2014 for a forceful argument that realist metaphysics should be informed by cognitive science, and related arguments in Schaffer 2016 and Paul 2015.

which enables her to learn about her preferences and make intuitive judgments and predictions about her self-involving possibilities. We then introduce the notion of modal prospection, and discuss connections between the ideas we are exploring and some contemporary philosophical debates in metaphysics, mind, and epistemology. In section 9.2, we consider two sample vignettes in which participants are asked to make a potentially transformative decision, and explore the philosophical and methodological reasoning behind our surveys, with special attention to our use of far-fetched, fantastical examples, the stock-in-trade of the metaphysician. Our farfetched possibilities involve metaphysically possible scenarios that highlight meaningful or fundamental elements of how people think about themselves, and we explain why we think these scenarios are especially apt for our purposes.[2] Section 9.3 presents the empirical part of our project, and discusses the ten self-involving possibilities in the surveys we conducted. We present the details of our surveys, and discuss the empirical and theoretical questions that devolve from our results, laying the groundwork for further work on this rich interdisciplinary topic. Our appendix lists the vignettes we used.

9.1.1 Background I: The Metaphysics of Counterfactuals and the Cognitive Science of Intuitive Physics Judgments

On a standard approach (Lewis 1986; Stalnaker 1976), reasoning about possibilities involves reasoning about possible worlds. If Finbarr drops a glass on the sidewalk, it will (likely) break: at time t in a world with laws just like the actual world, just like our world but for the initial state at t in which Finbarr drops his glass, it falls to the sidewalk and breaks. The metaphysics of counterfactuals, when understood using this possible worlds framework, involves the notion of similarity: we assess various modal claims in virtue of similarity relations between worlds. Very roughly, to evaluate counterfactuals such as "if C had not occurred, then E would not have occurred," we move to the closest possible world where C does not occur. If E does not occur in that world, the counterfactual is true. Closeness of worlds, here, is based on relevance and similarity: the closest possible world is the world that is most similar to the world of evaluation (usually the actual world) in salient respects. More generally, to determine whether S is possible in world W, we need to know which possible worlds are most similar to W in the relevant ways. Different similarity relations will define different kinds of possibilities.

Cognitive scientists are interested in how the mind represents the world when people make fast, natural, and intuitive judgments about ordinary goings-on in their environment.[3] How am I representing the world when I quickly, intuitively,

[2] One of our scenarios, Swap, is (at least arguably) not metaphysically possible. We ignore this complication in what follows.

[3] For related work on implicit representations of possibility in cognition, see, e.g., Phillips and Cushman 2017.

and naturally judge that, if Finbarr were to drop the glass on the sidewalk, it would (likely) break? Recent research on intuitive physics judgments (e.g. Gerstenberg et al. 2012; Battaglia et al. 2013; Sanborn et al. 2013; Smith and Vul 2013; Hamrick et al. 2016; Ullman et al. 2018) frames this sort of understanding of the dynamics of the world as people having a *world simulator* in their heads where this world simulator is a "mental physics engine," in analogy to the physics engines software that powers modern animations and computer games (although this is not the only way people may be making intuitive physics judgments; see e.g. Forbus 1988; Marcus and Davis 2013). A physics engine is software that generates a simulation of a dynamic physical system, such as simulations of collisions, explosions, or the movement of fluids. In particular, the analogy is to physics engines that support relatively fast and approximate simulations like those that power games, rather than engines underlying scientific simulations such as galaxy formation or protein folding (Ullman et al. 2017). As well as using their mental physics engine to evolve the world forward and predict what will happen, people can also use it to reason about the physical properties of a scene. For example, by observing objects collide, people can make inferences about the mass of the objects or the friction of the surface on which they were moving.

According to this account, the mind can use a quasi-Newtonian simulation to predict how a physical scene would unfold over a short time-span, in the same way that a real physics engine can quickly simulate the results of acting in a game world. My intuitive, commonsensical judgment that if Finbarr drops his glass it will probably break is based on the underlying computations of a mental physics engine with an approximate understanding of bodies and the forces acting on them.

We can now see a structural parallel between the metaphysics and the cognitive science. From the metaphysics perspective, we can consider a counterfactual claim about some ordinary goings-on in the local environment: "If Finbarr were to drop the glass, it would break." To evaluate this counterfactual, we move to the closest possible world where Finbarr drops the glass (and all other relevant features are the same), and evolve that world forward. If the glass breaks in that world, then we judge the counterfactual as true. If the glass doesn't break, then we judge the counterfactual as false. Assume, in the actual world, that the counterfactual is true, and I know that it is true. Then I know something about the nature of this part of my environment; for example, I know something about the laws of this world, and I know something about the disposition of Finbarr's glass to break. I know that the world is such that the laws make these sorts of ordinary counterfactuals true. But since my knowledge didn't require any sort of sophisticated physics knowledge, we can also say that part of what I know concerns manifest physics, the physics of ordinary goings-on at the level of human experience.

The manifest image concerns the world as it appears to us. Manifest physics can be thought of as concerning manifest laws of nature, that is, as concerning an implicit representation of what's generating events like the breaking of the glass. We can think of it in terms of a representation of a machine running the appearances, a machine that is implicitly represented by us when we understand and predict the ordinary goings-on in our environment. If I can successfully assert and know ordinary counterfactuals like "If Finbarr were to drop the glass, it would break," then I understand enough about the manifest physics of the actual world to make such predictions. Importantly, this means I understand enough about which features of the actual world, including its manifest features, I need to hold fixed when I determine which possible world is closest, as I assess the truth value of the counterfactual. Note that while we've framed our discussion in terms of the manifest, our account can support a realist approach to counterfactuals. The fact that our simulations are largely successful can be taken as evidence that we are getting something right about the nature of the world: we are in fact grasping counterfactual truths.[4]

Blending the cognitive science with the metaphysics, we can think of my counterfactual claim ("If Finbarr were to drop the glass, it would break") as a judgment based on my underlying computation involving a quasi-Newtonian simulation derived from my approximate understanding of the appearances involving the dynamics, that is, from my approximate understanding of the manifest physics. This simulation is, in effect, a representation of the possible world W1 where Finbarr drops the glass and it breaks: in some (admittedly implicit sense) the simulation evolves W1 forward from the dropping of the glass to the breaking of the glass. If my simulation correctly predicts the approximate result (the breaking of the glass), then we can interpret this in terms of my selecting the closest possible world, and by extension, correctly identifying the relevant similarity relation: the world I chose for my simulation (W1) was in fact the closest possible world in the relevant respects.

9.1.2 Background II: Indexicals, Self-Locating Attitudes, and Theory of Mind

Philosophers have discovered that reasoning about certain kinds of possibilities can require a more fine-grained modal semantics. In particular, reasoning about possibilities targeted to particular individuals, places, and times requires a semantics that can handle these types of targeted possibilities. Revising our sample counterfactual to include more indexicality, I can assert: "If I were to drop this glass right now, on this sidewalk, it would break." To evaluate the truth value of this counterfactual, I need to assess a possible world where I (or my counterpart) drop my glass at the designated time and place. One popular approach for such assessment uses a semantics that distinguishes the context of utterance from the circumstance of evaluation (Kaplan 1989) in order to determine truth values for claims that are sensitive to indexical content. Our interest in indexical claims can be expanded to self-locating claims, such as "I am the person spilling the sugar," and beliefs based on this, such

[4] Paul identifies as a metaphysical realist and endorses this realist approach. Thanks to Ross Cameron for discussion.

as my believing that "I am spilling the sugar." Some argue that reasoning about self-locating (or *de se*) attitudes requires a semantics involving centered worlds, where, to determine the content of the attitude, in addition to specifying the world of evaluation, we must also specify the relevant individual and time. Whether we endorse a semantics that includes *de se* content, or endorse this sort of centered worlds reasoning, or even whether we endorse a particular treatment of indexicality, it seems clear that indexical, self-involving judgments require a distinctive semantics, and this extends to indexical, self-involving counterfactual judgments.

We can make a related claim with respect to the cognitive science about the need for a distinctive model for intuitive, targeted judgments of particular types. One such relevant class of judgments involves judgments about the minds and actions of people. Just as psychologists have studied how people represent the world and simulate it forward, psychologists have also studied how people represent the minds of other people, and use this representation to simulate the actions of other people, given their beliefs and desires. These *intuitive psychology* judgments happen quickly, automatically, and consistently.[5] One influential account of the representations underlying these judgments hypothesizes that these representation takes the form of a "theory of mind" about other people (Wellman and Gelman 1992; Dennett 1987). According to this view, people assume that other people have minds, which include hidden unobservable mental variables (such as beliefs and desires), which cause observable actions.[6]

On our version of this approach, when I intuitively make the prediction that Finbarr will go to the movie theater today, we frame my understanding of the dynamics of Finbarr's mind as involving a planning algorithm, which takes in mental variables (such as Finbarr's beliefs about the particular movies that are playing today, Finbarr's relative preference for movies over opera, and so on), and generates the next likely action. To capture this idea, we'll describe people as having *agent simulators*. When I predict what Finbarr will do, I employ my agent simulator to approximately predict the likely actions of other people. The notion of an agent simulator is akin to the world simulator of intuitive physics: the way we understand others using a mental agent simulator is like the way we understand the manifest world using a mental physics engine. An approximate world simulator can be used to reason from observed data to hidden dynamic variables (such as mass and friction) or to predict a subsequent state of the world; analogously, the agent simulator can be used to learn about other people's hidden mental variables (such as their desires and beliefs) by observing their actions or to predict their subsequent actions. Mental agent simulators have been the topic of recent influential research in theory

[5] People interpret even impoverished visual stimuli such as simple geometric shapes moving around with this intuitive psychology perspective (Heider and Simmel 1944).

[6] We are focusing on adults. See Starmans 2017 for work on children's theories of the self.

of mind, including quantitative action prediction and understanding in adults and children (e.g. Baker et al. 2009, 2014, 2017; Jara-Ettinger et al. 2016).

Just as my intuitive, commonsensical judgment that "if Finbarr drops the glass it will probably break" is based on the underlying computations of a mental physics engine (my world simulator) with an approximate understanding of bodies and the forces acting on them, my intuitive commonsensical judgment that Finbarr will go to the movies today is based on the underlying computations of my mental planning engine (my agent simulator) with an approximate understanding of agents' beliefs, desires, and planning—perhaps more specifically my approximate understanding of Finbarr's beliefs, desires, and planning architecture.

Before we develop our argument further, an important note on terminology is in order: In cognitive science, theory of mind has often been contrasted with accounts that use the term "simulation." Both accounts concern the explanation and prediction of actions, on the part of both other people and ourselves. As mentioned, theory of mind sees people as constructing intuitive theories of themselves and others, positing indirect hidden variables such as belief and desire. Simulation accounts (e.g., Goldman 1989, 2006), by contrast, see people as having direct access to their decision-making apparatus.[7] On this account, people can explain and predict the actions of others by running a direct simulation of those people on the same neural and cognitive machinery they use to make their own decisions (Nichols and Stich 2003). However, theory of mind is also used to predict the actions of others in a way that is well described by the term *simulation* (in the sense of evolving the world forward in time from a given initial state). A better distinction between the accounts for our purposes would be to say that one account involves direct access

[7] The most influential philosophical theory of simulation is Goldman's mind-reading theory (1989, 2006). (Another important approach is that of Carruthers 2006, which is closer to the view we defend in this chapter.) On Goldman's mind-reading approach, in contrast to the approach we adopt here, we understand the minds of others by first, mentally projecting ourselves into their minds or first-personal perspective, and then performing a simulation. In a context where we are assessing our future or possible selves, on this view, we'd simulate ourselves in a future or merely possible scenario by, first, generating the appropriate initial mental state representing the preferences that this self would start with, and then simulating the mental process that this self would undergo in the relevant scenario. In a decision-making context, we'd then use the results of our simulation in order to determine how to act. Those that prefer Goldman-style approaches to simulation may prefer this account to ours. (Here, the relevant issue concerns the structure of the simulation process and what we know, and when. Our view, recall, is that often we simulate first, before or perhaps even as we discover our preferences.) To the extent that the way we simulate is an open empirical question, there are empirical issues here that need further exploration, and so our official position here is that it is premature to pronounce on which kind of simulation our transformative, self-involving possibilities involve. (And it may well be that there is more than one kind of simulation involved.) Our central point is that whether we use theory of mind when we assess our self-involving counterfactuals, or whether we use Goldman-style mind reading, or whether we use some other approach, an empirically informed discussion of how we think of self-involving possibilities in the context of transformative decision-making raises a host of fruitful possibilities for interdisciplinary work in metaphysics and mind.

to decision-making mechanisms, while the other account involves intuitive theories (Saxe 2009). Here, we are primarily concerned with intuitive theories, and so the term *simulation* should be understood as referring to evolving the world forward based on such approximate theories, whether in the physical or psychological domain.

Once we have this model in place, we can think about reasoning involving the sorts of self-involving possibilities and preferences that we are interested in. Just as I can reason about what would happen if I dropped the glass, using my world simulator, I can reason about what you might do if you were faced with a particular choice, using my understanding of your mind given by my agent simulator. But there's a further kind of judgment that I can make, a self-involving one. I can reason about what *I* might do if *I* were faced with the same choice. We will explore the possibility that I use something akin to my agent simulator to reason about myself as well as other agents. We'll refer to this as a *self simulator*.

Recall how believing that "if Finbarr were to drop the glass, it would break" can be interpreted as based on a judgment based on my underlying computation involving a quasi-Newtonian simulation, where the simulation is a representation of the possible world W1 where Finbarr (or his counterpart) drops the glass and it breaks. In a parallel fashion, we can treat my self-involving belief that "if I were given the choice, I would pick the red pill" as a judgment based on my underlying computation involving a self simulation, where the simulation is a representation of the possible world W2 in which I am given the choice and I pick the red pill. This self simulation won't necessarily activate the actual decision-making apparatus that would lead me to a given choice, were I actually faced with taking a red or blue pill. Rather, it's a prediction based on my beliefs about my own values and planning apparatus. Our approach investigates a range of possibilities for how simulating others (agent simulators) could relate to simulating oneself (self simulators) and the world (world simulators).

We can extend this to our philosophical treatment of modality, taking the simulations to, in effect, encode a decision about which world to move to. That is, the nature of the simulation generated by the engine of the mind can be interpreted, philosophically, as an implicit specification of the relevant similarity relation. It can be understood as encoding an implicit specification of which ways of understanding one's self matter (and which don't), which in turn define the relevant respects of the similarity relation, the relation that is used to identify the properties and ways of understanding oneself in the relevant possible world. In the semantics of counterfactuals, the properties that matter determine the relevant world for the assessment of the counterfactual. In our example with the red pill, the relevant properties define a similarity relation that takes W2 to be the closest possible world for the assessment of the counterfactual. Since in W2 my counterpart chooses the red pill, if W2 is indeed the closest world, the claim "If I were given the choice, I would pick the red pill" is true.

9.1.3 Reasoning about Self-Involving Possibilities

The way we've blended a philosophical account of reasoning about oneself with an empirical model of self simulation gives us a framework for exploring philosophical questions about modal reasoning from an empirical perspective, in particular, a framework for exploring the way we assess self-involving possibilities.

In the work we present and discuss below, we explore empirical results collected from participants who were asked to imagine and evaluate a series of life-defining, self-involving possibilities. We surveyed a wide range of individuals, including philosophers, and asked them to consider big decisions involving life-changing possibilities of various kinds. Our scenarios involved possibilities such as becoming a vampire, exploring the universe with aliens, freezing time, knowing the future, consulting all-knowing oracles, and other fantastical situations. (See the appendix for our ten vignettes.)

In presenting people with these vignettes, our primary interest was in exploring the way that individuals reflect on and decide about whether to undergo various sorts of transformative experiences (Paul 2014). These kinds of decisions are focused on self-involving possibilities, possibilities concerning significant changes in one's self. (Of course, such possibilities also concern others, especially the selves of others, and while we don't focus on this element, we take it to be implicitly represented in most of our discussion, and we discuss it explicitly in places later in the chapter.) When we consider these kinds of self-involving possibilities, we think of what it would be like to be in various hypothetical situations. As we will discuss subsequently, this is an important way to learn about ourselves, and then use what we've learned to decide how to act. This approach to decision-making is based on how we imaginatively represent or simulate ourselves in different scenarios, which the metaphysician can understand as representing oneself at future times in different possible worlds. As we'll define it, *temporal prospection* is the act of representing or assessing one's own future (or present, or past) experiences. *Modal prospection* is the act of representing or assessing one's own possible experiences. Such prospection, more generally, is an act involving the assessment of various sorts of self-involving possibilities.

Why is it philosophically interesting to explore the nature of modal prospection from an empirically informed perspective?[8] One important reason stems from the importance of understanding the way we intuitively evaluate such self-involving possibilities in order to gain insight into how we reason and learn about ourselves.

As we discussed previously, work on theory of mind suggests that we learn about others from the decisions they make. We think this point extends to ourselves: we might think we have some sort of privileged insight into our own preferences,

[8] Our main focus here is on modal prospection. For more on temporal prospection and subjective temporal experience, see Paul 2016 on subjective endurance.

but in fact, appearances here are often deceptive. In the sorts of vignettes that our participants consider, we don't think the empirical research supports the view that the best way, or the ordinary way, or even the most natural way, for people to approach these types of decisions is through having direct access to the relevant mental states. That is, we don't think the most natural approach to our ordinary decision-making process for self-involving possibilities uses a formal or deductive reasoning process where we start with our preferences and reason our way through a series of steps to determine how to act.

In support of this point, much evidence from psychology shows that we know less about our own minds or reasons for acting than we normally assume (Epley 2014; Wegner 2002). For example, in one of the best-known experiments in social psychology, participants chose between four stockings that were actually identical (Nisbett and Wilson 1977). The later an item was considered, the more likely a participant was to choose it. However, none of the eighty reasons people gave for their choice mentioned the order of viewing, and when explicitly asked whether order could have affected their judgment, only one participant responded affirmatively. People have all kinds of false beliefs about themselves, from overestimating their positive mental and personality attributes relative to others (Kruger and Dunning 1999), to incorrectly picking out an image of themselves merged with an attractive target as their own face (Epley and Whitchurch 2008). Order and framing effects also seem to influence the moral judgments of people with academic expertise and professional training, just as much as they influence the general population (Schwitzgebel and Cushman 2012, 2015).

Given the evidence that we have limited access to our own preferences, we prefer an alternative where we learn about our own preferences from observing or recalling how we responded in the past to that type of situation, or, if we lack the relevant experience, through simulating or imagining the responses we would have in this type of situation. That is, we may take a certain action, or imagine taking the action, and then think, "Why did I do that? I guess I like X more than I thought." We are not claiming that we have less direct access to our own mental states than we have to the mental states of others. Rather, in both cases, it can be important to infer information about underlying beliefs and desires by observing what decision is made.

Our approach to intuitive judgments suggests that, if we want to decide how to act in a novel situation, we may start by imaginatively assessing ourselves in one hypothetical future as opposed to another, and use that to understand the value and desirability of these different possibilities for ourselves. A very natural way (and perhaps, the most natural way) for us to approach a novel decision is to start by simulating ourselves in the proposed scenario(s), seeing how we'd respond, and reverse-engineering our preferences from this response.[9] We don't have to start by

[9] We are setting aside, for the moment, the question of whether we know enough to veridically simulate the novel context. This, of course, is at issue when epistemic transformations are involved.

attempting to identify or list our preferences and then deciding how to act. Indeed, in the survey we present in detail subsequently, 75% of people sampled from the standard US population and 53% of the philosophers who took our survey indicated that they had learned something about themselves as a result of taking the survey, which suggests that they had discovered something about their preferences by thinking through the novel scenarios. People predominantly reported learning something about their personality, their current satisfaction with their life, and their attitudes towards family, friends, and relations.

Exploring how we respond in various novel scenarios, then, can be an important way of discovering what we value, and thus of discovering various truths about ourselves.[10] That is, it can be an important way to discover our preferences when faced with a choice between these possibilities. Once we know our preferences, we can decide how to act.

We see many connections to questions of philosophical interest. One obvious connection is with the debate about the nature of decision-making in cases involving transformative experiences (e.g., Pettigrew 2015; Dougherty et al. 2015; Harman 2015). In that debate, a key issue concerns the epistemically transformative nature of the experience under consideration. If an individual lacks the ability to perform an *informed* simulation, she may lack the ability to determine the relevant similarity relation when assessing an important counterfactual. This can mean that she is impeded in her ability to learn about or form her preferences in the way she needs or wants to, with implications for how she is to approach life-defining transformative choices. See Paul 2018 for an account of how these ideas draw together questions about preference formation, perceptual modes of presentation, and the role of experience in determining salient *de se* truths about oneself.

Understanding the evaluation of self-involving counterfactuals under our preferred framework raises additional interesting questions about the metaphysics of mind. If we can understand self-involving judgments computationally in terms of people relying on their mental self simulators, we can ask: what sort of self simulator is the individual using to make her intuitive judgment? What sort of engine is she using to reason about herself? When making intuitive physics judgments, there are—at least conceivably—many different types of physics engines a philosopher could imagine the mind using, including ones that don't simulate the specific reality people inhabit. A physics engine may be used to simulate worlds with different laws of nature, such as worlds of different dimensions, worlds with odd time-dependent forces, without gravity, and so on. In a similar manner, there are, at least conceivably,

[10] See Paul 2018 for a companion, albeit highly a priori, philosophical discussion of these issues, with a framing of the discovery of such truths and preferences as the discovery of *de se* truths through experience.

many different types of self simulators we could use to make predictions about what we would think and choose in various scenarios.[11]

That is, the question about the metaphysics of self simulators is analogous to a metaphysical question about laws of nature, in particular, it is analogous to asking, "What are the laws of nature that govern the evolution of the world from t_1 to t_2?" If different kinds of people use different kinds of self simulators when assessing self-involving counterfactuals, they may be, intuitively speaking, using different kinds of mental machines to generate their intuitive judgments. Recall that we can (metaphorically) characterize a self simulator as a machine that your mind uses to generate intuitive judgments about yourself. We are here asking about the nature of this machine: what kind of machine is it? That is, what kinds of simulations does it generate? Just as the mental physics engine in adults generates noisy Newtonian simulations with forces of gravity and collision, rather than, say, noiseless simulations with random time-dependent forces, a mental engine for generating self simulations could generate one kind of simulation rather than another.[12] Our hope is that, just as work in cognitive science supports the thesis that our intuitive physics judgments are grounded by noisy Newtonian simulations and rules out, say, noiseless simulations or other ways of evolving the world forward, empirical work may guide us to a clearer understanding of the actual types of self simulations we use to make intuitive judgments and learn about ourselves.

Speculating further, in the context of our framework, we see several different (nonexclusive) possible models for how we might simulate self-involving possibilities. We might simulate (1) agentially, by modeling the evolution of a (centered) possible world from our own first-personal perspective, that is, from the first-personal, conscious perspective of the individual who is the indexical center. We might describe this as using a subjective mode of presentation when we simulate. Metaphorically, as we simulate, we "occupy our own shoes." Or (2), we could simulate observationally, by modeling the evolution of a centered world using an impersonal or third-person perspective. On this approach, we simulate using a more "objective" epistemic perspective, something akin to taking a "bird's-eye" view on ourselves in a situation. Metaphorically, we simulate as though we were watching ourselves respond in the counterfactual scenario. (For more on the contrast between agential and observational perspectives on oneself, see Pronin and Ross 2006; Paul 2016, 2018.) Other, quite different, models are also possible. For example, we

[11] Of course, as we noted previously, we are focusing on self simulation here, but we aren't ruling out other options. People may use different kinds of simulators or non- simulation algorithms in different situations. That is, just as people may use different techniques to judge a physical situation (using logical deduction rather than summoning the mental physics engine to evaluate "The glass is fragile and so it will break"), people may use different techniques to make or judge a psychological prediction. We are simply focusing on a natural and important way people make these judgments, with special attention to judgments for novel situations.

[12] Metaphysically speaking, they might be selecting different similarity relations, and thus moving to different kinds of worlds, to determine the truth value of their counterfactuals.

might (3) simply judge these counterfactuals retrospectively, in a "model-free" sense (Crockett 2013). Further research is needed to disentangle these possibilities and explore whether different kinds of simulations lead to different judgments about the relevant counterfactuals.[13]

Another philosophical issue concerns the way we reason about others. Prospection can be something we try to do for others when we try to determine what others might prefer in various situations. Such information is important for projects in decision theory, practical reasoning, ethics, and medical ethics, among other topics. (For example, see Bykvist 2006; Carel et al. 2016; Pettigrew 2018; Shupe 2016; Barnes 2015; Briggs 2015; Collins 2015; Dougherty et al. 2015.)

More generally, thinking about the different ways we can make intuitive judgments about ourselves, and understanding more about how we discover and evaluate our preferences and values, should inform philosophical work on metaphysical theories of the self as well as epistemological theories of how we reason about possibilities and understand self-locating attitudes.[14] Understanding the way we self-simulate might be especially fruitful when connected to work on narrative theories of selves and personal identity (e.g., Schechtman 2011; Strawson 2017; Parfit 1984). A better understanding of the psychology of self-involving judgments could also inform current debates about whether we need a distinctive notion of first-personal or *de se* content (e.g., Cappelen and Dever 2013; Magidor 2014),[15] could help us to understand the structure of indexicality in first-personal thought (Recanati 2012) and whether our understanding of the world and the nature of our thought is fundamentally perspectival (McGinn 1983), can enrich our understanding of the relationship of first-personal thought to motivation and action (Perry 1979), and may even be important for various topics in formal epistemology.[16]

[13] There are also other theories of simulation on offer, such as Goldman 2006. The empirical debate here is ongoing.

[14] For example, Paul 2018 argues that we need experience to discover certain kinds of *de se* truths about ourselves, which then inform or create *de se* preferences. Such experience can come from actually engaging in the experience, or from (correctly) simulating oneself having the experience.

[15] Consider this remark from Cappelen and Dever (2013): "There is no attempt in the arguments [in defense of a need for first-personal indexical beliefs to explain action] to study in detail the underlying physical structure of humans and their ability to act. That would require arguments and evidence of a completely different kind from what we find in the philosophical tradition we engage with in this work" (40). Our focus on the computational basis for first-personal judgments is a start at just this sort of study.

[16] For a sample connection to formal epistemology, consider the debate between fans of norms of diachronic rationality and time-slice epistemologists (e.g., Moss 2015; Heddon 2015). Moss 2015 characterizes "time-slice epistemology" as "the combination of two claims. The first claim: what is rationally permissible or obligatory for you at some time is entirely determined by what mental states you are in at that time. This supervenience claim governs facts about the rationality of your actions, as well as the rationality of your full beliefs and your degreed belief states. The second claim: the fundamental facts about rationality are exhausted by these temporally local facts. There may be some fact about whether you are a rational person, for instance. But this fact is a derivative fact, one that just depends on whether your actions and opinions at various times are rational for you at those times" (172). Once we have a clearer understanding of just which mental states define

We can expand on the connections to ethics and practical reasoning by considering a related philosophical project. Railton (2017) explores the idea that prospection is a natural outgrowth of the Humean insight that we understand the world through a combination of experience and cognitive projection.[17] Railton describes how, just as we can project causal structure onto a sequence of actual events, we can project modal structure onto hypothetical events, mentally extending reality in various ways in order to understand possibility. Railton suggests that imaginative projection can inform moral learning via moral prospection and the empathetic assessment of others.[18] We agree, and take the possibilities here to be expansive: imaginative projection can inform a wide variety of prospective assessments, especially, for our purposes, modal prospection concerning transformative changes in one's own self and others.[19] Prospection and simulation (or imaginative projection) informs learning about and empathy for others and ourselves, with moral learning and morally informed empathy as a special case. In his paper, Railton argues that imaginative prospection teaches us to evaluate moral possibilities from a "non-egocentric" perspective. Again, we take the possibilities here to be expansive. Imaginative prospection may be understood in more than one way. Agents may evaluate possibilities from a variety of perspectives, egocentric and non-egocentric, and different kinds of evaluations may lead to different assessments of the possibilities.[20]

Finally, we hope it is clear that our chapter provides a more general foundation for understanding new connections between topics in metaphysics and mind and cognitive science. For example, metaphysical discussions involving modality often concern our judgments about what is necessary, possible, and impossible, and the nature and structure of our modal intuitions play a key role in these discussions. An important debate in the metaphysics of mind explores the relationship between metaphysical possibility and various sorts of thought experiments designed to show

a person's preferences, and especially if it is the mental states that *result* from one's simulations that determine one's preferences, we can see that the ontological structure of human rationality may require a temporal and causal structure that makes this sort of debate, at the very least, more complex. (Since time-slice epistemologists are opposed to Bayesian conditionalization, the connection is unsurprising. This brings out how there can be empirical work on reasoning processes that formal epistemologists may want to engage with.)

[17] We are not endorsing projectivism here, merely discussing how Railton's approach relates to our project.

[18] For related work, see Graham et al. 2017.

[19] We note again that our view (in contrast to Railton's projectivist stance) makes no commitment to antirealism. Our imaginative projections may well be representing real modal structure.

[20] Railton (in conversation) notes that he sees psychological projection as part of a learning process about actual modal features, but does not think of the metaphysics projectively. Further, he agrees that prospection goes well beyond the moral case, can involve various kinds of modal and egocentric modeling and simulation as well as moral and nonegocentric, and can work with various kinds of evaluation and counterfactual suppositions. See Seligman et al. 2013 and Seligman et al. 2016 for further discussion.

that S is indeed possible in the relevant sense. Explorations of whether S is metaphysically possible often involve questions about whether S is conceivable, whether S is imaginable, what the relationship is between conceivability and imaginability, and whether conceivability entails metaphysical possibility. Some have argued that the way in which we imaginatively represent possibilities needs to be clearly and critically assessed in order to determine whether the structure of what we seem to conceive implies the type of possibility we seem to be discovering (Hill 1997; Hill and McLaughlin 1999). Others argue for a range of ways that imagination relates to assessments of possibilities (e.g., Gendler 2010; Nanay 2016; Ninan 2016, Kind 2016; Williamson 2016). The connections our work draws between empirical research concerning the details of the way the mind predictively represents possibilities and the evaluation of self-involving counterfactuals should be of significant interest to those who explore the relationships between possibility and imaginability. Finally, as we noted previously, there are important debates over the nature of simulation theory that span philosophy of mind and cognitive science (e.g., Goldman 2006; Carruthers 2006; Nichols and Stich 2003). Our work explicitly ties together theories of simulation with work in the metaphysics and epistemology of transformative experience, counterfactual semantics, and *de se* reasoning. This opens new avenues for enquiry and collaboration about the nature of simulation, showing how a topic that is central to the philosophy of mind may also be central to debates in metaphysics and epistemology.

The case is clear. Our philosophical understanding of self-involving possibilities, particularly in the context of modal prospection about transformative decision-making, is enriched by engaging with the relevant work in cognitive science. We turn now to a discussion of the empirical part of our project. In the next section, we introduce the sorts of self-involving possibilities we asked people to make choices about, and discuss the features of these decision tasks that we take to be especially probative.

9.2 MODAL PROSPECTION

Consider the following scenario.

Imagine that aliens come down to Earth, and give you the option to go with them on their travels throughout the universe. The aliens are friendly and honest, and tell you that you would see amazing things on your travels with them if you decide to go with them.

If you decide to go, you will have a week to say goodbye to your family and friends. Once you leave, you will never again return to Earth, nor be able to communicate with people on Earth.

Do you go?

What percentage of other people do you think will choose to go?

Now consider another scenario.

Imagine that there is a magical hourglass. If you flip the hourglass, the following happens:

Every person on Earth stops moving, but you are free to move around as you please. You do not age during this frozen time, but you can be hurt and you can die. For example, if you jump off a tall cliff, you will die. The internet, electricity, and so on carry on working. You will remember everything you did during the frozen time.

If you decide to flip the hourglass, you have to decide in advance how long to freeze time for. You cannot change your mind and unfreeze time in the middle.

If you wish to flip the hourglass it must be done now—you will not be able to flip it at a later time.

Would you like to flip the hourglass? For how long?

What percentage of other people do you think will choose to flip the hourglass?

These kinds of questions ask you to perform a distinctive sort of task: they ask you to deliberate about what you might do in a merely possible scenario involving a transformative experience.[21] Previously, we defined the assessment of one's possible experiences as *modal prospection*. Our interest is in "big decisions," that is, in tasks involving modal prospection about life-defining choices. But our scenarios are highly artificial. Why have we chosen such far-fetched, bizarre scenarios, rather than more everyday sorts of scenarios with transformative experiences such as choosing to have a child or taking a job in a foreign country?

First, because we want to force our participants to imagine and reflect on the situation in order to decide what they prefer. That is, we want participants to think through the situation, ideally through simulating themselves in it, and so we used epistemically new, imaginatively far-flung scenarios, including scenarios that are not physically possible (for example, scenarios where you could do things like freeze time). A familiar scenario could allow a participant to respond simply by drawing on previously stored responses to similar situations, or by using a pre-established convention or known scientific fact. While it's true that our participants may have had late-night dorm room conversations about vampires or debates in the bar about whether the government is covering up the aliens creating crop circles, we expect that, prior to our survey, most of our participants had never considered the explicit, self-involving possibilities we raise.[22]

[21] For related work on addiction, transformative experience, and distancing from the past and future self, see Iskiwitch et al. 2017.

[22] We say "most," because our far-fetched scenarios did not, in fact, guarantee novelty. Some participants reported having learned nothing from the vignettes, exactly because they "already thought about this in deep detail before," as one participant put it. This reinforces the idea that actually performing the simulation is important for learning and discovery.

Why is it important to have our participants reason about what they would actually do in each scenario? Because our project focuses on the way agents simulate themselves, and we are interested in gathering data on this feature of our mental lives. We are interested in knowing more about this feature of our mental lives because it is psychologically, philosophically, and practically important. Philosophers and psychologists care about the nature of selves, and about discovering fundamental human values, and understanding self simulators may give us knowledge of the nature and structure of selves and human values. More generally, Kappes and Morewedge (2016) argue that mental simulation can substitute for experience with respect to its evidentiary value and benefits for practice. In situations where we lack the relevant experience, such as novel or transformative decision contexts, simulation may play an especially important role. We think that gaining a better understanding of how we simulate our self in potentially transformative futures is relevant to getting a deeper understanding of agent deliberation and choice that is applicable in a wide range of practical and theoretical contexts.

The second reason we use fantastical thought experiments is that we want to isolate key properties of the self, and our thought experiments give us an excellent way to do this. That is, we are interested in identifying properties that are fundamental to the way people think of themselves and the way they live their lives. These scenarios, if they were real, would bring about life-defining changes: if you were actually to leave with the aliens or freeze time, your life would change in a dramatic way. Speaking philosophically, we are interested in properties and values that are fundamental to one's self conception, perhaps even metaphysically fundamental to one's self, or at least, we are interested in properties and values that determine central or core features of one's lived experience and self-understanding.[23]

Another reason we use fantastical examples is that we want to isolate particular properties or values for our participants to consider (the "key" properties and values we describe in the previous paragraph). To do this, we need to abstract away from the irrelevant and potentially distracting features of ordinary, familiar scenarios. We also need to idealize in order to properly isolate the relevant concept. The need for abstraction and idealization to isolate the concept of interest is familiar from work on the theory and practice of scientific theorizing and discovery. Consider the important role of the Maxwell's Demon thought experiment in the development of our understanding of thermodynamics, and its continued relevance today in teaching and learning about entropy.[24]

Our fantastical examples, then, are designed to isolate particularly important properties and values, and to highlight what we really care about when making high-stakes, forced choices between contrasting values. For example, the vignette

[23] In the sense of Paul (2014), these situations involve *personally transformative experiences.*

[24] See Paul 2012 for more discussion of how this works, and for a fuller explanation of the role and importance of abstraction and idealization in philosophical thought experiments.

involving friendly aliens offers you a stark choice: which do you value more, discovery and novelty, or your attachments here on Earth? Our magical hourglass vignette frames a different type of question: would you be willing to isolate yourself from the rest of humanity to gain freedom from the relentless pace of everyday life? To have time to explore and reflect, uninterrupted? Reflecting on what you'd do in these cases is a way of thinking imaginatively, yet precisely, about who you are and what you care about. Would you choose to go with the aliens? Or would you elect to stay on Earth? Would you freeze time, in the process freezing all of your family and friends, and temporarily cutting yourself off from all human interaction? If so, for how long?

So while these situations are far-flung and removed from reality, they connect back to things that many people care about. Their underlying structure concerns deep and fundamental questions about who we are, the sorts of questions we grapple with when making decisions to undergo big life changes. Metaphysically speaking, reflecting on these cases is a way of thinking about the person or self that you are in terms of the values and desires that structure your preferences and define your psychological profile. Arguably, if you'd choose to go with the aliens, you are someone who values exploration and discovery over the status quo. If you choose to stay on Earth, you might be someone who values family and friends over discovery of the unknown. If you choose to freeze time, you might value the freedom this gives you over having contact with friends and loved ones and the rest of humanity, and the "duration" of the freezing you choose may reflect something further about your values.

An additional benefit of our fantastical contexts is that they showcase the particular challenges people can face when they find themselves in high-stakes, highly unusual situations. In such situations, even in the real world, we can find ourselves effectively without guidance, with little anecdotal or scientific information available. Compare the way we'd use our mental physics engine in a high-stakes, novel, but potentially real-world situation where we lacked background experience or detailed scientific guidance. Imagine being marooned on a desert island, where your only chance of survival was to build a seaworthy boat out of some cardboard that had washed up on the beach. On the assumption that you've never had to do this before, and that you lack detailed knowledge of the engineering required to build such a boat, your best option is to rely on your intuitive physics judgments in order to construct your craft. Similarly, in a high-stakes, novel situation involving a big life decision, your best option may be to rely on your simulations when making your decision, for they may be your only guide. (Alas, for most of us, our intuitive understanding of physics is probably as bad at helping us construct a seaworthy cardboard boat as our intuitive understanding of ourselves is at helping us make good decisions in novel situations. And yet, in some circumstances, it may be all we have.)

We can now address a final issue. Subjectively, to us (the authors), it feels like the scenarios are capturing something important when we deliberate. Why? Why does it seem so relevant, and important, and interesting to imagine these kinds of highly speculative, fantastical scenarios, as opposed to reflecting on more mundane, ordinary cases? Why does reflecting on these types of decisions seem so meaningful?

The answer can be drawn out from what we've established already.

We care about understanding ourselves and our values. However, as we've suggested, it may be impossible to divine these fundamental values merely by introspecting and trying to perceive them directly. We may have no more access to these mental constructs than we do to the mental constructs of other people, for whom we construct a theory of mind (Saxe 2009). Our fantastical thought experiments give us another way to discover ourselves. They engage the theory-of-mind module and have us assess fundamental or core values, allowing us to discover our responses and make inferences about what we really care about.[25] That is, discovering our responses in the scenarios allows us to make an inference about who we are. As we might put it, when you consider your response to one of our scenarios, you infer something about the nature of your self simulator.

Perhaps your response surprises you. You discover that, even while you think of yourself as adventurous and free-spirited, you'd refuse to go with the aliens. You care too much about your attachments to other people here on Earth. That is, when you actually perform your simulation, you discover that you value your relationships with other people more than adventure and novelty. Here, you discover something about your nature, and so you improve your understanding of yourself.

Or perhaps your response doesn't surprise you. You think of yourself as adventurous and free-spirited, and, consistent with this, you jump at the chance to go with the aliens. You value adventure and amazing discovery over the familiar things you have here on Earth. In this case, you partially confirm your understanding of yourself. Perhaps you still feel you've learned something, but that knowledge came mainly as reinforcing your existing beliefs about yourself.

If people had direct access to the mental inputs of their self simulator, they wouldn't be capable of this learning process, neither the surprise nor the reinforcement of existing beliefs. To show the intuition behind this, consider a case in which you have to predict the actions of a friend. Suppose you hold certain beliefs about your friend's values and beliefs that lead you to predict your friend will go with the aliens. Suppose these beliefs are rather firm, but not absolute. But as it turns out, your friend decides not to go with the aliens. This surprises you, and you radically revise your beliefs accordingly. Suppose instead your friend decides to go with the aliens, as you predicted. This doesn't surprise you much, but it reinforces your beliefs. This belief updating towards growing certainty is also a form of learning. The only case in which you won't learn anything is if your beliefs weren't just firm, but absolute. If you are absolutely certain your friend will go with the aliens, and they indeed go, you learn nothing. Now, replace your friend with yourself. If you had direct

[25] Thanks here to Josh Tenenbaum for discussion.

access to the mental variables that go into planning your actions, this would be akin to being absolutely certain of their value, and your responses would not surprise you nor reinforce your existing notions in the slightest.

Thus, unlike with direct introspection, our fantastical scenarios recruit the theory-of-mind module, leading us to update our beliefs about our own selves. Moreover, the scenarios are constructed to make us focus on distinctive, core human values, and so our deliberations involve the assessment and updating of central, self-defining beliefs. As a result, we find these scenarios engaging and meaningful: reflecting on them can teach us about ourselves.

Finally, the idea that we can learn something from modal prospection may also explain why considering our actions in the kinds of situations described by the opening vignettes is often enjoyable. The majority of people sampled from the general US population in our survey reported that they enjoyed it a great deal more than the average study. While this is a low bar, people's free-form comments also indicated how unusually fun the survey was. This sentiment is at odds with how big real-life decisions can be difficult and painful to contemplate (Ullmann-Margalit 2006). Unlike real-world scenarios that involve big decisions, they don't come with the baggage of facing real regret and closing off of opportunity. At the same time, the information gained from considering such situations may lead to positive feelings, reflecting the intrinsic reward that accompanies information gain and exploration in general (Schmidhuber 2010; Gottlieb et al. 2011).

9.3 LEARNING FROM TEN VIGNETTES

In the following, we describe in more detail the methods and data analysis that informed the preceding discussions.

The empirical studies discussed here involve participants choosing whether to take a life-altering choice in an unusual, hypothetical situation, and then reflecting on how they had made their choice and what they learned from it. It is possible that the actions participants indicate that they will take in the presented scenarios sometimes differ from those that they would actually take. However, even if people are mistaken about the actions that they would actually take in the scenarios, the reasons they give for their actions may shed light on the factors that they consider when making such decisions. Moreover, people can learn something about themselves (or think they learned something about themselves) from the actions that they believe they would take, even if their predictions about these actions are inaccurate. In this setting, people's stated actions may differ from their actual actions for interesting reasons. For example, people may use a hypothetical answer to deceive themselves into thinking they have various characteristics that they value, such as a sense of adventure, even though in the actual situation they would choose to stay safely home, as this signal is far more costly.

We recruited two groups of participants independently. One group was a sample of American adults recruited through Amazon's Mechanical Turk service (hereafter "Turkers"). The other group was recruited by soliciting professional philosophers using social media (hereafter "philosophers"). The Turkers were financially compensated for their participation. Both groups were directed to an online survey, described in what follows. The survey was largely identical for both groups, although the philosophers were additionally asked to identify themselves as either graduate students, postdocs, or faculty members in philosophy or cognate departments (or as nonphilosophers). Those who did not identify as graduate students, postdocs, or faculty members in philosophy were excluded from the following analysis.

In the survey, participants were presented with a series of ten vignettes in random order (see full list in appendix). For each vignette, participants were asked to make a yes/no decision concerning their own action in the vignette. The yes answer always corresponded to choosing to transform in some way. Several questions had immediate follow-up decisions. For example, in the magical hourglass scenario described in the introduction, participants were also asked for how long they would choose to freeze time. For every participant, three of the ten vignettes were randomly selected for additional follow-up questions. These included asking participants how they made their decision (free-form text response), how confident they were in their answer (sliding scale), what percentage of other people they predicted would say yes to that question (textbox), how difficult it was to make their choice (sliding scale), what if anything they learned about themselves from their decision (free-form text response), and how much they think they would change as a result of their decision (sliding scale).

Following these ten vignettes, participants were asked to provide their age, gender, degree of education, relationship status, number of children, and any additional comments they may have about the survey. Participants were also invited to invent their own vignette.

We analyze data from 500 Turkers, ranging in age from eighteen to seventy-four with a median age of thirty-two. About a third of the Turkers reported having at least one child, and about 60% reported being in a relationship. We analyze data from 365 philosophers, ranging in age from twenty to seventy-seven with a median age of thirty-three. A fifth of the philosophers reported having at least one child, and about 80% reported being in a relationship.

We present below eight claims arising from our analysis of the empirical data.

9.3.1 People Do Choose Transformative Experiences, with Turkers and Philosophers Differing in Their Choices

It's perhaps striking that people actually choose yes in these vignettes, despite their transformative nature, encountering them for the first time, and with much

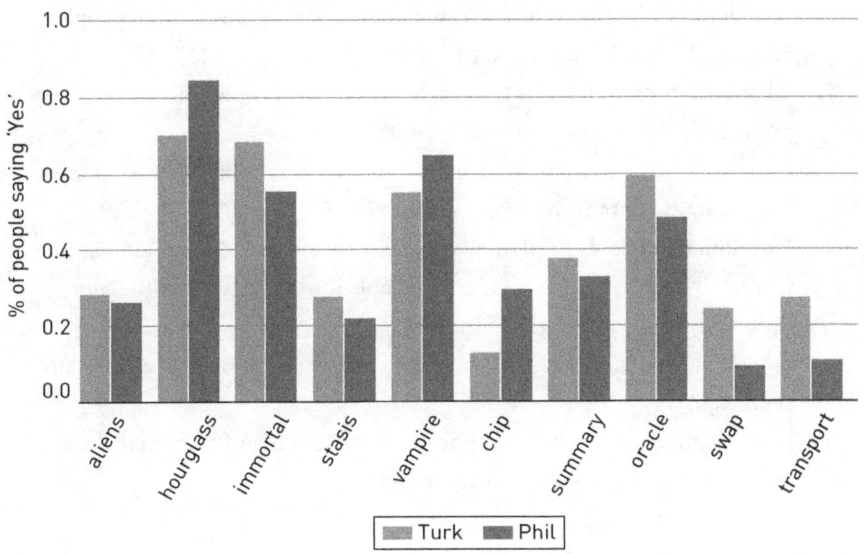

FIGURE 9.1 Percentage of philosophers and Turkers selecting yes in each scenario

uncertainty about the results. Out of ten vignettes, Turkers said yes an average of 4.1 times (sd 2.25), and philosophers 3.8 times (sd 1.6). However, the philosophers we survey are not simply more reluctant to transform than Turkers across all scenarios, the philosophers also make different choices. For example, the philosophers are about three times more likely to take the chip than Turkers. See figure 9.1 for the choices of the two groups in each scenario.

While the exact percentages of people saying yes to a particular vignette is not important for our discussion in this chapter, it is interesting that the average choice differs between philosophers and Turkers for some of the vignettes, specifically the Hourglass, Immortal, Chip, Oracle, Swap, and Transport scenarios.[26]

9.3.2 Demographics and Personality Affect Choices in These Vignettes, but Are Fairly Small Effects

A number of different factors influence the pattern of people's choice across questions. We analyzed the relationship between answers to personality and demographic questions, and the number of scenarios for which somebody responded yes. The correlations for Turkers are always given first in the following:

(A) Age matters to a small degree: the older people are, the less likely they are to transform ($r = -.26$, $r = -.18$)

[26] This is after controlling for age, gender, happiness, number of children, and whether someone is single.

(B) Number of children matters to a small degree: the more children people have, the less likely they are to transform ($r = -.24$, $r = -.13$)

(C) Relationship status matters to a moderate degree: people who are single are more likely to transform ($d = .30$, $d = .18$)

(D) Happiness matters to a small degree: the less happy people are, the more likely they are to transform ($r = -.26$, $r = -.16$)

(E) How different people think they will be in 10, 20, and 30 years all matter to a small degree: the more different people think they will be, the more likely they are to transform (Turkers 10, 20, 30 years: $r = .18$, $r = .21$, $r = .23$; philosophers 10, 20, 30 years: $r = .15$, $r = .12$, $r = .08$; the last is not significant)

(F) How accurate people think they are at predicting their future state does *not* matter ($r = -.04$, $r = -.07$, not significant)

The preceding analysis was done across all vignettes, but as one might expect, some demographic variables were more relevant to particular scenarios, rather than the vignettes generally.

Some of the demographic variables involve people's relationships to other people, either as partners or as parents. That these variables predict choice suggests that when people imagine such acts of transformation, they consider not simply how they themselves will change, but also how their relationships will change. Many of the comments explicitly indicated that despite a no answer, responders would have chosen otherwise if they were not in a relationship or if they did not have children. Some of the comments further reflect on what people would have done prior to being in a relationship (e.g., "I have a wife and children. Primarily because of them (but also friends), I wouldn't want to abandon my life for adventuring around the world"). At least from the comments, some factors seemed to matter in opposite directions ("I'm too old for these things; I want to stay with my family" vs. "Since I am older now, and will probably at best only live ten to fifteen more years . . . I feel it would be morally permissible").

The relatively small effect of these demographic and personality variables points to the complexity of imagining these experiences. Imagining what you will be like after a transformative experience thus depends on far more factors than captured by these simple measures.

9.3.3 People Are Highly Confident in Their Responses

The median confidence of Turkers saying no was 96.0 (out of a maximal score of 100.0), and for philosophers it was 80.0. Both Turkers and philosophers tended to be less confident when they said yes to a vignette, although the median remains high (figure 9.2).

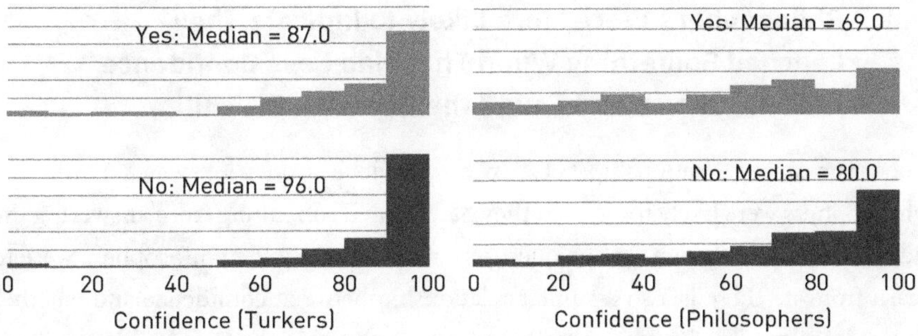

FIGURE 9.2 The distribution of reported confidences by Turkers and philosophers, broken

Given that these are complex transformative experiences where the outcome is difficult to imagine, this result is somewhat surprising. One possibility is that after the decision is made, people may become confident that they are the sort of person who would give such an answer—after all, they just gave it!

9.3.4 People Learn Things about Themselves from Their Choices

Recall that people were asked what they had learned about themselves for three vignettes during the survey (randomly chosen for each person), as well as at the end of the survey for the survey as a whole. People from both groups self-report learning about themselves. They do this both in their answers to individual vignettes, and for the survey overall. Approximately 75% of Turkers indicated that they learned something from the survey overall, as did 53% of philosophers.[27] People predominantly reported that they learned about their personality (e.g. "That I am more risk averse when faced with transformative decisions"), how much they were satisfied or dissatisfied with their current life or self (e.g. "That I am attached to my actual, earthly existence more than I might have thought otherwise"), and their feelings towards their family or other people they had relationships with (e.g. "I've learned that I value personal relationships more than I thought").

[27] These percentages are based on our hand-coding of the free-form text responses. This coding proved challenging at times. Responses that include text such as "Learned nothing" or "N/A" are easy to code as no, but other responses are more difficult. We tended to err on the side of coding responses as not having learned anything. For example, for individual scenarios in which people simply said they had learned how they would react to the very specific scenario (e.g. "I learned that I would flip over a magic hourglass"), we did not code this as a yes.

9.3.5 Philosophers Were More Likely to Indicate They Learned Something When They Had Less Confidence in Their Choice and Found the Question Difficult

There is a strong relationship between how difficult a question was and whether philosophers were likely to say that they had learned something from answering the question. That is, the harder the question, the more likely the philosophers were to learn from it. There is also a similar relationship between confidence and whether or not a philosopher learned something from a question—the less confident in their answer, the more likely the philosophers were to learn from it. This relationship does not exist for Turkers, although it's possible that this is because hand-coding whether Turkers learned something was more difficult than for philosophers.

9.3.6 People Learn from Choices Both Taken and Refused

Turkers and philosophers were as likely to say that they learned something about themselves when they said yes to a particular transformative experience as when they said no to a transformative experience.

This is somewhat surprising. Assuming that no is the conservative option, the decision not to change (even in theory) does not seem to involve much new knowledge. On the other hand, people also seemed to be surprised by how conservative they were, even in responses to hypothetical situations ("I'm more conservative than I expected"; "I'm more conservative than I used to be, yet also quite touchy about this. I feel a sense of nagging guilt at abandoning my lone wanderer ways"; "I used to think of myself as an explorer keen for any chance to explore new possibilities. No more, it seems"). This suggests that there is new knowledge to be gained about the self from reflecting on possible futures even if one decides not to take them, whether it is knowledge about the self one didn't know about, or an updating of an outdated self-perception held over from younger days.

9.3.7 People Are Inaccurate at Predicting Other People

On average, people are about 20% inaccurate in predicting the percentage of other people saying yes to a scenario, with philosophers and Turkers about equally inaccurate (a bootstrap analysis shows no difference between the two groups). Figure 9.3 shows the percentage of philosophers and Turkers saying yes for each vignette, as well as the average prediction of people saying yes made by each group (gray dots for Turkers and black dots for philosophers).

In line with the standard "false consensus effect" (Ross et al. 1977; Dawes 1990; Robbins and Krueger 2005), both Turkers and philosophers who said yes to a vignette are more likely to think other people will say yes to the same vignette. For example, Turkers who would themselves go with the aliens thought that approximately

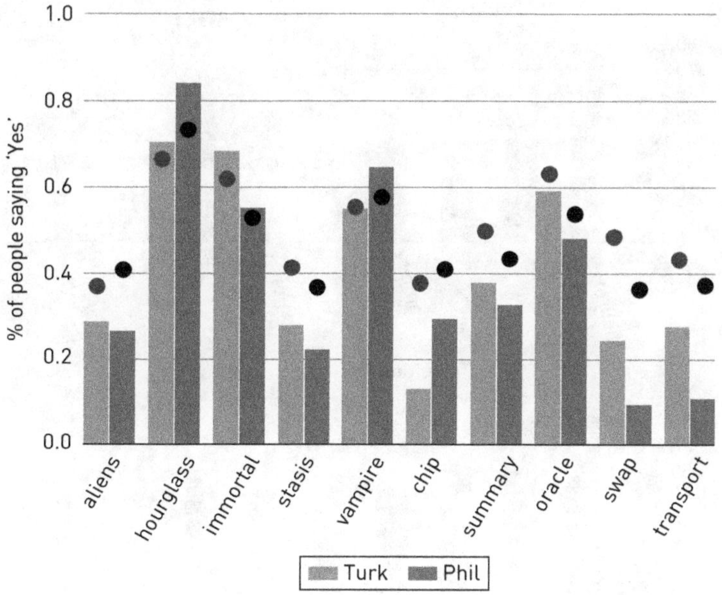

FIGURE 9.3 Predictions made by Turkers and philosophers of the percentage of people

50% of other people would go with the aliens, while Turkers who would not themselves go with the aliens thought only approximately 30% of other people would go with the aliens.[28] Philosophers showed a smaller false consensus effect than Turkers. In figure 9.4, we display the false consensus effect for each group by splitting the predictions of group members who said yes and no.

9.3.8 People Enjoy Deciding about Such Vignettes

People seem to enjoy considering their choices in the kinds of vignettes that we describe in this chapter. Figure 9.5 provides some quantitative evidence for this claim, with Turkers indicating that they enjoyed this survey considerably more than other studies on Mechanical Turk, although admittedly this is a fairly low bar.

As mentioned in section 9.2, in real life people agonize over big decisions and try to avoid them, raising the question of why this task is enjoyable. We suggested that it both removes the negative aspects of big-decision-making, and simultaneously provides reward by giving decision-makers new information about themselves.

9.4 CONCLUSION

In this chapter, we explored how people engage in modal prospection, that is, how they understand and reason about self-involving possibilities. Drawing parallels

[28] The actual number was 28%.

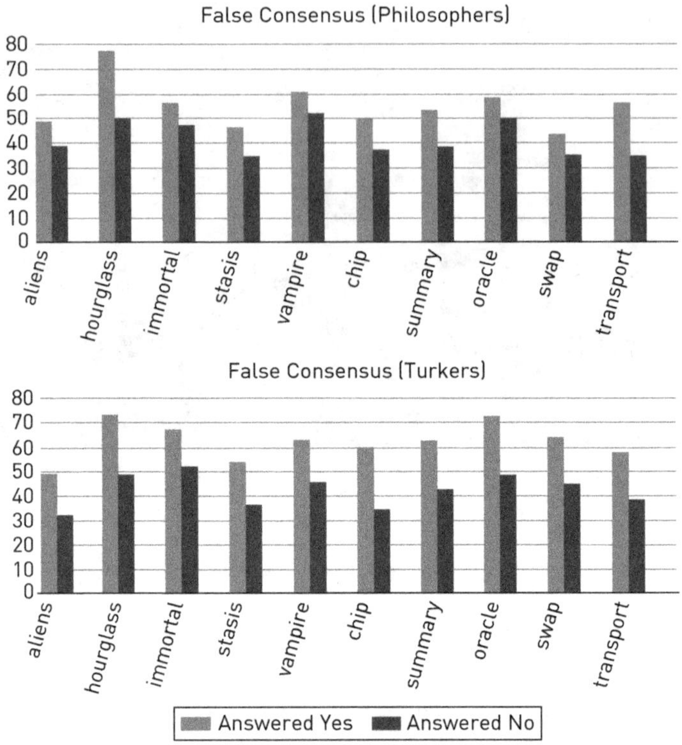

FIGURE 9.4 Average prediction by Turkers and philosophers, split up by how the person making the prediction had responded to the yes/no question. Both Turkers and philosophers display a false consensus effect.

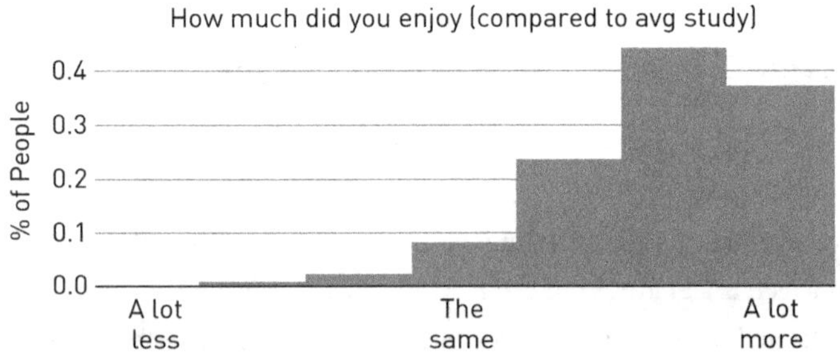

FIGURE 9.5 Distribution showing how much Turkers enjoyed answering the vignettes, compared to the average study on Mechanical Turk

between the metaphysics of counterfactuals and work from cognitive science on people's mental physics engine, we sketched an account of how we might evaluate self-involving possibilities, with particular attention to the way we'd evaluate them in transformative contexts. Our account suggests that people have an agent simulator, specifically a self simulator, that informs the way we assess our preferences and counterfactuals. Through observing or simulating their own choices, people understand themselves through a process similar to that by which they understand other people. In this way, they learn about their own beliefs and desires.

Our theoretical presentation was informed by empirical data from surveys we conducted, where lay people and philosophers decided what action they would take in unusual vignettes, and we developed a theoretical framework that interpreted and discussed people's reactions to these vignettes. While the empirical data bore out some of our theoretical speculations, for example about the possibility of learning about oneself from considering such choices, we've raised many questions that we don't yet have the empirical data to answer.

For example, when and how do people use self simulation to discover their preferences in these scenarios? Are there differences in the types of self simulators people use, and do these differences stem from different modes of presentation, different contextual factors, or other features of the decision process? How do we represent the inputs to our imaginative simulation, and what properties govern the richness of our simulations?

Our conceptual framework raises many questions, both philosophical and psychological, about the process of modal prospection and our understanding of possibilities. Moreover, our vignettes encouraged participants to discover important, self-defining truths about themselves, which suggests that the questions we are raising here may also be relevant to the emerging psychological literature on "true selves." The concept of a "true self" concerns the possibility that certain properties or dispositions of a person define, in some essential or important sense, who that person really is. Speaking philosophically, we might say that these properties and dispositions somehow ground who the person is, and provide a framework for identifying the deep or inner nature of a person. Much attention has been paid to the role that moral properties seem to play in constituting the true self (Strohminger and Nichols 2014; Prinz and Nichols 2016; Strohminger et al. 2017). Our research suggests that we need to look at this issue through a wider lens. Just as we think that moral prospection is likely to be a species of modal prospection, moral properties are likely to be a subset of the properties that are important to who we are, or who we take ourselves to be. Our self-defining values may extend past moral values. And just as we might use prospection to think about the true self from a non-egocentric view, we might use it to think about the true self from an egocentric view. We encourage those who are interested in empirical approaches to philosophical questions of personal identity, persistence, and selves to explore these possibilities further.

More generally, we hope that our conceptual framework and empirical data will encourage philosophers and psychologists to grapple with some of the many open questions about how people conceive of self-involving possibilities and engage in modal prospection, particularly in the context of intuitive judgments about transformative experiences.

9.A APPENDIX: VIGNETTES

9.A.1 THE ALIENS

Imagine that aliens come down to Earth, and give you the option to go with them on their travels throughout the universe. The aliens are friendly and honest, and tell you that you would see amazing things on your travels with them if you decide to go with them.

If you decide to go, you will have a week to say goodbye to your family and friends. Once you leave, you will never again return to Earth, nor be able to communicate with people on Earth.

Do you go?

9.A.2 THE HOURGLASS

Imagine that there is a magical hourglass. If you flip the hourglass, the following happens:

Every person on Earth stops moving, but you are free to move around as you please. You do not age during this frozen time, but you can be hurt and you can die. For example, if you jump off a tall cliff, you will die. The internet, electricity, and so on carry on working. You will remember everything you did during the frozen time.

If you decide to flip the hourglass, you have to decide in advance how long to freeze time for. You cannot change your mind and unfreeze time in the middle.

If you wish to flip the hourglass it must be done now—you will not be able to flip it at a later time.

Would you like to flip the hourglass?

(Follow-up: How long would you freeze time for?)

9.A.3 THE HIGHLANDER

Would you like to live forever, assuming good health and a youthful physique?

(Follow-up: Assuming you cannot live forever, how long do you want to live for, assuming good health and a youthful physique?)

9.A.4 THE CHAMBER

Imagine that scientists offer you the immediate, one-time opportunity to go into a chamber that works in the following way:

Once in the chamber, you will fall into a dreamless sleep for as long as the chamber is running. While you are in the chamber, you will not age. When the chamber opens, you will wake up without any side effects.

If you choose to go into the chamber, you need to decide in advance how long to stay in the chamber. You can choose any length of time. The chamber is completely impervious to tampering or damage, and is guaranteed to fully function indefinitely.

Do you go in the chamber?

(Follow-up: for how many years do you go into the chamber?)

9.A.5 THE VAMPIRE

Imagine that you have the chance to become a vampire. With one swift, painless bite, you'll be permanently transformed into an elegant and fabulous creature of the night. As a member of the undead, your life will be completely different. You'll experience a range of intense, revelatory new sense experiences, you'll gain immortal strength, speed, and power, and you'll look fantastic in everything you wear. You'll also need to drink blood and avoid sunlight.

Suppose that all of your friends, people whose interests, views, and lives were similar to yours, have already decided to become vampires. And all of them tell you that they love it. They describe their new lives with unbridled enthusiasm, and encourage you to become a vampire too. They assuage your fears and explain that modern vampires don't kill humans; they drink the blood of cows and chickens. They say things like: "I'd never go back, even if I could. Life has meaning and a sense of purpose now that it never had when I was human. I understand Reality in a way I just couldn't before. It's amazing. But I can't really explain it to you, a mere human—you have to be a vampire to know what it's like." Suppose that you also know that if you pass up this opportunity up, you'll never have another chance.

Would you do it?

9.A.6 THE CHIP

Imagine that scientists have developed a chip that can be painlessly implanted in your head with a simple procedure. If you choose to have the chip implanted, you will gain an entirely new sense (completely different from taste, touch, smell, sight, and hearing). However, you will also lose your sense of taste. The procedure is irreversible.

Do you want to have the chip implanted?

9.A.7 THE SUMMARY

Imagine an honest time-traveler from the future comes to you and says:

"I have written a paragraph summarizing your entire life from start to finish. Whether or not you choose to read this summary, the events of your life will unfold as it says."

Would you like to read this summary?

9.A.8 THE ORACLE

Imagine that there exists an advanced machine called The Oracle, which can answer the question "What should I do with my life to be as happy as possible?"

To do this, The Oracle scans your brain and accurately understands what makes you happy, what you like and dislike, what you value, what you hope and dream for.

The Oracle is completely accurate at predicting the future. The Oracle is honest, and error-free.

A condition of consulting with The Oracle is that you must do whatever it tells you to do.

Everyone who has asked The Oracle what to do with their lives report that they are extremely happy.

Do you ask the Oracle what to do with your life?

9.A.9 THE SWAP

Imagine that you can push a button that works in the following way:

If you push the button, you immediately swap lives with a person of your choosing. You will completely swap bodies, memories, personalities, abilities, current locations, and so on. Neither you nor the person you choose will remember that the swap occurred. Nobody else will know the swap happened.

Do you push the button?

(Follow-up: Who would you switch with?)

9.A.10 THE TRANSPORTER

Imagine that the world is divided up into a "transporter grid" of ten-mile by ten-mile nonoverlapping blocks, so that every point on Earth is in one of these blocks.

You are offered the only key to this transporter grid. If you choose to use the key, the following happens:

You will gain the ability to transport yourself instantly to any block you choose in the world, with no side effects. Nothing transports with you, except the clothes on your body, a wallet, and a phone.

You must stay within the block you chose for exactly thirty days, no more and no less. After thirty days you must use the transporter key again and transport to a new block that you have never visited before. This will continue for the rest of your life.

If you do not take the key now, it will disappear.

Do you choose to use the transporter key?

ACKNOWLEDGMENTS

Authors are listed in alphabetical order, in accordance with philosophy conventions, and contributed equally to this work. We are indebted to discussion with Ross Cameron, Jessica John Collins, Alvin Goldman, Kris McDaniel, Brian McLaughlin, Daniel Nolan, David Rose, Josh Tenenbaum, and participants at the 2017 Metaphysics Ranch Workshop.

REFERENCES

Baker, C. L., Jara-Ettinger, J., Saxe, R., and Tenenbaum, J. B. (2017). Rational quantitative attribution of beliefs, desires and percepts in human mentalizing. *Nature Human Behaviour 1*: 0064.

Baker, C. L., Saxe, R., and Tenenbaum, J. B. (2009). Action understanding as inverse planning. *Cognition 113* (3): 329–49.

Baker, C. L., and Tenenbaum, J. B. (2014). Modeling human plan recognition using Bayesian theory of mind. In Gita Sukthankar, Christopher Geib, Hung Hai Bui, David V. Pynadath, and Robert P. Goldman, eds., *Plan, Activity, and Intent Recognition*. Burlington, MA: Morgan Kaufmann, 177–204.

Barnes, E. 2015. Social identities and transformative experience. *Res Philosophica 92* (2): 171–87.

Battaglia, P. W., Hamrick, J. B., and Tenenbaum, J. B. (2013). Simulation as an engine of physical scene understanding. *Proceedings of the National Academy of Sciences 110* (45): 18327–32.

Briggs, R. 2015. Transformative experience and interpersonal utility comparisons. *Res Philosophica 92* (2): 189–216.

Bykvist, K. (2006). Prudence for changing selves. *Utilitas 18* (3): 264–83.

Cappelen, H., and Dever, J. (2013). *The Inessential Indexical: On the Philosophical Insignificance of Perspective and the First Person*. Oxford: Oxford University Press.

Carel, H., Kidd I., and Pettigrew, P. (2016). Illness as transformative experience. *The Lancet 388* (10050): 1152–53.

Carruthers, P. (2006). *The Architecture of the Mind: Massive Modularity and the Flexibility of Thought*. Oxford: Oxford University Press.

Collins, J. 2015. Neophobia. *Res Philosophica 92* (2): 283–300.

Crockett, M. J. (2013). Models of morality. *Trends in Cognitive Sciences 17* (8): 363–66.

Dawes, R. M. (1990). False consensus effect. In R. M. Hogarth, ed., *Insights in Decision Making: A Tribute to Hillel J. Einhorn*. Chicago: University of Chicago Press, 179–99.

Dennett, D. C. (1987). The Intentional Stance. Cambridge, MA: MIT Press.

Dougherty, T., Horwitz, S., and Sliwa, P. 2015. Expecting the unexpected. *Res Philosophica 92* (2): 301–21.

Epley, N. (2014). *Mindwise: Why We Misunderstand What Others Think, Believe, Feel, and Want*. New York: Vintage.

Epley, N., and Whitchurch, E. (2008). Mirror, mirror on the wall: Enhancement in self-recognition. *Personality and Social Psychology Bulletin 34* (9): 1159–70.

Forbus, K. D. (1988). Qualitative physics: Past, present and future. In Howard E. Shrobe, ed., *Exploring Artificial Intelligence*. Burlington, MA: Morgan Kaufmann, 239–96.

Gendler, T. (2010). *Intuition, Imagination, and Philosophical Methodology*. New York: Oxford University Press.

Gerstenberg, T., Goodman, N., Lagnado, D. A., and Tenenbaum, J. B. (2012). Noisy Newtons: Unifying process and dependency accounts of causal attribution. In *Proceedings of the 34th Annual Conference of the Cognitive Science Society*. Austin, TX: Cognitive Science Society.

Goldman, A. (1989). Interpretation psychologized. *Mind & Language 4*: 161–85.

Goldman, A. (2006). *Simulating Minds: The Philosophy, Psychology, and Neuroscience of Mindreading*. Oxford: Oxford University Press.

Goldman, A. (2015). Naturalizing metaphysics with the help of cognitive science. In K. Bennett and D. Zimmerman, eds., *Oxford Studies in Metaphysics*, vol. 9. New York: Oxford University Press, 171–216.

Gottlieb, J., Oudeyer, P. Y., Lopes, M., and Baranes, A. (2013). Information-seeking, curiosity, and attention: Computational and neural mechanisms. *Trends in Cognitive Sciences 17* (11): 585–93.

Graham, J., Waytz, A., Meindl, P., Iyer, R., and Young, L. (2017). Centripetal and centrifugal forces in the moral circle: Competing constraints on moral learning. *Cognition 167*: 58–65.

Hamrick, J. B., Battaglia, P. W., Griffiths, T. L., and Tenenbaum, J. B. (2016). Inferring mass in complex scenes by mental simulation. *Cognition 157*: 61–76.

Harman, E. 2015. Transformative experience and reliance on moral testimony. *Res Philosophica 92* (2): 323–39.

Heddon, B. (2015). Time-slice rationality. *Mind 124* (494): 449–91.

Heider, F., and Simmel, M. (1944). An experimental study of apparent behavior. *American Journal of Psychology 57* (2): 243.

Iskiwitch, C., Carden, L., Meindl, P., Dehghani, M., Monterosso, J., Doris, J. M., and Graham, J. (2017). Moral purity and self-distancing. Manuscript in preparation.

Jara-Ettinger, J., Gweon, H., Schulz, L. E., and Tenenbaum, J. B. (2016). The naïve utility calculus: Computational principles underlying commonsense psychology. *Trends in Cognitive Sciences 20* (8): 589–604.

Kaplan, David (1989). Demonstratives. In J. Almog, J. Perry, and H. Wettstein, eds., Themes from Kaplan. New York: Oxford University Press, 481–563.

Kappes, H. B., and Morewedge, C. (2016). Mental simulation as substitute for experience. *Social and Personality Psychology Compass 10* (7): 405–20.

Kind, A. (2016). Imagining under constraints. In A. Kind and P. Kung, eds., *Knowledge Through Imagination*. Oxford: Oxford University Press, 145–59.

Kruger, J., and Dunning, D. (1999). Unskilled and unaware of it: How difficulties in recognizing one's own incompetence lead to inflated self-assessments. *Journal of Personality and Social Psychology 77* (6): 1121–34.

Lewis, D. (1986). *On the Plurality of Worlds*. Oxford: Blackwell.

Magidor, O. (2015). The myth of the de se. *Philosophical Perspectives 29* (1): 249–83.

Marcus, G. F., and Davis, E. (2013). How robust are probabilistic models of higher-level cognition? *Psychological science 24* (12): 2351–60.

McGinn, C. (1983). *The Subjective View: Secondary Qualities and Indexical Thoughts*. Oxford: Clarendon Press.

Moss, S. (2015). Time-slice epistemology and action under indeterminacy. In Tamar Szabó Gendler and John Hawthorne, eds., *Oxford Studies in Epistemology*, vol. 5. New York: Oxford University Press, 172–94.

Nanay, B. (2016). The role of imagination in decision-making. *Mind & Language 31* (1): 127–43.

Nichols, S., and Stich, S. (2003). Mindreading: An Integrated Account of Pretence, Self-Awareness, and Understanding Other Minds. New York: Oxford University Press.

Ninan, D. (2016). Imagination and the self. In A. Kind, ed., The Routledge Handbook of the Philosophy of Imagination. New York: Routledge, 274–85.

Nisbett, R. E., and Wilson, T. D. (1977). Telling more than we can know: Verbal reports on mental processes. *Psychological Review 84* (3): 231.

Parfit, Derek (1984). Reasons and Persons. New York: Oxford University Press.

Paul, L. A. (2012). Metaphysics as modeling: The handmaiden's tale. *Philosophical Studies 160* (1): 1–29.

Paul, L. A. (2014). *Transformative Experience*. Oxford: Oxford University Press.

Paul, L. A. (2018, forthcoming). De se preferences and empathy for future selves. In John Hawthorne and Jason Turner, eds., *Philosophical Perspectives: Philosophy of Mind*.

Paul, L. A. and Healy (2017, forthcoming). Transformative treatments. *Noûs*.

Perry, J. (1979). The problem of the essential indexical. *Noûs 13*: 3–21.

Pettigrew, R. (2015). Transformative experience and decision theory. *Philosophy and Phenomenological Research 91* (3): 766–74.

Pettigrew, R. (2018). Choosing for changing selves. July 10. https://drive.google.com/file/d/1FPKDxYu7JyxIVGd_qW491djJDMBpSQe4/view.

Phillips, J., and Cushman, F. (2017). Morality constrains the default representation of what is possible. *PNAS 114* (18): 4649–54.

Prinz, J., and Nichols, S. (2016). Diachronic identity and the moral self. In J. Kiverstein, ed., *Handbook of the Social Mind*. London: Routledge, 449–64.

Pronin, E., and Ross, L. (2006). Temporal differences in trait self-ascription: When the self is seen as an other. *Journal of Personality and Social Psychology 90* (2): 197–209.

Railton, P. (2017). Moral learning: Conceptual foundations and normative relevance. Cognition *167*: 172–90.

Recanati (2012). Mental Files. New York: Oxford University Press.

Robbins, J. M., and Krueger, J. I. (2005). Social projection to ingroups and outgroups: A review and meta-analysis. *Personality and Social Psychology Review 9* (1): 32–47.

Ross, L., Greene, D., and House, P. (1977). The "false consensus effect": An egocentric bias in social perception and attribution processes. *Journal of Experimental Social Psychology 13* (3): 279–301.

Sanborn, A. N., Mansinghka, V. K., and Griffiths, T. L. (2013). Reconciling intuitive physics and Newtonian mechanics for colliding objects. *Psychological Review 120* (2): 411–37.

Saxe, R. (2009). The happiness of the fish: Evidence for a common theory of one's own and others' actions. In Keith D. Markman, William M. P. Klein, and Julie A. Suhr, eds., *The Handbook of Imagination and Mental Simulation*. New York, NY: Psychology Press, 257–66.

Schaffer, J. (2016). Cognitive science and metaphysics: Partners in debunking. In B. McLaughlin and H. Kornblith, eds., *Goldman and His Critics*. Malden, MA: Wiley-Blackwell, 337–65.

Schechtman, M. (2011). The narrative self. In Shaun Gallagher, ed., The Oxford Handbook of the Self. New York: Oxford University Press, 394–418.

Schmidhuber, J. (2010). Formal theory of creativity, fun, and intrinsic motivation (1990–2010). *IEEE Transactions on Autonomous Mental Development 2* (3): 230–47.

Schwitzgebel, E., and Cushman, F. (2012). Expertise in moral reasoning? Order effects on moral judgment in professional philosophers and non-philosophers. *Mind & Language 27* (2): 135–53.

Schwitzgebel, E., and Cushman, F. (2015). Philosophers' biased judgments persist despite training, expertise and reflection. *Cognition 141*: 127–37.

Seligman M., Railton P., Baumeister R., and Sripada, C. (2013). Navigating into the future or driven by the past. *Perspectives on Psychological Science 8* (2) 119–41.

Seligman M., Railton, P., Baumeister, R., and Sripada, C. (2016). *Homo Prospectus*. New York: Oxford University Press.

Shupe, E. (2016). Transformative experience and the limits of revelation. *Philosophical Studies 173* (11): 3119–32.

Smith, K. A., and Vul, E. (2013). Sources of uncertainty in intuitive physics. *Topics in Cognitive Science 5* (1): 185–99.

Stalnaker, R. 1976. Possible worlds. *Noûs 10* (1): 65–75.

Starmans, C. (2017). Children's theories of the self. *Child Development 88* (6): 1774–85.

Strawson, G. (2017). *The Subject of Experience*. New York: Oxford University Press.

Strohminger N., Newman, G., and Knobe, J. (2017). The true self: A psychological concept distinct from the self. *Perspectives on Psychological Science 12* (4): 551–60.

Strohminger, N., and Nichols, S. (2014). The essential moral self. *Cognition 131* (1): 159–71.

Ullman, T. D., Spelke, E. S., Battaglia, P., and Tenenbaum, J. B. (2017). Mind Games: Game Engines as an Architecture for Intuitive Physics. *Trends in Cognitive Science* 21(9), 649–665.

Ullman, T. D., Stuhlmüller, A., Goodman, N. D., and Tenenbaum, J. B. (2018). Learning physical parameters from dynamic scenes. *Cognitive Psychology 104*: 57–82.

Ullmann-Margalit, E. (2006). Big decisions: Opting, converting, drifting. *Royal Institute of Philosophy Supplement 58*: 157–72.

Wegner, D. M. (2002). The illusion of conscious will. Cambridge, MA: MIT Press.

Wellman, H., and Gelman, S. (1992). Cognitive development: Foundational theories of core domains. *Annual Review of Psychology 43* (1): 337–75.

Williamson, T. (2016). Knowing by imagining. In A. Kind and P. Kung, eds., *Knowledge Through Imagination*. Oxford: Oxford University Press, 113–23.

10 Against Phenomenal Parsimony

A Plea for Bodily Feelings

Frédérique de Vignemont

10.1 INTRODUCTION

According to Wu (forthcoming), one should not posit additional phenomenal properties in one's mental ontology when one can account for them by appealing to other properties. In particular, Wu argues that feelings of mineness can be explained away both in the case of bodily ownership and in the case of agency. Likewise, several authors have recently denied that the sense of agency and the sense of bodily ownership have a distinctive phenomenology (Alsmith 2015; Bermúdez 2010, 2011, 2015; Martin 1992, 1995; Mylopoulos 2015). On their view, one knows that one is at the origin of one's actions and that this body is one's own but there is no need to assume the burden of extra phenomenal properties such as agency and bodily ownership: one does not *experience* those facts, and they are not manifested to the subject in a more primitive form than judgments.

At first sight, the principle of phenomenal parsimony seems hardly controversial. There is indeed a risk of an unwarranted multiplication of phenomenal properties. On the other hand, one should not neglect the opposite risk of greatly impoverishing our mental ontology by denying phenomenology even when it is the most intuitive. The crucial question is then to determine what reasons there can be to enrich our phenomenal ontology. Interestingly, this debate has recently turned to cognitive science to find answers. Those who defend a liberal or rich view of phenomenology have taken pathological disorders and illusions as evidence in favor of the existence of feelings of agency and of bodily ownership (Bayne 2011; Vignemont 2017). In a nutshell, the argument is that if these feelings can go missing under pathological conditions, then they must exist under normal conditions. The appeal of these borderline cases can be easily explained. To some extent, they are simply thought experiments that happen to be actual. Like thought experiments, they involve situations brought to their extreme, such as situations in which one experiences one's movements controlled by an external force (i.e.

delusion of control) or situations in which one no longer experiences one's body as one's own (i.e. somatoparaphrenia). Like thought experiments too, they allow evaluating necessary and sufficient conditions, although only in our world and not in all possible worlds. As we shall see, sometimes what is originally conceived of as a mere thought experiment can actually be found in neurology and psychiatry textbooks. Conceptual analysis ultimately rests on intuitions, but sometimes empirical facts contradict one's intuitions. The difficulty, however, is that even in the case of disorders and illusions there is room for interpretation, and where the proponents of the liberal view see feelings, the proponents of a conservative view of phenomenology see cognitive attitudes (Alsmith 2015; Bermúdez 2015; Mylopoulos 2015; Wu, forthcoming). The debate over our phenomenal toolkit has thus become partly a debate about how to best understand this or that syndrome or illusion.

In this chapter, I shall focus exclusively on the existence of what I call bodily feelings, that is, feelings of high-level features of the body, such as feelings of bodily presence, of bodily ownership, and of bodily capacities. I shall review the main objections that have been put forward against the liberal view, both a priori and empirical, and conclude that none of them is conclusive.

10.2 THE DEBATE

We are aware of our bodily posture, of its temperature, of its physiological balance, of the pressure exerted on it, and so forth. Insofar as these properties are detected by a range of inner sensory receptors in the skin, joints, muscles, tendons, inner ear, and internal organs, one may conceive of bodily awareness on the model of perceptual awareness. However, there are other features of the body that are of a higher level, which characterize what may be described as the fundamental state of the body, that is, the enduring relation of the body with the world and with the self. We are aware that this body is here in the external world, that it belongs to us, and that it has two arms and two legs that can do some things but not others. The question is then: do we actually *feel* the body that way, or do we merely *know* those facts? Such high-level bodily properties cannot be directly extracted at the sensory level. In short, there is no inner receptor for ownership, and its detection must involve more complex computations. Furthermore, low-level bodily properties fluctuate all the time. You feel cold and then you do not and then you do again; you are thirsty, you drink, you no longer feel thirsty until next time, and so forth. But the fundamental core of bodily awareness is relatively permanent, and it is thus rarely at the forefront of consciousness: it does not attract attention because it normally does not change. Consequently, we cannot simply turn inside by an act of introspection and decide whether there are specific feelings that correspond to these high-level properties of our body. There is thus an ongoing debate over their existence.

On the one hand, advocates of the liberal view—also sometimes called the inflationary view (Bermúdez 2011)—claim that there are such things as feelings of high-level bodily features that go beyond the awareness of low-level bodily features (Bayne 2011; Billon 2017; Peacocke 2014; Vignemont 2017). On this view, we have a primitive nonconceptual awareness of bodily presence, ownership, and capacities, which is over and above the experience of pressure, temperature, position, balance, movement, and so forth. These feelings do not have to be conceived as purely "raw feels" without representational properties. Instead, the feeling of bodily presence can consist in experiencing the body as being here, the feeling of bodily ownership in experiencing the body as being one's own, and the feeling of bodily capacities as being able to perform such movements.

On the other hand, advocates of the conservative view—sometimes also called deflationary—reject a distinctive experiential signature for the sense of bodily ownership and for the sense of agency (Alsmith 2015; Bermúdez 2010, 2011; Martin 1992, 1995; Mylopoulos 2015; Wu, forthcoming). Instead, one judges that the high-level bodily features are instantiated thanks to the experience of low-level bodily features. For example, Martin (1992, 201–2) claims: "What marks out a felt limb as one's own is not some special quality that it has." Still, he grants that there is a phenomenal contrast between cases in which one is aware of one's hand as one's own and cases in which one is not. He does not explain the difference in terms of ownership feelings but instead in terms of the phenomenology of other bodily features that always accompany ownership, and more specifically in terms of the spatial phenomenology of bodily sensations. He proposes that the distinction between what is one's own body and what is not one's own body consists in the distinction between inside and outside bodily boundaries in which one can experience bodily sensations. Consequently, according to Martin (1995), it is sufficient to feel sensations as being located in a body part to experience this body part as one's own. The phenomenology of ownership is then inherited, so to speak, from the spatial phenomenology of bodily sensations.

The crucial question is how to settle the debate between the liberal and the conservative views. How can we determine whether high-level bodily properties are part of the experiential content? Two kinds of arguments have been proposed. The first type of argument is based on a priori considerations about the nature of feelings and their role for knowledge. However, as we shall see, it ultimately relies on phenomenological intuitions. By contrast, the second type of argument is based on the analysis of a series of borderline cases, in which the awareness of high-level bodily features is disturbed one way or the other, following pathological disorders or experimental illusions. These empirical facts provide a common ground for discussion between the liberals and the conservatives. The argument then becomes an inference to the best explanation. But whose side offers the best explanation?

10.3 A PRIORI ARGUMENTS: THE NATURE OF SENSATIONS

Let us start with some general a priori considerations. One might, for instance, reject the liberal view because there is phenomenology only for nonconceptual content, and high-level bodily features can only be represented conceptually. One might also reject the liberal view because only lower-level properties (colors, shapes, sounds, etc.) can be part of the content of experiences, but not high-level properties. These replies, however, beg the question at several levels. It is first controversial whether there cannot be nonconceptual content that represent high-level bodily features. Peacocke (2014), for instance, analyzes the sense of bodily ownership in terms of nonconceptual *de se* content, what he describes as the degree 1 of self-representation. Siegel (2010) proposes that perceptual experiences also include certain high-level properties (natural or artifactual kinds, meanings, expressions, values, gestalt properties, causal or modal properties, etc.). Why not agency and ownership too? Second, even if one grants that the content can be only conceptual, it does not follow that there cannot be any phenomenology. Advocates of cognitive phenomenology at least would reject such a conclusion (Strawson 1994; Siewert 1998): even if bodily presence, ownership, and capacities could be represented only conceptually, it would not preclude them for being experienced. Nor is the complexity of the underlying mechanisms an obstacle. It has been recently suggested that we have a range of noetic feelings, which are not based on simple sensory processing but rather on metacognitive monitoring (Dokic 2012). Suppose I have a déjà vu experience in a certain spot although I perfectly know that I have never been there before. In this case, it seems legitimate to assume that I *feel* something specific when I have this experience. Arguably, this feeling is not sensory and its grounds are relatively complex, involving metacognitive monitoring of visual processing. Still, it has been argued that there is a distinct phenomenology associated with the feeling of déjà vu and one should not discount it just because its grounds are metacognitive. One can also mention the feeling of familiarity. It cannot be reduced to the sensory recognition of the visual features of the face, but involves autonomic responses, which result in increased arousal in front of familiar faces (Ellis and Lewis 2001). But again it is characterized by a specific phenomenology, which can be conceived of in affective terms (Dokic and Martin 2015). If we accept that there are feelings of familiarity, confidence, déjà vu, and so forth, then why not feelings of ownership, agency, or presence? Bodily feeling might actually be conceived of in the same way as these nonsensory types of feelings.

Another type of a priori argument finds its origin in Anscombe. Anscombe (1962) notes that the use of the term "sensation" can be misleading. In a nutshell, it is not because we talk of sensations of a specific feature that we do have them. Instead, what we may really have are only beliefs or judgments about this feature. How do we decide then whether we truly have a sensation or not? Anscombe (1957) provides an epistemological criterion. She claims that sensations must be able to ground knowledge, and in order to do so, their internal content must be "independently describable": there is a sensation of *x* if its description has a different content than *x* and this content is taken as a sign that indicates *x*. For example, there is a sensation of going down in a lift since one can provide an independent description of its internal content in terms of lightness and of one's stomach lurching upward. By contrast, it is not legitimate to talk of sensation of sitting cross-legged, Anscombe claims, because there is no such independent description that can be given.

> [A] man usually knows the position of his limbs without observation. It is without observation, because nothing shews him the position of his limbs; it is not as if he were going by a tingle in his knee, which is the sign that it is bent and not straight. Where we can speak of separately describable sensations, having which is in some sense our criterion for saying something, then we can speak of observing that thing. (Anscombe 1957, 13)

On her view, the confusion between sensations and judgments of bodily posture arises because in some rare cases it seems to us that our legs are crossed when they are not. Such errors do not show that we feel that they are crossed. They only show that we are entitled to talk of knowledge when they are crossed.[1] This knowledge, however, is knowledge without observation: it is not epistemically grounded in sensations. We can reconstruct her argument as follows:

(1) A state qualifies as a sensation only if its content can be described independently of what it refers to.
(2) Only independent content can be used as the epistemic basis for judgments.
(3) One cannot provide an independent description in the case of bodily position.
(4) Thus, there are no sensations of bodily position.
(5) Thus, bodily knowledge is knowledge without observation.

Bermúdez (2010, 2011, 2015) has recently given a second life to Anscombe's argument in discussions about the sense of agency and the sense of bodily ownership. On his view, sensations that can be used as an epistemic basis of knowledge must meet two criteria. On the one hand, they must be "focused," that is, their content must provide information that is precise and specific. If they are too vague, they fail to justify the specific judgment that one makes. On the other hand, they must also be independent, that is, they cannot simply duplicate the content of the judgment. As Bermúdez (2015) notes, one cannot justify an assertion by simply repeating it.

[1] Following Wittgenstein, Anscombe assumes that there is knowledge only if there is the possibility of being mistaken. For instance, she claims that we cannot know that we are in pain because we can never be wrong.

He claims that neither the feeling of agency nor the feeling of ownership can meet these two criteria.

> There is no determinate "quale" associated with agency that can be identified and considered independently of our knowledge without observation that we are acting. And, by the same token, there is no distinctive experience of agency that can ground and underwrite our beliefs about our own agency. (Bermúdez 2010, 588)
>
> And yet a feeling of myness that can only be described in those very terms is not sufficiently independent of the judgment of ownership that it is claimed to justify. So the postulated non-conceptual intuitive awareness of ownership falls foul of Anscombe's dilemma. (Bermúdez 2015, 39)

What is surprising is that Anscombe and Bermúdez reach their metaphysical conclusion on the basis of epistemological considerations. On their view, we can decide whether or not there are bodily feelings by determining the epistemic ground of bodily knowledge. One may thus be tempted to immediately rule out their objection because it assumes the priority of epistemology over metaphysics. For many indeed, whether or not there are bodily feelings is a matter independent of epistemic issues. But even if one accepts Anscombe's epistemology-first strategy, one may still challenge her specific epistemological theory: why do contents have to be independent to ground judgments? And how to characterize independent contents? The distinction between sensations whose content is, and sensations whose content is not, separately describable is problematic, and Anscombe's (1962, 57) own examples are not helpful. For instance, she claims that "the visual impression of a blue expanse" has an independent content and can thus ground the judgment that the sky is blue. Since she defends the view that sensations can be captured in intentional terms (Anscombe 1965), her point is not that there is a qualitative raw feel of blueness independently of the visual property of blue. But then it is not clear in what sense this qualifies as an independent description, and it may sometimes be difficult to see why perceptual awareness does not fall into the same category as bodily awareness. Finally, the whole argument rests on the assumption (3) according to which one cannot provide an independent description in the case of bodily position: "no question of any appearance of the position to me, of any sensations which give me the position" (Anscombe 1962, 58). Similarly, Bermúdez (2015, 39) assumes that one cannot describe the sense of ownership without referring to the fact that this is one's own body (i.e., myness):

> It is highly *implausible* that there is a determinate quale of ownership that can be identified, described and considered independently of the myness that it is supposed to be communicating. (Bermúdez 2015, 39, my emphasis)

But why is it so? Neither Anscombe nor Bermúdez gives arguments. They merely state that there cannot be independent descriptions. Bermúdez (2015, 44) further adds that ownership is "a phenomenological given" and that it is impossible to ground it in further nonconceptual content. But this seems to be simply begging the question. By offering no principled reasons for their third assumption, the opponents of the liberal conception undermine their own objection. It may then seem that their epistemological argument eventually relies on phenomenological intuitions.

10.4 EMPIRICAL ARGUMENTS: AN INFERENCE TO THE BEST EXPLANATION

Let us thus go back to our phenomenology. As said earlier, one cannot settle the debate by a direct use of introspection in everyday life because bodily feelings, if they do exist, are recessive in normal situations. What is more interesting is to consider borderline cases, such as those that are found in pathologies and illusions of bodily awareness.

Consider first Mike Martin's thought experiment about the awareness of one's body as one's own.

> If the sense of ownership is a positive quality over and above the felt quality of the sensation and the location—that there is hurt in an ankle for example—then it should be conceivable that some sensations lack this extra quality while continuing to possess the other features. Just as we conceive of cold as the converse quality of warmth, could we not also conceive of a converse quality of sensation location such that one might feel pain in an ankle not positively felt to belong to one's own body. If O'Shaughnessy is right, we can make no sense of either possibility. (Martin 1995, 270)

As seen earlier, Martin can make no sense of sensations in a limb that is disowned because he assumes that it is sufficient to feel sensations as being located in a body part to experience this body part as one's own. However, the problem for Martin is that what he claims to be unintelligible can actually happen in some neurological and psychiatric disorders. For instance, after a lesion in the right parietal lobe, some patients suffer from what is known as somatoparaphrenia: they deny that their limbs belong to them. Surprisingly, they report feeling their hand as alien although they can still feel touch and pain located in their "alien" hand (e.g., Moro et al. 2004). What these patients seem to experience is exactly what Martin could not make sense of, and if we follow him, then we should conclude that the sense of ownership is "a positive quality over and above" bodily sensations (Vignemont 2018). In other words, we cannot explain away the phenomenal property of ownership and we should thus add it to our ontology.

Yet proponents of the conservative view resist this conclusion (Bermúdez 2015; Wu, forthcoming). Their objection can be more or less strong. The milder form consists in denying the relevance of pathological disorders for the debate. The hypothesis is that these patients experience a feeling of disownership, but disownership feelings do not entail that there are normally ownership feelings that are lacking in these patients (Bermúdez 2015).[2] It is true that feelings of disownership cannot be reduced to the lack of feelings of ownership and that one can account for them independently of any reference to them. For instance, one might argue that the mechanism responsible for ownership involves some kind of comparator between multisensory signals that, in case of mismatch, gives rise to an error signal that is experienced as a feeling of disownership. But if everything goes well, then no signal is sent. This explanation may be plausible in the case of somatoparaphrenia, but it cannot account for other pathological disorders. Consider for instance the better-known phenomenon of phantom limbs:

> To anyone looking at me, I have no arm. But I can feel the entirety of my phantom hand and arm. (Mezue and Makin 2017, 34)

First described by the French surgeon Ambroise Paré in 1551, this sensation has been reported many times by amputees who say that they still experience their missing limbs. It is commonly accepted that amputees can feel kinesthetic, tactile, thermal, and painful sensations as being located in their phantom limb. But do these sensations exhaust the content of what they experience? According to the neurologist Melzack (1992), they also experience their phantom limb as being *present* to such an extent that they can forget that it has been amputated. This sense of reality, he claims, can be "reinforced" by phantom sensations but goes beyond them. In addition, the amputees experience their phantom limb *as being part of their own body*. Again, this feeling of ownership should be distinguished from phantom sensations and from the feeling of presence.

One may still be tempted to discount these reports and argue that they do not show that there are feelings of presence and of ownership. But to do so, one would have to reject the phenomenological interpretation of the patients' subjective reports. According to Wu (forthcoming), the mistake that proponents of the liberal view commit is that they assume what he calls a spotlight model of introspection: they believe that the patient expresses what she feels instead of what she judges because they take introspection to provide a direct read off of phenomenal properties. However, on Wu's hypothesis, introspection involves more complex processing than that and we should not take the patients' reports at face value when the somatoparaphrenic patient asserts that she no longer experiences her hand as being her own or when the amputee asserts that she experiences her amputated limb as present.[3] The patients merely *judge* that it is not their own hand or that the limb is still there.

[2] Likewise, it has been argued that the syndrome of delusion of control in schizophrenia should not be taken as evidence that one normally experience agentive feelings (Mylopoulos 2015).

[3] This kind of doubt is actually what motivated the American surgeon Mitchell, who introduced the term "phantom limb" in 1866, to do so in an unsigned short story. He feared that the patients'

At first sight, this cognitive interpretation may have some plausibility as far as somatoparaphrenia is concerned because the patients are delusional: they are convinced that the hand is not their own. However, their delusional belief appears to be strongly anchored in their abnormal feelings:

> [M]y eyes and my feelings don't agree, and I must believe my feelings. I know they look like mine, but I can feel they are not, and I can't believe my eyes. (Nielsen 1938, 555)

According to an influential theory of delusion, the two-factor model, one needs to distinguish between the factors that trigger the initial implausible thought (and thus contribute to explaining the thematic content of a particular delusion), and the factors that explain the uncritical adoption of the implausible thought as a delusional belief (Langdon and Coltheart 2000). Abnormal rationality can then account for the feeling of confidence in the delusional beliefs, but not for their content. Instead, on this theory the delusions result from sensory or motor impairment leading to *abnormal experiences* that the patient tries to account for.

There is a further and more fundamental problem for the cognitive interpretation. Consider again the case of phantom limb. Amputees are not delusional. They are aware that the limb is just a phantom: they know that it has been amputated and they can see that the limb is missing. If asked, they would not claim that the limb is really present. It merely seems to them as if it were present. According to the liberal conception, amputees cannot help but experience the limb as being present despite having the belief that the limb is absent. Conversely, some patients who feel that their body has disappeared or that it is no longer their own are still aware that their body is still here and that it belongs to them. This is the case in the psychiatric condition of depersonalization, in which patients report that it only seems to them *as if* it were not their body. A patient, for example, reported: "I can sit looking at my foot or my hand and not feel like they are mine" (Sierra 2009, 27). She correctly self-ascribes her foot and her hand but still experiences as if they were not hers.

What the patients experience then appears as equivalent to what subjects experience in classic perceptual illusions, such as the Müller-Lyer illusion: one cannot help but experience the two lines as being different despite having the belief that they are actually of the same size. Interestingly, one can actually induce belief-immune illusions of bodily ownership in healthy participants (Botvinick and Cohen 1998). In the rubber hand illusion (RHI) participants sit with their arm hidden behind a screen, while fixating on a rubber hand presented in their bodily alignment; the rubber hand can then be touched either in synchrony or in asynchrony with their hand. In the synchronous condition, participants report that it seems as if the rubber

reports would not be believed. For a review of the history of the notion of phantom limbs, see Halligan 2002.

hand were their own hand although they are fully aware that this is a mere piece of rubber. Again, an advocate of the conservative view may deny that the subjects *feel* the rubber hand as being their own. But if the amputee, the depersonalized patient, and the subject in the RHI experiments do not express what they feel, then what kind of state do they express?

There are attitudes other than feelings and sensations that can be encapsulated and immune to the influence of beliefs and judgments (Alsmith 2015; McDowell 2011; Bermúdez 2015; Wu, forthcoming; Mylopoulos 2015). For instance, Wu (forthcoming) proposes that participants in the RHI express "a gradable attitude of agreement towards a proposition," although they do not believe it. To some extent Wu's proposal is hardly controversial: level of agreement is precisely what is asked for in the RHI questionnaires.[4] The question is what grounds the participants' replies. According to Wu, participants draw an inference on the basis of evidence that points to features that they would expect if the rubber hand were their own hand. The problem, however, is that there can be strong defeaters, including differences in skin color, size, and laterality between the rubber hand and the real hand. How can participants even partly agree that this is their own hand when it does not look at all like their hand? To counterbalance the defeaters there must be strong evidence that this is their own hand, and what more powerful evidence than the participants' feelings?

Alternatively, proponents of the conservative view may go for something weaker than partial agreement and suggest that participants only *suppose* that the hand is their own while they know that this is not true. As far as this attitude is not driven by evidence, it does not matter whether there are strong defeaters. However, it does not seem plausible in the pathological cases. I doubt that even the advocates of the conservative view would claim that the amputee *supposes* her hand to be still there, while believing it not to be there or that the patient suffering from depersonalization simply *supposes* her hand to have disappeared, while believing that it has not. One problem indeed is that the depersonalized patient is not simply in a cold cognitive state about the disappearance of her body. Her introspective reports are associated with strong distress. Why are depersonalized patients greatly disturbed? They know that their body is still here, that it only seems to them as if it were not. How can mere suppositions have such an emotional impact?

Imagination may then be a better candidate because it can be affectively loaded. One may suggest that the depersonalized patient *imagines* her body not to be her own and that imagining suffices to make her feel bad. Likewise, Alsmith (2015) argues that in the RHI one imagines that the rubber hand is one's hand although one believes that it is not. However, imagination is generally under voluntary control,

[4] Participants are asked to rate to what extent they agreed with ownership statements such as "It seems to me as if the rubber hand were my own hand" on a scale that goes from −3 (strongly disagree) to +3 (strongly agree) (Longo et al. 2008).

whereas the patients cannot stop feeling the disappearance of their body. Nor can amputees stop feeling the presence of their phantom limb. Even if imagination can escape control, it can be affectively loaded because it can be experiential. On one interpretation indeed, imagination involves mentally recreating selected experiences (Goldman 2006). If one does imagine the phantom hand as being there, it means that one imagines *experiencing* it as being there. The imaginative proposal is thus not purely cognitive; it also includes a phenomenal component with it, that is, the experience that is re-enacted in imagination.

To recapitulate, the argument from cognitive impenetrability does not suffice to show that there are bodily feelings, but this is not to say that the liberal view is false. It only means that there may be alternative interpretations of these illusions and disorders. What needs to be done, then, is to compare the strength and the weakness of each interpretation. It seems that the liberal conception provides a simple and unified explanation for a wide range of borderline cases, from the rubber hand illusion to phantom limbs, from depersonalization to somatoparaphrenia. Its only cost is to posit new phenomenal properties. By contrast, opponents of the liberal conception avoid the burden of adding extra feelings. But it is unclear how their alternative interpretation can account for these borderline cases. Indeed the alleged cognitive attitude that the conservatives posit must not only be insensitive to the influence of beliefs; it must also be beyond one's control and affectively loaded. So far, imaginative states may be the best candidates, but only because they consist in experiential states. Finally, the conservatives must find those other properties that can explain away bodily feelings. This is actually a difficult task because of the variety of cases in which the awareness of high-level features of the body and the awareness of low-level features can come apart. We have seen that patients with depersonalization can feel as if their body does not belong to them and as if it had disappeared. What is interesting is that they display no sensory alteration (Leroy 1901; Janet 1928; Billon 2017): "Even if I touch my face I feel or sense something but my face is not there" (Sierra 2009, 29). Hence, one cannot simply appeal to the lack of bodily sensations to explain why they report the lack of presence and of bodily ownership. The reverse dissociations can be found: one can be aware of one's body as being here in the absence of bodily sensations. For instance, it has been shown that the complete absence of somatosensory inputs does not eradicate the feeling of presence (Gandevia and Phegan 1999). Subjects had their thumb anaesthetized so that they could no longer detect light touch and pain. They were then asked to evaluate the size of the anaesthetized body part. As noted by the authors, no subject reported that the thumb was missing. Instead, they found a dramatic increase in perceived size: the thumb felt bigger. Such an illusion is frequent after dental surgery: your cheek often feels huge following anesthesia, but when you look at yourself in the mirror, you see that it has actually not inflated. Thus, blocking bodily sensations does not preclude feelings of bodily presence. Each of these cases might be open to interpretation, but taken all together they provide strong support for a rich conception of bodily phenomenology, which can offer a unified and relatively simple explanation of all these cases in terms of bodily feelings, which are above and beyond the phenomenology of low-level bodily properties.

10.5 CONCLUSION: REVISITING THE THEORETICAL LANDSCAPE

We have seen that no principled arguments have been provided so far that would make the liberal prospect simply a dead end. There may be alternative interpretations of the evidence provided in support of the liberal conception, but it remains to be shown that they have more explanatory power than the liberal one. This may not suffice to show that the amputees do *feel* their phantom hand as being there. But at least we have shown that there is no reason to discount their subjective reports when they claim that they do. I now want to conclude by showing that the boundaries between the views may be more blurred than expected and that views that claim to reject bodily feelings are actually not incompatible with their existence.

To recapitulate, we started with the dichotomy between the liberal view, which defends a rich phenomenal ontology, and the conservative view, which defends a more parsimonious one. They can both be characterized as follows, with B, a high-level bodily feature such as bodily presence, bodily ownership, or bodily capacities.

Liberal conception: There is a phenomenal property of B, on the basis of which one can report that B.

Conservative conception: There are phenomenal properties of low-level bodily features, on the basis of which one can report that B.

The difference between these two conceptions is well illustrated in the case of phantom limb. According to the liberal conception, amputees still experience their amputated limb as being there. According to the conservative conception, amputees experience tactile, kinesthetic, and painful sensations as located in the phantom limb and thus report that the limb is there. Their reports express their cognitive attitude instead of their phenomenology.

However, there may be another way to draw the line between the views. Let us start with the basic question: is there a phenomenology of high-level bodily properties? The liberal view clearly replies positively, but it is not the only one to do so. Consider first an intermediary position between the liberal and the conservative views. On this position, amputees do express a cognitive attitude, but this attitude is associated with a distinctive phenomenology. To some extent, this view is liberal because it assumes that there are agentive and bodily feelings, but it consists in a cognitive version of the liberal view: these feelings are cognitive, or even metacognitive.

Liberal cognitive conception: there are subpersonal computations that lead to the cognitive attitude that B and there is a phenomenology associated with this cognitive attitude.

Some of the opponents of the liberal conception actually reject only the hypothesis of a *sensory* phenomenology of ownership and agency and may be willing to accept such a cognitive phenomenology (Alsmith 2015; Mylopoulos 2015). This suggests that one should clearly distinguish the following two questions: (1) is there a phenomenology of high-level bodily properties? And (2) is the awareness of high-level bodily properties cognitive? A positive reply to the second question does not entail a negative answer to the first one.

Now consider the conservative view. It assumes that one reports that B on the basis of the awareness of low-level bodily features. The question is: what is the nature of the relationship between the phenomenology of low-level bodily features and judgments of high-level bodily features? One possibility is that one can infer B because B consists in these other bodily features.

Reductionist conception: There are phenomenal properties on the basis of which one can report that B because of their constitutive relation to B.

Such an approach is defended by Martin (1995) about the sense of bodily ownership. As seen earlier, he argues that the sense of bodily ownership consists in the spatial phenomenology of bodily sensations. To defend his view, he makes the following metaphysical assumption, which he calls the Sole Object view: there is an identity between one's own body and the body in which one locates bodily experiences. In other words, the fact of ownership (whose body is mine) consists in the body in which I can feel sensations to be located. This identity, he claims, enables the spatial content of bodily experiences to ground bodily self-ascriptions. Hence, on Martin's view, there is a phenomenology of ownership, but it consists in spatial phenomenology.[5] His project is thus reductionist, and reducing is not eliminating.

However, there are other versions of the conservative conception that are eliminativist. Consider the following interpretation. On this view, one can infer B on the basis of the awareness of low-level bodily features because they are features that are associated with B, what Anscombe (1962, 56) herself describes as "what produces or always goes with it." For instance, one might claim that if one feels that one controls a limb, then one experiences the body part as one's own. It does not entail that bodily ownership consists in agentive phenomenology but only that bodily control is either what causes it or a side effect of it. There may then be a range of causes and effects normally associated with ownership and on the basis of which one can report that this is one's own body. One might possibly use the notion of sense of

[5] One might even suggest that the reductionist version provides an independent description of the feeling of ownership in Anscombe's terms.

ownership to cover this bundle of various sensations, but none of them is conceived of as being necessary to the sense of ownership. There is then literally no feeling of ownership.

Bundle conception: There are phenomenal properties on the basis of which one can report that B because of their causal connection to B.

One might go even further into eliminativism and claim that the properties that are causally associated to B do not need to be consciously available. On this fully eliminativist view, amputees do not report that their missing hand is still there on the basis of their tactile and painful phenomenology. Instead, their report directly results from some neural computations. The eliminativist does not attempt to reduce the phenomenology of ownership and agency to other phenomenal properties; she simply denies it. For instance, there would be no phenomenal contrast between cases in which one is aware of one's hand as one's own and cases in which one is not. Instead, it directly results from complex neural computations, which do not give rise to a feeling, but directly inform our cognitive attitude about ownership.

Eliminativist conception: There are subpersonal computations that lead to the report that B.

To conclude, when considering bodily awareness, one needs to answer the following questions. First, is there a phenomenology of B? Another way to put it is whether it feels the same when one is aware that B and one is aware that non-B. Most conceptions easily grant that it makes a phenomenological difference, and it is only the most extreme version of the eliminativist conception that denies it. Now if one grants that there is a phenomenology of B, then one must ask about its nature. Is it sensory, affective, cognitive, or even metacognitive? Can one reduce it to other properties? Or is it simply associated with a bundle of properties that are causally associated with B? There is thus a large range of theoretical options, and even the conservatives can be advocates of bodily feelings if their main objective is to reduce them. Cognitive science may not be able to adjudicate the debate among these different views, but at least empirical cases constrain the shape that the theory must take by providing an agenda of what it has to explain.

REFERENCES

Alsmith, A. J. T. (2015). Mental activity and the sense of ownership. *Review of Philosophy and Psychology* 6 (4): 881–96.

Anscombe, G. E. M. (1957). *Intentions*. Cambridge, MA: Harvard University Press.

Anscombe, G. E. M. (1962). On sensations of position. *Analysis* 22 (3): 55–58.

Anscombe, G. E. M. (1965). The intentionality of sensation: A grammatical feature. In R. J. Butler, ed., *Analytic Philosophy*. Oxford: Blackwell, 158–80.

Bayne, T. (2011). The sense of agency. In F. Macpherson, ed., *The Senses*. Oxford: Oxford University Press, 355–74.

Bermúdez, J. L. (2010). Action and awareness of agency: Comments on Chris Frith. *Pragmatics and Cognition* 18: 576–88.

Bermúdez, J. L. (2011). Bodily awareness and self-consciousness. In S. Gallagher, ed., *Oxford Handbook of the Self*. Oxford: Oxford University Press, 157–79.

Bermúdez, J. L. (2015). Bodily ownership, bodily awareness, and knowledge without observation. *Analysis* 75 (1): 37–45.

Billon, A. (2017). Mineness first. In F. de Vignemont and A. Alsmith, eds., *The Subject's Matter: Self-Consciousness and the Body*. Cambridge, MA: MIT Press, 189–216.

Botvinick, M., and Cohen, J. (1998). Rubber hands "feel" touch that eyes see. *Nature* 391: 756.

Dokic, J. (2012). Seeds of self-knowledge: Noetic feelings and metacognition. In M. J. Beran, J. L. Brandl, J. Perner, and J. Proust, eds., *Foundations of Metacognition* Oxford: Oxford University Press, 302–21.

Dokic, J., and Martin, J.-R. (2015). "Looks the same but feels different": A metacognitive approach to cognitive penetrability. In A. Raftopoulos, and J. Zeimbekis, eds., *Cognitive Effects on Perception: New Philosophical Perspectives*. Oxford: Oxford University Press, 241–67.

Ellis, H. D., and Lewis, M. B. (2001). Capgras delusion: A window on face recognition. *Trends in Cognitive Sciences* 5 (4): 149–56.

Gandevia, S. C., and Phegan, C. M. (1999). Perceptual distortions of the human body image produced by local anaesthesia, pain and cutaneous stimulation. *Journal of Physiology* 514: 609–16.

Goldman, A. I. (2006). *Simulating Minds*. Oxford: Oxford University Press.

Janet, P. (1928). *De l'angoisse à l'extase*. Vol. 2. Paris: Alcan.

Langdon, R., and Coltheart, M. (2000). The cognitive neuropsychology of delusions. *Mind & Language* 15: 183–216.

Leroy, E.-B. (1901). Sur l'illusion dite dépersonnalisation. *L'année Psychologique*: 519–22.

Martin, M. G. F. (1992). Sight and touch. In T. Crane, ed., *The Content of Experience*. Cambridge: Cambridge University Press, 199–201.

Martin, M. G. F. (1995). Bodily awareness: A sense of ownership. In J. L. Bermúdez, T. Marcel, and N. Eilan, eds., *The Body and the Self*. Cambridge, MA: MIT Press, 267–89.

McDowell, J. (2011). Anscombe on bodily self-knowledge. In A. Ford, J. Hornsby, and F. Stoutland, eds., *Essays on Anscombe's Intention*. Cambridge MA: MIT Press, 128–46.

Melzack, R. (1992). Phantom limbs. *Scientific American* 266 (4): 120–26.

Mezue, M., and Makin, T. (2017). Immutable body representations: Lessons from phantoms in amputees. In F. de Vignemont and A. Alsmith, eds., *The Subject's Matter: Self-Consciousness and the Body*. Cambridge, MA: MIT Press, 33–50.

Mitchell, S. W. (1871). Phantom limbs. *Lippincott's Magazine of Popular Literature and Science* 8: 563–69.

Moro, V., Zampini, M., and Aglioti, S. M. (2004). Changes in spatial position of hands modify tactile extinction but not disownership of contralesional hand in two right brain-damaged patients. *Neurocase* 10–16: 437–43.

Mylopoulos, M. (2015). Agentive awareness is not sensory awareness. *Philosophical Studies* 172 (3): 761–80.

Nielsen, M. (1938). Gerstmann syndrome: Finger agnosia, agraphia, confusion of right and left and acalculia. Comparison of this syndrome with disturbance of body scheme resulting from lesions of the right side of the brain. *Archives of Neurology & Psychiatry* 39 (3): 536–60.

Peacocke, C. (2014). *The Mirror of the World: Subjects, Consciousness, and Self-Consciousness.* Oxford: Oxford University Press.

Siegel, S. (2009). The visual experience of causation. *Philosophical Quarterly* 59 (236): 519–40.

Siegel, S. (2010). *The Contents of Visual Experience.* New York: Oxford University Press.

Sierra, M. (2009). *Depersonalization: A New Look at a Neglected Syndrome.* New York: Cambridge University Press.

Siewert, C. (1998). *The Significance of Consciousness.* Princeton, NJ: Princeton University Press.

Strawson, G. (1994). *Mental Reality*. Cambridge, MA: MIT Press.

Vignemont, F. de (2018). *Mind the Body*. Oxford: Oxford University Press.

Wu, W. (forthcoming). Mineness and introspective data. In M. Guillot and M. Garcia-Carpintero, eds., *The Sense of Mineness.* New York: Oxford: Oxford University Press.

PART VI
SORTALS AND NATURAL KINDS

11 Does the Identity of an Object Depend on Its Category?

The Role of Sortals in Thought

Lance J. Rips and Nick Leonard

Individual members of a natural category tend to remain members throughout their lifetime. Fido's existence, for example, tends to coincide with his membership in the dog category, and a particular apple's existence with its membership in the apple category. Of course, this tidy temporal relationship between existence and membership fails to hold for all categories. Fido outlived his membership in the puppy category, for example, and he may outlive his membership in the pet category if he goes on the loose. But for the type of categories that psychologists call "basic level"—dog, apple, oak, trout, and so on—members tend to stay put (see Rosch et al. 1976, for the special status of basic-level categories). For instance, we don't expect Fido to be able to undergo an extreme makeover that would cause him to become a cat.

A similar relation between basic categories and their instances holds in the case of artifacts, although matters are less clear-cut. You can easily imagine someone reshaping a coffee pot into a bird feeder or an umbrella into a lampshade (Keil 1989; Leonard and Rips 2015; Rips 1989), and it seems possible that these transformations don't disrupt the identity of the original object. The same thing that was formerly a coffee pot is now a bird feeder, and the same thing that was a formerly an umbrella is now a lampshade. Still, the underlying point remains: Many artifacts belong to the same basic-level category throughout their entire existence.

These mundane facts about everyday objects may encourage a strong thesis about the relation between an individual object and the basic-level category to which it belongs. Perhaps the lifetime of an object necessarily coincides with its membership in a basic-level category because the category dictates the lifetime. For instance, perhaps it is by virtue of being a dog that Fido comes to have an identity—one that begins at the time of his birth, ends at the time of his death, and traces a spatial and temporal path between these two extremes. Because his status as a dog determines his identity, Fido can't stop being a dog without ceasing to exist entirely. No metamorphosis or reincarnation can turn Fido into a cat, since such a purported transformation would in fact put an end to Fido and bring into existence something entirely new.

Let's call metaphysical theories along these lines *sortal* theories. According to such views, sortals are nouns, such as *dog* or *coffee pot*, that name categories with identity-conferring properties. The key idea is that sortals have, as *essential* parts of their meaning, both criteria of identity and criteria of individuation (more on this in the next section). Of course, not all nouns qualify as sortals. For example, *pet* doesn't for the reasons we've seen. So how do sortalists determine which nouns are sortals? Sortal theories (e.g., Wiggins 2001) typically hold that sortal nouns are those that give the most fundamental answer to the question "What is it?" for the items to which they apply. In these terms, the sortal for Fido would presumably be *dog*, and the sortal for a coffee pot, *coffee pot*. Because people overwhelmingly use basic-level terms in answering "What is it?" for ordinary objects (Rosch et al. 1976), basic-level terms and sortals tend to coincide.

Our aim in this chapter is to examine what happens when sortal theories are transferred from their home in metaphysics to play a new role in cognitive science. A number of developmental psychologists have taken philosophers' arguments in favor of sortals to show that people must possess concepts—mental representations—of sortals in order to be able to individuate objects and to trace the identity of a single object over time. According to these theories, people must have the sortal concept DOG in order to tell Fido from Rover and in order to reidentify Fido from one moment to the next.[1] What we'll try to show here is that transplanting sortal theories in this way has not been successful. Arguments based on purely metaphysical grounds don't secure a role for sortals as parts of people's representational apparatus. And arguments based on empirical considerations in psychology similarly fail to establish that people have sortal concepts that provide conditions of identity and individuation. Backing up these claims will occupy most of this chapter in sections 11.3 and 11.4. But first we need to say a bit more about the role of sortals in metaphysics (section 11.1) and in cognition (section 11.2).

11.1 SORTALS IN METAPHYSICS

One way to motivate sortal theories is to consider the way we count objects. Following Frege ([1884] 1968), sortal theorists claim that the number of objects that occupy a given region isn't a determinate quantity. A particular region may contain playing cards that we could count as one stack, two decks, four suits, 104 cards, and

[1] Following common practice, we use words with all caps to refer to concepts (i.e., mental representations), and we will continue to use italics to name words. So we use *duck* to refer to the sortal (noun) and DUCK to refer to the sortal concept for ducks.

so on. In order to pin down a definite number, a sortal (e.g., *card, deck*, etc.) must specify the appropriate unit. Sortals perform this task by including, as part of their meaning, *criteria of individuation* that carve out the units from the category named by the sortal. A sortal like *dog*, for example, provides individuation criteria that determine, at least in principle, how many dogs occupy Grant Park at a particular time.

In addition to individuation criteria, sortal theories often assume that sortals provide *criteria of identity* that determine which object at a later time (if any) is identical to a given object at an earlier time. The sortal *dog* dictates which dog on Thursday is identical to Fido on Tuesday. Count noun sortals, like *dog* and *coffee pot*, provide both criteria of individuation and identity, and some sortal theories maintain that these count noun sortals are the only sortals there are. Other sortal theories, however, grant that some mass nouns (e.g., *water, air*, or *gold*) can also be sortals: Although they do not come with criteria of individuation, they do supply criteria of identity (Gupta 1980). According to these theories, for instance, the meaning of *water* specifies when the water in a particular glass is the same water that was earlier in a pitcher. But *water* cannot determine how many quantities of water exist in the glass, since any quantity of water comprises many overlapping subquantities, down to the molecular level.

On either view, all sortals carry criteria of identity, which according to Lowe (1998) and Williamson (1990) can take the following form:

$$(\forall x, y)\ [S(x) \& S(y)] \rightarrow [x = y \leftrightarrow R_S(x, y)], \qquad (11.1)$$

where *S* is a sortal category and R_S is an equivalence relation (a relation that is reflexive, symmetric, and transitive) that is specific to sortal *S* (see Blok et al. 2007). To take the standard example, we can spell out the identity criterion for sets by saying that if *x* and *y* are sets then they are identical if and only if they have exactly the same members. Here, *set* is the sortal and *has exactly the same members as* is the equivalence relation (i.e., *x* has exactly the same members as *x*; if *x* has exactly the same members as *y*, then *y* has exactly the same members as *x*; and if *x* has exactly the same members as *y* and *y* exactly the same members as *z*, then *x* has exactly the same members as *z*). One point to note about (11.1) is that, although it is not a definition for the sortal itself (nor for identity in general), it does provide both necessary and sufficient conditions for two instances of the sortal to be identical. The biconditional (↔) enforces this two-way relationship. For example, having exactly the same members is both necessary and sufficient for one set to be equal to another.[2]

[2] Debate concerns whether all identity criteria can be reduced to the form of (11.1) or whether, instead, some identity conditions are irreducibly "two level" (see Lowe 1998; Williamson 1990). Two-level criteria spell out the identity of the sortal instances in terms of equivalence relations between items to which the instances are functionally related rather than in terms of equivalence relations between the items themselves, as in (11.1). In both cases, however, the criteria are necessary and sufficient conditions, which is all that will concern us here.

Another point to note is that the relation R_S must be noncircular. Identity is itself an equivalence relation, but replacing $R_S(x,y)$ in (11.1) with $x = y$ would not produce an informative criterion of identity. It doesn't help in specifying identity conditions for dogs to know that two dogs are identical if and only if they are identical.

Philosophers have recruited sortals to deal with a variety of problems, such as the nature of counting, individuation, identity over time, and essence. While these sortalist views are controversial, let's suppose for now that there are nouns that furnish identity and individuation conditions and consider the ways in which psychologists have put these sortals to work.[3]

11.2 SORTALS IN COGNITION

This section traces the use of sortals in theories of cognitive development. Initial theories of this type stuck closely to the philosophical accounts we've just glimpsed. But pressure from experimental findings produced modifications that sometimes put the psychological views into conflict with philosophical ones. In what follows, we'll use *sortalism* for the philosophical versions and *psychosortalism* for the psychological versions of these theories (following Blok et al. 2005).

11.2.1 Psychosortalism's Beginnings

To our knowledge, John Macnamara (1986) was the first to use sortals in the context of explanations in cognitive psychology. Macnamara's project was to understand children's learning of proper names by considering the theoretical constraints that the semantics of these names impose. Taking a leaf from philosophical work on sortals, he claimed that children could not correctly understand a proper name unless they already possessed a sortal concept that applied to the name's referent. In order to learn the meaning of the name *Fido*, for example, children would have to possess the concept DOG (or some similar concept), since DOG provides the identity and individuation conditions for Fido.[4] Without those conditions, children

[3] The notion of a sortals can conflict, the notion of a sortal itself may be obscure (as Feldman 1973 and Grandy 2014 contend). In addition, sortals' ability to handle some of these demands is not always clear cut. For example, *picture frame* arguably provides conditions for identity over time (i.e., conditions under which a later picture frame is identical to an earlier one). But does it properly individuate objects? For example, if Patricia fastens together six small picture frames in a three-by-two arrangement and gives the arrangement to her friend Paul, then the number of picture frames that Paul receives is unclear to many college students (Leonard and Rips 2015). Most think that "*both* one and six" are correct answers (as opposed to: just "one" is correct, just "six" is correct, or "neither one nor six" is correct). So *picture frame* does not appear to be a sortal in the sense of providing determinate conditions for individuation—at least, not conditions to which students have easy access. We will return to problems like these in sections 11.3 and 11.4.

[4] Macnamara (1986, 132–33) uses the term *principle* to denote a metaphysical relation and *criterion* to denote a (possibly fallible) test for that relation. So a principle of identity, in these terms, is the actual relation that specifies identity for individuals (in the way R_S does in (11.1)), but a

would have little idea that *Fido* applies to the whole dog rather than to one of its parts or to its matter. And they would have little idea that *Fido* applies at each moment of its life rather than to some scattered moments or temporal segments.

Macnamara (1986) is clear that children do not need to formulate explicitly the identity conditions they employ. Nor do they have to use in speech the sortal that supplies these conditions before they can acquire the name for a member of that sortal category. For example, they don't have to use the word *dog* before learning *Fido*. What they *do* have to possess, according to Macnamara, is a concept—a mental representation—corresponding to the sortal, a concept that connects to the relevant identity conditions. Macnamara believed that the representation is, at least initially, what he called a *visual gestalt type* of the sortal category. Although the description of these gestalt types is very brief, they seem to contain visual information that is general enough to categorize members of the type. In learning the name *Fido*, for example, children would recognize Fido's "gestalt as being of a familiar type," namely, the dog type (Macnamara 1986, 73). Either the visual gestalt itself or a more abstract symbol corresponding to it (Macnamara suggests something like "Sortal_{17}") becomes part of the language of thought, where it denotes dogs. In learning the word *dog* at a possibly later stage, children come to understand the equivalence between the new term for dogs and the one they have already acquired as a visual gestalt. They learn that all and only dogs are members of the category denoted by the gestalt (or by Sortal_{17}).

As Macnamara (1986, 148) points out, the referent of a sortal such as *dog* is not determined by "a set of perceptually given distinctive features, whether in the form of a list or an image." Similarly, criteria of identity are not principally visual (Macnamara 1986, 70): "Our intuitions of identity run deeper than perception." We certainly allow Fido as a puppy to be identical with Fido as an older dog, despite dramatic changes in his appearance. This suggests that the visual gestalt type does not itself determine the referent of a sortal or its identity conditions. Instead, people take the referent to be whatever it is in nature that's responsible for the sortal category, even though they usually can't describe this "whatever it is." The visual gestalt type provides no more than a rough guide to what's in this category. This idea seems consistent with more recent psychological essentialist theories in which people have an unspecified "placeholder"—that is, a variable—to stand for whatever it is inside an organism that makes it a member of its sortal category (Gelman 2003; Medin 1989). Similarly, according to Macnamara, "it seems unlikely that children have to

criterion of identity tests whether that relation is in place. Standard philosophical use of *criterion*, however, is closer to what Macnamara calls a principle. See, for example, Merricks (1998, 107 and n. 2): "criteria of identity over time are *informative* necessary and sufficient conditions for identity over time. Here . . . 'necessary and sufficient' means broadly logically, or metaphysically, necessary and sufficient." We will stick to the latter, more standard meaning of *criterion* in this chapter and will use *conditions* as a synonym for *criteria*. When we have to refer to a test or a procedure, we'll label it explicitly as such.

represent the rule for tracing identity under the sortal *dog*. . . . It may perhaps be applied by the interpreters and implicators that handle sortals and apply them to individuals" (1986, 140). But for Macnamara, "interpreters" are "devices that place the mind in intentional contact with the objects that form the semantic content of some linguistic expression," and he uses "'interpreter' in its semantic function to speak of what I see largely as a mystery" (1986, 36).

This indirect connection between people's representation of a category and the relevant sortal puts some pressure on the psychosortal theory. Recall that the justification for the theory comes from logical constraints. No counting of a category's members and no tracking of a member across time is possible without a sortal to supply criteria of individuation and identity. For example, it's impossible to count dogs or to follow them across time without the sortal *dog* to supply these criteria. But if the psychological representation—for example, the visual gestalt type—corresponding to the sortal provides only rough-and-ready tests or procedures, then it is no longer clear why people need to have a representation of the sortal prior to learning names for members of the category.

To see this, suppose for the sake of the argument that, as a purely metaphysical or logical matter, the identity and individuation of Fido depend on the sortal *dog* supplying criteria of identity and individuation. Suppose, too, that the psychological representation of dogs connects to these criteria only indirectly through mysterious "interpreters" and does not itself follow these metaphysical rules (or follows them only imperfectly). Then it is no longer clear why children need to have a representation of dogs prior to learning *Fido*. The representation (e.g., visual gestalt type) doesn't supply the criteria. So why not assume that *Fido* refers to Fido without the intermediate psychological representation of the sortal? Any metaphysical carving out of Fido for purposes of individuation or identity could be done by whatever metaphysical criteria attach to the sortal *dog*, but without children having to represent this sortal.[5]

If we're right about these difficulties, their source is the split between the metaphysical work and the psychological work that sortals are supposed to perform. The metaphysical demands seem compatible with very light commitments on the psychology side. In fact, Macnamara himself (1986, 148–49) makes a similar point in connection with reference: "The strength of reference from the psychological point of view is that so little knowledge of the referent is required in the user." Because the motivation Macnamara provides for sortals is almost entirely metaphysical, psychologists could get by with theories that short-circuit representation of sortals (and their criteria of individuation and identity), allowing the metaphysics to take

[5] Having gone this far, we might wonder whether we can account for the semantics and learning of proper names without the need for sortals even as part of the metaphysical side of the story. Why couldn't *Fido* refer to Fido directly without the intermediary of a sortal to individuate and identify him? This last possibility seems more in line with direct reference theories of proper names (e.g., Kripke 1980, 1981), but we don't pursue it here.

care of itself, as in externalist theories of reference (e.g., Burge 1979; Kripke 1980; Putnam 1975). So to justify the need for sortals in a psychological theory, we need empirical evidence that implicates these sortals in thought. Let's look at some possible evidence of just this kind.

11.2.2 Psychosortalism's Evolution

In commenting on Macnamara's theory, Carey and Xu (1999, 318) note "a logical problem" in his proposal, related to those we've just discussed. They agree with Macnamara that children need sortal concepts in order to learn proper names for objects that fall under those concepts. But if a child's sortal concept is a visual gestalt, how could the gestalt provide the appropriate guidance for the object's individuation and identity?

Carey and Xu propose instead that an infant's initial sortal concept is the concept of a *Spelke object:* "any entity that is three-dimensional, is bounded, and retains its boundedness as it moves through space and time" (1999, 327).[6] Infants have the concept of a Spelke object either innately or within the first two or three months of life. They begin to gain concepts for basic-level sortals, like *dog* or *cup*, only around their first birthday, as they acquire the use of these words. Unlike Macnamara's visual gestalt types, the concept of Spelke objects (and the later concepts of basic-level sortals) does provide criteria for identity and individuation.

Carey and Xu's (1999) theory is based on evidence from experiments that attempt to determine whether infants can distinguish objects that they see intermittently. In one version of these experiments (Xu and Carey 1996), infants watch objects moving behind a screen on a puppet stage. Figure 11.1 illustrates schematically the events that occur during an experimental trial. The infants see an object of one kind (here a toy duck) emerge from one side of the screen and then retreat behind it. They then see an object of a different kind (a cup) emerge from the other side of the screen and retreat. After a number of repetitions of these events, the experimenter removes the screen to reveal to the infants either one object (say, the duck) or both objects (the duck and the cup). Experimenters record the amount of time the infants look at the scene after the screen is removed. These looking-time results show that infants at twelve months look longer if one object is revealed than if two are revealed. The infants act as if they were expecting to see both objects they had witnessed earlier and are surprised (and therefore look longer) if only one is on

[6] Carey and Xu (1999) use *physical object* as a synonym for *Spelke object*. But *physical object* is probably not the best choice of terms, because many things that people consider physical objects (e.g., houses, trees, roads) are not Spelke objects since they don't move (as Hirsch 1997 and Wiggins 1997, have pointed out). We'll therefore stick with *Spelke object* in what follows as a technical term for the type of entity just defined. See Spelke (1990) for an account of the research and theory behind Spelke objects.

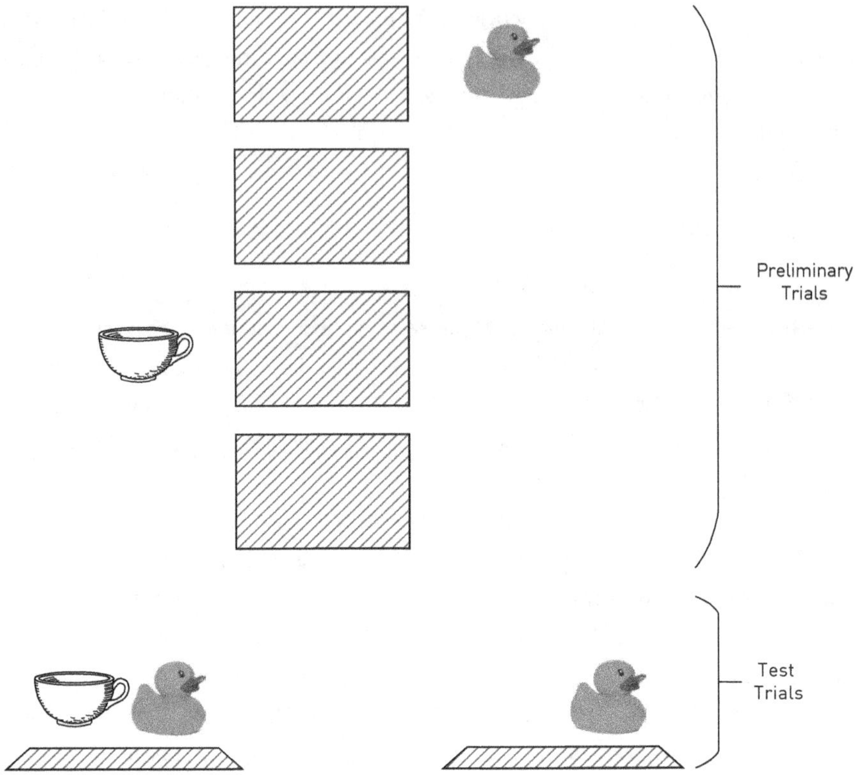

FIGURE 11.1 The events in Xu and Carey's (1996) Is-it-one-or-two task. See the text for an explanation of these events.

hand. Infants at ten months, however, look no longer at the one-object scene than at the two-object scene.

One variation on this experiment is relevant to its interpretation in terms of sortal concepts: If infants see both objects simultaneously in full view before these objects begin their back-and-forth traversals, then both ten-month-olds and twelve-month-olds succeed in looking longer when one object appears than when two objects appear at the end of the trial.

Carey and Xu's (1999) interpretation of this experiment follows from their ideas about when infants gain sortal concepts. According to Carey and Xu, infants at ten months have acquired the concept of a Spelke object, but not the concepts of sortals for lower-level categories like ducks or cups. If they see a simultaneous preview of both objects, they can deploy their Spelke-object concept to discriminate them. The Spelke-object concept comes along with the information that the same object can't be in two different places at the same time, and this information is sufficient to alert the infant to the presence of two items. But without the preview, infants cannot use the spatiotemporal cues associated with Spelke objects to disambiguate the number of objects behind the screen. Lacking sortal concepts for ducks and cups, ten-month-olds are also unable to deduce that a duck-like object can't turn into a cup-like object, and so they don't know how many objects to expect. By contrast, twelve-month-olds do have sortal concepts for ducks and cups that provide identity conditions, and so they do know that two objects should be behind the screen. They look longer at the unexpected one-object tableau, both with a preview and without a preview of these items.

Our purpose in reviewing these findings is to see whether they implicate infants' use of sortal concepts, as Carey and Xu (1999) contend. When infants succeed in these tasks, is it because their knowledge of sortal concepts gives them cognitive access to criteria of identity for the objects that appear in the displays? To answer this question, we need to look at two claims: (*a*) that older infants use concepts of sortals at the basic level (e.g., DUCK, CUP) to figure out the number of objects in conditions without a preview of these items, and (*b*) that younger infants use a sortal concept for Spelke objects to figure out the number of objects in conditions with the preview. Let's start with the performance of the older infants.

11.3 BASIC-LEVEL SORTALS

The aim of this section is to probe the motivation for positing basic-level sortal concepts. More specifically, psychosortalists and nonpsychosortalists now seem to agree that infants and adults use properties (e.g., color, size, and shape) to identify and individuate objects. If appealing to the use of this property information can adequately account for all the relevant experimental data—that is, the data that is taken to support the existence of basic-level sortals like DOG and CUP—then positing the existence of basic-level sortal concepts is unnecessary.

To begin, consider this question: Did the twelve-month-olds' success in the figure 11.1 task depend on the identity or individuation conditions of sortals such as *duck* and *cup*? As Carey and Xu (1999) note, an immediate alternative explanation is that infants' success could be due to their taking the properties of the objects to signal their distinctness. The twelve-month-olds might guess that something that's yellow, rubbery, and duck-shaped is probably not the same object as one that's white, ceramic, and cup-shaped. The ten-month-olds may lack the cognitive sophistication to perform this inference and so fail the task. "Although concepts lexicalized as properties do not provide criteria for individuation (we can't count the red in this room, the big in this room, the striped in this room), in the situations we have described above [e.g., those of figure 11.1], the property differences among the objects certainly do provide relevant information about number" (Carey and Xu 1999, 324). Let's call this alternative the *property account* to contrast it with the *sortal account* of the findings.

Carey and Xu (1999) criticize the property account based on further findings that show that even twelve-month-olds fail the Is-it-one-or-two task in figure 11.1 if the objects have different properties but belong to the same sortal category (Xu et al. 2004). For example, if the infants see a big red plastic cup with colored dots come from one end of the screen and a small blue ceramic cup from the other, they do not look longer at the one-object scene than at the two-object scene at the end of the trial. This finding suggests that twelve-month-olds' success in the original task was not due to their use of property differences (e.g., yellow vs. white) but depended instead on the objects coming from distinct sortal categories (e.g., duck vs. cup).

The worry about this objection to the property account is that later studies have found that even infants as young as four months can use properties to succeed in simple variants of the Is-it-one-or-two task (e.g., Wilcox 1999; Wilcox and Baillargeon 1998; Xu and Baker 2005). In one such experiment (Wilcox 1999), four-month-old infants saw a big green ball move behind the left side of a screen and, a few seconds later, a small green ball emerge from the right side, as figure 11.2 illustrates. The small ball then retreated behind the screen and, a few seconds later, the large ball emerged from the left side. The same sequence repeated until the trial ended. The screen remained in place throughout the trial. The crucial variable in this experiment was the width of the screen. In the wide-screen condition (left side of figure 11.2), the large and the small ball could both fit side by side behind the screen simultaneously. In the narrow-screen condition (right side of figure 11.2), however, the screen was too narrow to accommodate both balls. If infants in the narrow-screen condition take the two balls to be distinct solid objects, they should find it odd that both fit behind the screen at the same time during certain parts of the trial. But infants in the wide-screen condition should have no difficulty understanding

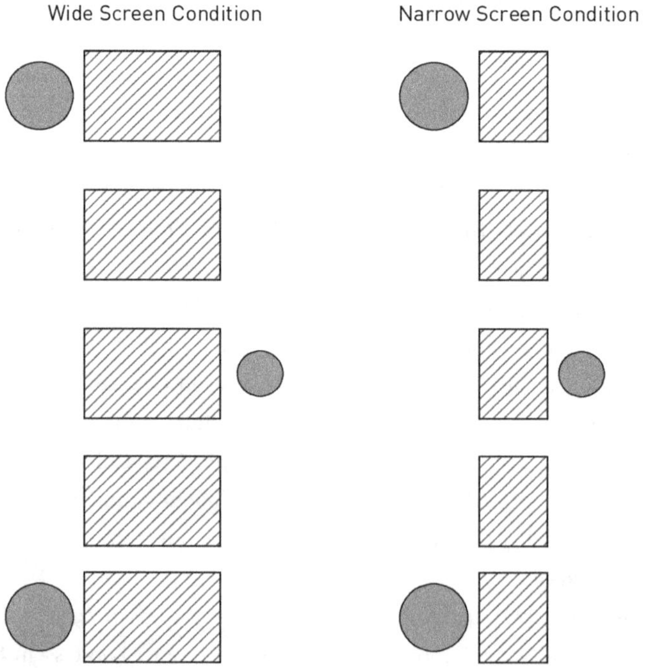

FIGURE 11.2 The events in Wilcox's (1999) study of object discrimination. See the text for an explanation of these events.

how both balls could fit behind the screen. In fact, infants in the narrow-screen condition did look reliably longer than did infants in the wide-screen condition. This suggests that the four-month-olds understood that the two balls were distinct, space-occupying objects that could not be in the same place at the same time. To support this interpretation, control conditions showed infants the same scenes but with another large ball replacing the small one to prevent discrimination based on size. In this situation, infants looked approximately the same amount of time in the narrow-screen as in the wide-screen condition, presumably taking the identical-looking large balls to be the same object. In short, infants seem to use the difference in size of the balls to infer the presence of two balls rather than one.

Investigators currently agree that infants can use property information—for example, size (big vs. small), shape (spherical vs. cubical), and pattern (dots vs. stripes)—to individuate objects in simplified versions of the Is-it-one-or-two task, such as that of figure 11.2. (Earlier philosophical discussions of object concepts in infants seem handicapped by failing to take these findings into account; see Sarnecki 2008). In these procedures, the property account seems correct. The failure of property differences to help infants in some earlier studies (e.g., Xu and Carey 1996) could be put down to the complexity of the task in these studies (i.e., the procedure of figure 11.1, to be discussed in a moment). The key question for psychosortalists, then, is whether *any* evidence exists to implicate concepts of basic-level sortals like *duck* or *cup* in infants' performance.[7]

Xu (2005) continues to maintain that sortals play a crucial role in tasks like that of figure 11.1. She concedes that younger infants (e.g., four-month-olds) can use properties to differentiate objects in simple procedures, such as that of figure 11.2. But ten-month-old infants are unable to use property information in more complex tasks to determine the correct number of objects because features of the task override that information, causing the ten-month-olds to fail. One such complex feature is the fact that seeing the objects move back and forth along the same trajectory could create the illusory impression of a single object changing its properties behind the screen (a "tunneling effect"; see Xu 2005). A second complex feature is that infants must draw on memory to compare the events they see in the first part of the trial to the test tableau at the end (as Wilcox and Baillargeon [1998] propose). According to Xu, however, twelve-month-olds gain access to sortal concepts that override these complexities, allowing the twelve-month-olds to succeed when

[7] You might try to argue that properties like size and pattern are effective in the simpler procedures, such as that of figure 11.2, because basic-level sortal concepts for the objects (BALL? SPHERE?) dictate their effectiveness. For some sortals, size individuates objects, whereas for other sortals, it does not (see Xu 2007, for a related suggestion). But this is not a tactic that psychosortalists can consistently use in explaining Wilcox's (1999) findings, while also maintaining that infants gain basic-level sortal concepts only at twelve months. Remember that infants in Wilcox's study were only four months old. Psychosortalists seem forced to say that properties alone are sufficient for differentiating things (perhaps in conjunction with the higher-level concept for Spelke objects).

objects cross sortal boundaries (duck vs. cup) but not when the objects fall within the same sortal (big red plastic cup vs. small blue ceramic cup).

Although this psychosortalist view seems consistent with the data, the property account can adequately explain these results too. The property account can go along with the idea that ten-months-olds can use property information to succeed on the simpler (figure 11.2) task and that task complexities inhibit their use of properties on the more difficult (figure 11.1) task. But the property account can go on to claim that, at twelve months, infants have gained enough know-how to overcome the complexities and use property information (rather than sortal information) to succeed at even the more difficult problem. This extra know-how might include an increase in working memory capacity (allowing the older infants to compare more easily the preliminary back-and-forth events with the later scene after the screen is removed; see Levine and Baillargeon 2016) or an increase in attentional skills (allowing the older infants to ignore the misleading trajectory cues). According to the property account, then, lower-level sortals play no role in explaining infants' performance at either age; rather, everything can be chalked up to the use of property information and varying degrees of know-how. So, if psychosortalists admit that infants can use property information and acquire more know-how as they get older, it is hard to see a need to posit basic-level sortal concepts to explain this data.

In our view, the biggest remaining worry with the property account is that, at first sight, it seems to fail to explain why twelve-month-olds don't perform correctly in the figure 11.1 task when the items are from the same sortal category but have different properties (e.g., big red plastic cup vs. small blue ceramic cup) in Xu et al.'s (2004) study. If twelve-month-olds can differentiate a duck from a cup on the basis of their distinctive properties, why not a cup from another, distinctive-looking cup?

But as psychosortalists themselves acknowledge, cross-kind items, such as a duck and a cup, differ more in shape than within-kind items, such as a red cup and a blue one. This is crucial, since shape is a highly salient cue for children. For example, it dominates color and texture in classification and generalization tasks (e.g., Baldwin 1989; Soja et al. 1991). Thus, if shape similarity can override the differences that distinguish the cups, the property account can explain all the relevant data without appealing to basic-level sortal concepts. Of course, additional empirical evidence is need to establish the relevant differences in salience in this experimental setup. But at the very least, we think that as things currently stand, there is no more reason to posit the existence of basic-level sortals than to prefer a more fully developed property account.

Before moving on, though, we think two further points are in order about the explanations for the Is-it-one-or-two tasks. First, in granting that properties are sometimes sufficient for differentiating objects in this task, Carey and Xu (1999) have backed away from a strong form of psychosortalism, such as Macnamara's. Recall that Macnamara's (1986) reason for introducing sortals in cognitive explanations was his conviction that sortals are logically necessary for individuating and identifying objects. This conviction in turn derived from the special role that sortals play in furnishing conditions for individuation and identity, a role that specifically distinguishes sortals from predicates for properties, like color and size. Only sortals yield individuation and identity conditions. This sortal/predicate distinction is at the very heart of the sortal theories that we glimpsed in section 11.1. But if infants can use properties for individuating objects in the Is-it-one-or-two task, this underlying logical motivation is undermined. The case for psychosortalism then rests on the ability of sortal concepts to earn their keep empirically as part of psychological explanations. So ambiguities about whether the property account or the sortal account is correct for the Is-it-one-or-two task raise critical difficulties for the claim that sortals play an important role in human thinking.

Second, basic-level concepts like DUCK and CUP could still play a role in infants' (and adults') decisions about whether one or two objects exist behind the screen, even if these concepts don't have the status of sortals—that is, aren't associated with criteria of identity and individuation. People may know as a simple matter of empirical fact that a toy duck is unlikely to turn into a cup while it's briefly behind a screen. But a simple empirical generalization of this sort does not require anything as strong as the criteria of individuation and identity that sortals are supposed to supply. One way to see this is to note that such generalizations are often true of objects in categories that don't qualify as sortals. For example, iPhones don't turn into Samsung Galaxies when they move behind screens, and many adults could use this fact to make correct predictions about the number of cell phones in an Is-it-one-or-two task. But this is presumably not because they know criteria of identity and individuation for iPhones and Galaxies. Instead, facts about how easily one object could change into another depend, not on criteria of individuation or identity, but instead on the physical processes governing these items and their properties. It is not at all metaphysically impossible that Apple and Samsung could agree to produce a transformer model that would quickly convert from an iPhone to a Galaxy and back. It's just that our beliefs about the current state of technology make it unlikely that we're dealing with such a model. The present point is that concepts like DUCK, CUP, and PHONE can aid in individuation by supplying information of an empirical kind even if they are not sortals.

To recap, we've argued that psychologists can't simply assume on purely logical or metaphysical grounds that if infants manage to differentiate two things, they must have a sortal concept to provide conditions of individuation. Whatever work sortals do by way of individuation, they don't seem to require people to represent them mentally, in the sense of representing the sortals' criteria of individuation and identity. In the case of basic-level concepts, such as DUCK or TOASTER, we've argued that such concepts don't have to possess the status of sortals to explain the empirical results on individuation (e.g., the results of the Is-it-one-or-two tasks of figures 11.1 and 11.2). Could the evidence for sortal status be any stronger in the case of SPELKE OBJECT? We turn to this issue in the next section.

11.4 THE SPELKE-OBJECT SORTAL

The aim of this section is to argue that although empirical evidence suggests the existence of a SPELKE OBJECT concept, there is no reason to maintain that it is a sortal concept. Because psychosortalism crucially rests on SPELKE OBJECT being a sortal concept, psychosortalism is in trouble.

To begin, recall that sortal concepts enter Carey and Xu's (1999) account in two ways. Basic-level sortals like *duck* and *cup* are supposed to explain twelve-month-olds' advantage over ten-month-olds in the figure 11.1 task. But a higher-level sortal concept, SPELKE OBJECT, explains why ten-month-olds can perform correctly in this task if they have seen a preview of the two objects. The same concept can explain how it is that people (both children and adults) can recognize something as an object even when they are unable to classify it as a member of a lower-level category. For example, suppose we look out over the lagoon and see a moving entity that might be a tractor or a horse or something else. We can still take the thing to be an object, despite not being able to identify it more precisely (see Campbell 2006 and Goodman 2012, for antisortalist treatments of such cases). The question for us is whether *Spelke object* is a sortal.

This issue has sparked debate between Xu (1997) and a number of philosopher critics, some accepting traditional sortal theories (e.g., Hirsch 1997; Wiggins 1997) and others not (Ayers 1997). From the orthodox sortalists' point of view, a concept at the level of SPELKE OBJECT is too abstract to provide a grip on an individual's identity or individuation. You can no more count the Spelke objects in a given region than count the yellow in that region. From the antisortalists' point of view, of course, sortals are unnecessary at any level of abstraction. In either case, the key question in this debate seems to be this one (Ayers 1997, 393):

Xu's argument for this common-sense conclusion ["that infants and adults can and do individuate and track unitary, coherent, bounded material objects without knowing on each occasion what kind of thing is before them"] is itself open to question, in particular the question of why she presents her own thesis in the terms of the theory she rejects, although there seems no particular reason for her to retain the philosophical assumptions embodied in those terms. Specifically, why does she interpret her own and others' observations as evidence that infants and adults employ a "sortal concept," *physical object* [i.e., SPELKE OBJECT], which is more general, certainly, than the concepts postulated as basic individuators by strong sortalists, but which is assigned the same kind of logical-cum-epistemological work as such concepts have been supposed to do—to "tell us what to count as one instance of something and whether something is the same one as what we have seen before," and to "provide the criteria" by reference to which we can decide when a thing has ceased to exist, and so forth? We have to ask why the capacity to discriminate physically unitary material objects cannot simply be regarded as a primitive function of the totality of the perceptual and agent mechanisms of the infant.

One way to take this comment is as a generalized version of the point we raised earlier with respect to basic-level sortals. Object tracking need not be the kind of skill that requires the individuation and identity criteria that define sortal concepts. Instead, infants and adults can use heuristic cues for tracking.

There is no evidence that prelinguistic infants require something to be a Spelke object before they can track it. As Green (2018) has pointed out, both infants and adults can track unfamiliar objects that don't happen to be "three-dimensional, . . . bounded, and [that] retain [their] boundedness as [they] move through space and time" (Carey and Xu 1999, 327). For instance, infants can track two-dimensional figures (non-3D: Cordes and Brannon 2009; Wynn et al. 2002) and swarms or ensembles of objects (unbounded: Feigenson and Halberda 2004; Wynn et al. 2002), at least under some conditions. These findings show that we have no serious reason to think that the SPELKE OBJECT concept is necessary for infants' successful tracking. What reason do sortalists have, then, for positing this concept at all?

Perhaps sortalists could try to explain these experimental facts by positing other mechanisms to supplement those associated with SPELKE OBJECT. For instance, perhaps they could maintain that we have special cognitive mechanisms for tracking two-dimensional (i.e., non-Spelke object) entities. (For a similar suggestion regarding nonsolid substances, such as sand, see Anderson et al. 2018 and Hespos et al. 2016.) Assuming that we do have these additional mechanisms, perhaps sortalists could say that infants really do track some objects (though not all) as Spelke objects. That is, the mechanisms we need in order to account for the results just cited don't also render SPELKE OBJECT otiose, in that the SPELKE OBJECT concept does play an important cognitive role in cases in which these other mechanisms are not employed.

Because this is a substantive, empirical assumption and the center of active debate (see, e.g., Green 2018), we won't try to resolve it here. Instead, let's grant for the sake of the argument that infants really do attend to 3D, bounded, moving entities as such, since this will allow us to focus on our main issue—whether SPELKE OBJECT, distinguished in this way, is a sortal concept. In the next two subsections, we consider two anti–SPELKE OBJECT arguments addressed to this issue, the first questioning whether SPELKE OBJECT is a concept and the second questioning whether it is a sortal.

11.4.1 Is SPELKE OBJECT a Concept?

Perhaps one could argue against psychosortalism in a simple way just by arguing that SPELKE OBJECT is not a concept (i.e., a mental representation for the category of Spelke objects—3D, bounded, moving things). Psychosortalism requires SPELKE OBJECT to be a sortal concept in order to explain the data we described earlier (e.g., ten-month-olds' success on the Is-it-one-or-two task with a preview of the relevant objects). But if SPELKE OBJECT is not a concept (in general), then it isn't a sortal concept (in particular). And if it isn't a sortal concept, then psychosortalism is false.

In our view, the best way to argue that SPELKE OBJECT isn't a concept is to show that it isn't fit to do any important psychological work. One possibility along these lines is the idea that people possess *preconceptual* mechanisms that could produce the same results that SPELKE OBJECT is supposed to—singling out Spelke objects and tracking them—without first bringing them under a concept.

To see the motivation for this idea, consider that some well-known proposals in perceptual psychology feature similar preconceptual tracking devices. For instance, object files (Kahneman et al. 1992) and visual indexes (Pylyshyn 2001) are supposed to be preconceptual representations of objects that record their spatial location as they move through the visual field. A perceptual mechanism of this type could then undercut the need for the SPELKE OBJECT concept, as Goodman (2012) argues. The evidence supporting a SPELKE OBJECT concept in infants can instead be taken as showing "that infants have a *perceptual ability* to attend to and successfully track objects on the basis of spatiotemporal information *before* they have developed a full-fledged system of concepts (among which are sortal concepts like *cup*, *dog*, etc., and concepts like *physical object* [i.e., SPELKE OBJECT; see Footnote 6])" (Goodman 2012, 95). This possibility also goes along with Ayers's (1997) comment that we quoted earlier and with similar objections by Casati (2004).

Of course, if you argue against psychosortalism in this way, then you must defend the existence of a notoriously disputed border between what's preconceptual and what's conceptual. One way of drawing this distinction relies on the notion of a *perceptual module* (from Fodor 1983): an information-processing mechanism that accepts sensory input and transforms the input into a mental representation, but whose workings are sealed off from the rest of the cognitive system. A perceptual module responds to input that it is specialized to detect, automatically processes the input through a series of stages, and finally outputs its final representation to central cognition. During its internal processing stages, though, representations in a module are "encapsulated"—they can't be accessed or affected by beliefs and other long-term information. Only after the module's final representation enters central cognition can other parts of cognition use it. So the module's encapsulation provides one way to draw the line between the preconceptual—the work done within the module—and the rest of cognition. In line with this division, visual indexes for tracking objects are supposed to constitute a perceptual module (or, at least, to be encapsulated in the way modules are).

Given this notion, we can sharpen the argument against the SPELKE OBJECT concept. If (preconceptual) perceptual modules can carry out the work of tracking Spelke objects, then we have no need for SPELKE OBJECT, understood as a general *concept*, for performing this task. That is, if what's doing the work of tracking objects are purely preconceptual, modularized mechanisms, then posting a SPELKE OBJECT concept is unnecessary. Because object files and visual indexes are devices for object tracking, we have no need for the SPELKE OBJECT concept, provided they can do whatever work the SPELKE OBJECT concept was supposed to. Or, as Fodor and Pylyshyn (2015, 116) put it, "since tracking is a reflex, which is to say that it doesn't involve the application of a concept, the concept OBJECT need not come into the process of visual perception *at all*. That's just as well, since, so far at least, no one has been able to provide a plausible account of what the concept OBJECT is."

But how much weight will this argument bear? To see one difficulty with it, notice that the argument used the distinction between perceptual modules and central cognition to drive a wedge between what's preconceptual (internal to a module) and what's conceptual (internal to central cognition). It's not so obvious, though, that these distinctions coincide. Why can't we think of concepts—in particular, the purported SPELKE OBJECT concept—as embedded in a module? Spelke herself believes that the process that detects 3D, bounded, moving entities—her core object system—is encapsulated (Spelke 2000) and "is a cognitive module in Fodor's sense" (Spelke 2003, 31), but at the same time makes use of "representations of objects as enduring bodies over occlusion" (Spelke 2000, 1234).[8] So even if we agree that perceptual modules mark off perception from cognition, the SPELKE OBJECT concept could inhabit the perceptual side of the tracks. Although modules can't draw on concepts from long-term memory (since they're encapsulated), they might be able to incorporate general concepts of their own, at least if we understand general concepts as mental representations that bring entities together under a single heading. By analogy, the syntactic categories—noun phrase, verb phrase, and so on—that appear in classical Fodorian language-processing modules would seem to be concepts in exactly this sense. Likewise, Fodor and Pylyshyn (2015, 123) assert that "processes inside the visual module would allow it to look up a particular shape type $\mathfrak{L}(x)$ in a catalog of shape types," which seems a fancy way of saying that the module sorts things into categories with similar shapes. So why not take SPELKE OBJECT as a "modularized type" or concept?

As it stands, then, we think that the argument for denying that there is a SPELKE OBJECT concept is uncertain. On the one hand, antipsychosortalists might be able to find a better way to distinguish preconceptual from conceptual representations (i.e., better than the module/central cognition split we've been considering) and then show that preconceptual representations can do the work that SPELKE OBJECT concepts do. On the other hand, psychosortalists could respond by denying either the distinction or the sufficiency of the preconceptual representations for tracking 3D, bounded, moving objects. Whether or not these responses will ultimately succeed is not something that we will address here (for reasons of space). Rather, our

[8] For further discussion of Spelke's views on this topic, see Rips (2017).

primary aim was just to draw out some of the commitments that psychosortalists will need to make in order to maintain that SPELKE OBJECT is indeed a concept.

However, even if psychosortalists are able to meet these commitments, they are not yet out of the woods. For in addition to showing that there is a SPELKE OBJECT concept, psychosortalists must also show that this concept is a sortal. In the next subsection we will argue that there is good reason to think that it is not.

11.4.2 Is SPELKE OBJECT a Sortal?

Even if we grant that Spelke objects are associated with a general concept, why should we think of it as a *sortal* concept—a concept that comes along with criteria of individuation and identity? In discussing basic-level concepts like DUCK and CUP, we noted that these concepts could still play a role in identifying things, even though they don't possess criteria for identity in the sense of (11.1). Seeing a duck disappear behind a screen and a cup emerge from the other side is a good reason to think the screen covered two different items. But that's because we can infer from our ordinary empirical knowledge that a duck turning into a cup is awfully unlikely. Our aim in this subsection is to persuade you that the same is true of SPELKE OBJECT. Any role it plays in individuation or identity is heuristic rather than necessary—that is, it doesn't have to provide criteria of individuation and identity to perform these functions.

If SPELKE OBJECT is a sortal concept, then the criterion of identity it imposes on objects is presumably occupying the same spatiotemporal path. This seems to be the criterion that Xu and Carey (1996) and Xu (2005) appeal to in explaining their findings. Showing infants the duck and the cup together before the start of the trial provides them with the fact that these two objects occupy different spatial positions and must therefore be different objects. *Not* showing them the preview leaves them thinking that the objects are on the same trajectory as they go back and forth behind the screen and are therefore the same object.[9]

Similarly, in a well-known experiment by Spelke, Kestenbaum, Simons, and Wein (1995), four-month-old infants sat in front of a stage, looking at two screens separated by empty space, as shown in figure 11.3. In the one-object condition, the infants saw a cylinder begin at one side of the stage (let's say the left side), move behind the left-hand screen, move between the left-hand and right-hand screens,

[9] Fodor and Pylyshyn (2015, 37) point to a circularity problem for Spelke objects: "We've tried hard, but without success, to convince ourselves that the concept of a TRAJECTORY is more basic than the concept of a PHYSICAL OBJECT; isn't it true by definition that a trajectory is the path of an (actual or possible) physical object through space?" In what follows, we take sameness of trajectory as the criterion of identity for Spelke objects, rather than as part of the definition of SPELKE OBJECT itself. But we note that Fodor and Pylyshyn's problem could still arise if movement-as-a-whole, which *is* part of the definition, presupposes movement of an object. Fodor and Pylyshyn's criticism is made very briefly (we've just quoted the entire argument), and we won't attempt to pursue it here.

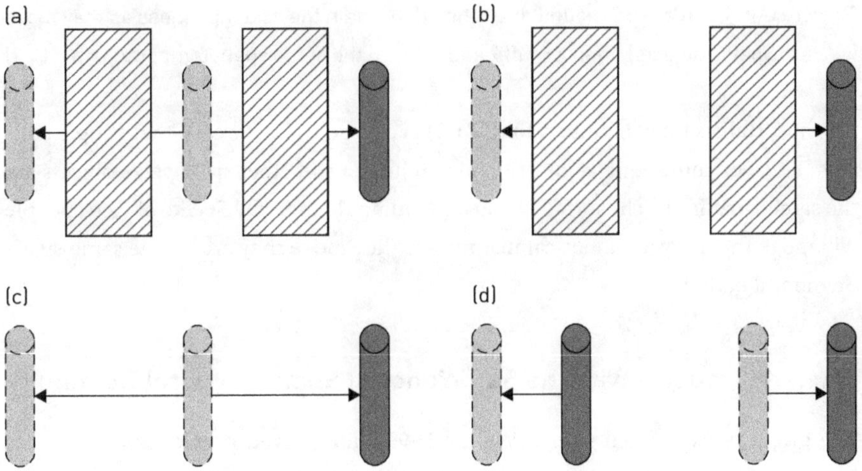

FIGURE 11.3 The events in Spelke et al.'s (1995) study of object discrimination. See the text for an explanation of these events.

move behind the right-hand screen, emerge at the far right, and then reverse direction (see figure 11.3a). This action was repeated for as long as the baby cared to look. In the *two-object* condition, everything was the same, except that the cylinder never appeared between the screens (see figure 11.3b). (The babies saw actual 3D objects, not the 2D images of the figure.) As an adult, you would probably infer during these habituation trials that one object was going back and forth behind the screens in the one-object condition, but two objects were going back and forth in the two-object condition, and the infants appeared to come to the same conclusion. Spelke and colleagues tested them in one of two kinds of test trials, both with the screens removed. In one test trial, the infants saw a single cylinder going back and forth, as in figure 11.3c. In a second test trial, the babies saw two objects, as in figure 11.3d. If the infants have adapted to the idea of one object going by during the habituation trials, they should be surprised to see two objects during the test trials, and they should look longer during the two-object test trials than during the one-object test trials. If the infants have adapted to the idea of two objects going back and forth, they should be surprised to see only one object during the test trials, and they should look longer during the one-object than during the two-object test trials. In other words, if the babies have used continuity of the spatiotemporal paths of the objects during habituation to make the right inference about how many objects are in play, they should be surprised if the *wrong* number of objects turns up in the test trials. And that's exactly what Spelke and company found: On the test trials, babies looked longer if the wrong number of objects appeared than if the correct number of objects appeared. So it looks as if four-month-olds can determine how many objects are around, just from the spatiotemporal paths of the items.

These results suggest that if SPELKE OBJECT is a sortal concept, its criterion of identity is spatiotemporal. Filling in the terms of the schema in (11.1) gives us:

For all *x* and *y* that are 3D, bounded entities that retain their boundedness as they move (i.e., are Spelke objects), then $x = y$ iff *x* and *y* are on the same spatiotemporal path. (11.2)

Psychosortalists must maintain that (11.2) is psychologically real—that people consult (11.2), or some equivalent principle, in judging whether two Spelke objects are the same. That is, psychosortalists must maintain that for any Spelke objects, people will judge that they are identical just in case they judge they are on the same spatiotemporal path.[10]

11.4.2.1 Troubles with the Sufficiency of Spatiotemporal Continuity

The property experiments (e.g., Wilcox 1999) that proved problematic for basic-level sortals also provide counterexamples to (11.2). Consider an experiment like that of figure 11.2, but in which the two spheres are replaced by a sphere with dots and a same-size sphere with stripes. At 7.5 months, infants are able to succeed at this task, looking longer in the narrow-screen condition than in the wide-screen condition. They infer from the markings that the spheres are distinct and so unable to fit behind the narrow screen at the same time. However, if the spheres have identical markings (e.g., both are striped), then infants look an equivalent amount of time in the two conditions. According Carey and Xu (1999), infants of this age don't have lower-level sortals to individuate the two spheres, and this means the only sortal available to them is that for Spelke objects. But if the criterion of individuation for Spelke objects is the one in (11.2), it fails to explain the infants' performance. In the narrow-screen condition, the spatiotemporal path is the same when the two spheres have different markings as when they have the same markings. So the spatiotemporal path can't explain why infants differentiate the spheres in the former case but not in the latter one.

A simple variation of this experiment is likely to yield the same conclusion for adults: Imagine, for example, two amorphous but identically shaped objects that you can't identify as members of any basic-level category (perhaps because they are too far away). If they move on the same figure 11.2 pathway, but one is dotted and the other is striped, you may be surprised when they both seem to be behind the narrow

[10] One qualification to (11.2) is that it fails in the forward direction for animate objects (Kuhlmeier et al. 2004). If *x* and *y* are people who have different trajectories, they aren't necessarily counted as distinct for the situation in figure 11.3b. Spelke and Kinzler (2007) think these cases are handled by a different core module for agents. But this means we should supplement (11.2) with a further restriction on Spelke objects, perhaps:

> For all *x* and *y* that are 3D, bounded entities that retain their boundedness as they move and *are not agents*, $x = y$ iff *x* and *y* have the same continuous spatiotemporal path. (11.2')

However, since this qualification does not affect our argument, we'll stick with the simpler formulation in (11.2).

screen at the same moment. Again, the criterion of identity for SPELKE OBJECT in (11.2) can't explain your reaction.

We take Wilcox's (1999) result to show that the Spelke-object concept isn't a sortal. Moving spheres qualify as Spelke objects in anyone's book; so the schema in (11.2) clearly applies to them. But (11.2) incorrectly predicts that infants should respond in the same way when confronted with spheres having different markings as with spheres having same markings. More specifically, even if people judge that the spheres have the same spatiotemporal path, they do not always think that those spheres are identical. And if this is the case, then the sufficiency claim in (11.2) is false. Thus, these results suggest that SPELKE OBJECT does not provide a criterion of identity and that it is, therefore, not a sortal concept.

Perhaps psychosortalists could respond by denying that "same spatiotemporal path" is the right criterion of identity for Spelke objects. But because this criterion is the one investigators have used to explain many earlier results—for example, the Spelke et al. (1995) finding in figure 11.3—we don't see that they can ignore it here. Alternatively, psychosortalists could try out the idea that infants believe the dotted and the striped spheres are on different spatiotemporal paths since the narrow screen interrupts infants' view of it. If so, then the infants could conclude from (11.2) that two spheres are in play. But this would again interfere with the explanations for infants' performance when the two spheres have the same markings. Here, infants seem to think only one sphere is in view.

A more reasonable explanation for Wilcox's (1999) finding, we think, is that infants believe that a single sphere is unlikely to pick up two different decorative patterns when it is briefly out of sight. So there are probably two spheres rather than one at hand. This explanation goes along with the suggestion by Carey and Xu (1999) and Xu (2005) that infants can sometimes use property information to determine the number of objects within a scene. As we mentioned in section 11.3, though, if infants can use properties to differentiate things, then there is no logical pressure to suppose that sortal concepts are required to explain these acts of individuation. Section 11.3 argued that basic-level concepts don't have to be sortals in order to explain why older infants think toy ducks don't change into cups while behind a screen. The present section suggests that the evidence for sortals is no stronger in the case of Spelke objects.

11.4.2.2 Troubles with Necessity of Spatiotemporal Continuity

Wilcox's (1999) study shows that the sufficiency half of (11.2) is incorrect: Sharing the same spatiotemporal path does not entail identity. But we also doubt that the necessity half is correct. If two objects are identical, do they necessarily share the same spatial-temporal pathway? A number of philosophers have offered hypothetical counterexamples directed against this possibility (e.g., Armstrong 1980; Hirsch 1982; Nozick 1981; Shoemaker 1979; see Rips et al. 2006, for discussion), but let's consider a more concrete case for which we can provide some experimental backing (Leonard and Rips 2015, Experiment 3).

Suppose that on Monday there is an assembled table, which we'll nickname "Timmy," in the dining room. On Tuesday the table is taken apart and the pieces are moved to the living room. On Wednesday, the pieces are resembled. Here are the key questions: Does Timmy exist on Tuesday and, if so, is Timmy a Spelke object? Does Timmy exist on Wednesday and, if so, is Timmy a Spelke object? With respect to the former, most participants said that Timmy did exist on Tuesday but was not a Spelke object (defined for them as "a bound, coherent, three-dimensional physical entity that moves as a whole"). And with respect to the latter, most participants said that Timmy did exist on Wednesday and was a Spelke object. So participants agreed that Timmy exists from Monday to Wednesday, but goes from being a Spelke object on Monday to a non-Spelke object on Tuesday, and then returns to being a Spelke object on Wednesday. (Of course, the participants in this experiment were adults and so had available the basic-level sortal *table*. But most participants also said that Timmy was not a table on Tuesday and went back to being a table on Wednesday. In other words, their judgments about whether Timmy was a table followed their judgments about whether Timmy was a Spelke object.)

Depending on what psychosortalists want to say about whether there is a spatiotemporal path that connects Timmy on Monday to Timmy on Wednesday, they seem to be faced with a dilemma. Psychosortalists can either say that people judge there to be a spatiotemporal path that connects the Monday Timmy to the Wednesday Timmy, or else they can deny that people make this judgment. If they deny that people make this judgment, then we have a counterexample to the necessity claim in (11.2): people judge that Monday Timmy and Wednesday Timmy are identical even though they do not judge that there is a spatiotemporal path that connects them.

But if psychosortalists say that people judge that there is a spatiotemporal path that connects the Monday Timmy and the Wednesday Timmy, then it is hard to see how they can explain the data in psychosortalist terms. That is, it is hard to see how they can appeal to the SPELKE OBJECT sortal to explain how Timmy can exist on Monday, Tuesday, and Wednesday (which they must maintain, given that on this horn of the dilemma we are supposing that people judge that there is a spatiotemporal path that connects Monday Timmy to Wednesday Timmy). To see why, notice that if people judge that there is a spatiotemporal path that connects Monday Timmy to Wednesday Timmy, then they must judge that this path also connects the Monday Timmy to the Tuesday Timmy. But people do not think that Tuesday Timmy is a Spelke object (rather, they think it is just a collection of unbounded table pieces). Thus, (11.2) does not apply, since (11.2) requires both x (Monday Timmy) and y (Tuesday Timmy) to be Spelke objects.

So (11.2) is silent about whether people will judge that x and y are identical, and psychosortalism is therefore too weak to explain all the relevant data.

Of course, as in the case of the concepts of ducks and cups, a SPELKE-OBJECT concept *can* provide information that allows people to distinguish objects. The discontinuous path of the items in figure 11.3b is likely to be a clue that two cylinders are present, in line with the notion that Spelke objects usually don't have discontinuous trajectories. What we are denying is that this kind of information amounts to a criterion for identity or individuation as in (11.2). So although a Spelke-object concept doesn't provide criteria (i.e., necessary and sufficient conditions) for identity and individuation, it can sometimes provide evidence in favor of identifying or individuating things. That's fine with us. Our quarrel is with sortal concepts, not with the SPELKE OBJECT concept.

Taken together, then, the arguments from sections 11.3 and 11.4 suggest that it is unnecessary to posit either basic-level sortal concepts or the SPELKE OBJECT sortal concept to accommodate and explain the relevant data. Thus, we take these arguments to undermine the main motivation for psychosortalism.

11.5 CONCLUSION

Metaphysics can inform cognitive science by providing possible models of how people think about things. In particular, metaphysics can suggest cognitive theories of object individuation and identity, part-whole constitution relations, relations between material substances and objects, and many others (see Rips and Hespos 2015, for a review of some connections of this kind). We wouldn't expect, however, that every metaphysical theory would translate into a successful cognitive one, and we have tried to make the case here that sortal theories are among metaphysics's less successful export commodities.

The initial reason why psychologists took up sortal theories was the conviction that singular thought about an individual was impossible unless people possessed a mental representation for a sortal that included that individual. We suspect this was a wrong turn at the start. As far as we can see, cognitive theories of reference to individuals only require a representation of the individual and the right kind of causal connection between the individual and the representation; they do not require the additional representation of a sortal to which the individual belongs. We can successfully refer to Fido (at least according to externalist theories) provided we have a Fido concept that's on the receiving end of a causal arrow from Fido. But we don't also need the concept DOG to mediate this connection. Indeterminacy arguments can suggest otherwise. What guarantees that our concept FIDO picks out Fido and not summed Fido parts? Doesn't *something* have to carve out Fido in a way that differentiates these possibilities? Maybe. But it's hard to see how this something could be another mental representation—the concept DOG—which presumably has much the same problem with indeterminacy of reference.

Despite the fact that sortal concepts aren't necessary for singular thought, some psychologists continue to assume that they *are* necessary and continue to suppose that when people correctly distinguish objects, sortal concepts are responsible. They interpret the results of experiments, such as those illustrated in figures 11.1 and 11.3, as evidence that children possess sortal concepts, either at the basic level (e.g., DOG) or at a higher level (that of SPELKE OBJECT). However, the findings from further studies (e.g., Wilcox and Baillargeon 1998; Wilcox 1999; Xu and Baker 2005) undercut this interpretation. The essential feature of sortal concepts is that they provide criteria for identity and individuation in the sense of (11.1) or (11.2). But children's use of property information to distinguish objects in tasks like that of figure 11.2 violates these criteria. Infants distinguish spheres with different patterns of markings, even though they are Spelke objects and are on the same spatiotemporal path, contrary to (11.2). Because psychosortalists believe that infants of this age do not possess sortal concepts other than SPELKE OBJECT, sortal concepts are unable to explain infants' ability to discriminate and identify things.

Here's our final list of disclaimers: We don't deny that basic-level concepts and the SPELKE OBJECT concept are relevant to identity and individuation or that they have distinctive and important roles to play in cognition. And we also don't deny that the psychological experiments motivated by the idea of sortal concepts have provided substantive information about the nature of infants' thinking. What we *do* deny is that there is any reason to believe that people have a distinguished class of concepts (basic-level ones, SPELKE OBJECT) that supply (necessary and sufficient) criteria for resolving questions of identity and individuation.

REFERENCES

Anderson, E. M., Hespos, S. J., and Rips, L. J. (2018). Five-month-old infants have expectations for the accumulation of nonsolid substances. *Cognition* 175: 1–10.

Armstrong, D. M. (1980). Identity through time. In P. van Inwagen, ed., *Time and Cause.* Dordrecht: Reidel, 67–78.

Ayers, M. (1997). Is physical object a sortal concept? A reply to Xu. *Mind & Language* 12: 393–405. doi: 10.1111/1468-0017.00053.

Baldwin, D. A. (1989). Priorities in children's expectations about object label reference: Form over color. *Child Development* 60: 1291–306. doi: 10.2307/1130921.

Blok, S. V., Newman, G. E., and Rips, L. J. (2005). Individuals and their concepts. In W.-K. Ahn, R. L. Goldstone, B. C. Love, A. B. Markman, and P. Wolff, eds., *Categorization inside and outside the Laboratory.* Washington, DC: American Psychological Association, 127–49.

Blok, S. V., Newman, G. E., and Rips, L. J. (2007). Out of sorts? Some remedies for theories of object concepts: A reply to Rhemtulla and Xu (2007). *Psychological Review* 114: 1096–102. doi: 10.1037/0033-295x.114.4.1096.

Burge, T. (1979). Individualism and the mental. *Midwest Studies in Philosophy* 4: 73–121.

Campbell, J. (2006). Sortals and the binding problem. In F. McBride, ed., *Identity and Modality.* Oxford: Oxford University Press, 203–18.

Carey, S., and Xu, F. (1999). Sortals and kinds: An appreciation of John Macnamara. In R. Jackendoff, P. Bloom, and K. Wynn, eds., *Language, Logic, and Concepts: Essays in Memory of John Macnamara.* Cambridge, MA: MIT Press, 311–35.

Casati, R. (2004). Is the object concept formal? *Dialectica* 58: 383–94.

Cordes, S., and Brannon, E. M. (2009). The relative salience of discrete and continuous quantity in young infants. *Developmental Science* 12: 453–63. doi: 10.1111/j.1467-7687.2008.00781.x.

Feigenson, L., and Halberda, J. (2004). Infants chunk object arrays into sets of individuals. *Cognition* 91: 173–90. doi: 10.1016/j.cognition.2003.09.003.

Feldman, F. (1973). Sortal predicates. *Noûs* 7: 268–82.

Fodor, J. A. (1983). *Modularity of Mind: An Essay on Faculty Psychology*. Cambridge, MA: MIT Press.

Fodor, J. A., and Pylyshyn, Z. W. (2015). *Minds without Meanings: An Essay on the Content of Concepts*. Cambridge, MA: MIT Press.

Frege, G. ([1884] 1968). *The Foundations of Arithmetic: A Logico-Mathematical Enquiry into the Concept of Number*. 2nd ed. Evanston, IL: Northwestern University Press.

Gelman, S. A. (2003). *The Essential Child: Origins of Essentialism in Everyday Thought*. Oxford: Oxford University Press.

Goodman, R. (2012). Why and how not to be a sortalist about thought. *Philosophical Perspectives* 26: 77–112.

Grandy, R. E. (2014). Sortals. In E. N. Zalta, ed., *The Stanford Encyclopedia of Philosophy*. Spring 2014 ed.

Green, E. J. (2018). What do object files pick out? *Philosophy of Science* 85: 177–200.

Gupta, A. (1980). *The Logic of Common Nouns: An Investigation in Quantified Modal Logic*. New Haven: Yale University Press.

Hespos, S. J., Ferry, A. L., Anderson, E. M., Hollenbeck, E. N., and Rips, L. J. (2016). Five-month-old infants have general knowledge of how nonsolid substances behave and interact. *Psychological Science* 27: 244–56. doi: 10.1177/0956797615617897.

Hirsch, E. (1982). *The Concept of Identity*. Oxford: Oxford University Press.

Hirsch, E. (1997). Basic objects: A reply to Xu. *Mind & Language* 12: 406–12. doi: 10.1111/1468-0017.00054.

Kahneman, D., Treisman, A., and Gibbs, B. J. (1992). The reviewing of object files: Object-specific integration of information. *Cognitive Psychology* 24: 175–219. doi: 10.1016/0010-0285(92)90007-o.

Keil, F. C. (1989). *Concepts, Kinds, and Cognitive Development*. Cambridge, MA: MIT Press.

Kripke, S. A. (1980). *Naming and Necessity*. Cambridge, MA: Harvard University Press.

Kripke, S. A. (1981). Two lectures on time and identity. *Patten Foundation Lectures* [sound recording]. December 1–2. Indiana University, Bloomington.

Kuhlmeier, V. A., Bloom, P., and Wynn, K. (2004). Do 5-month-old infants see humans as material objects? *Cognition* 94: 95–103. doi: 10.1016/j.cognition.2004.02.007.

Leonard, N., and Rips, L. J. (2015). Identifying and counting objects: The role of sortal concepts. *Cognition* 145: 89–103.

Levine, S. C., and Baillargeon, R. (2016). Different faces of language in numerical development: Exact number and individuation. In D. Barner and A. S. Baron, eds., *Core Knowledge and Conceptual Change*. Oxford: Oxford University Press, 127–50.

Lowe, E. J. (1998). *The Possibility of Metaphysics: Substance, Identity, and Time*. Oxford: Oxford University Press.

Macnamara, J. (1986). *A Border Dispute: The Place of Logic in Psychology*. Cambridge, MA: MIT Press.

Medin, D. L. (1989). Concepts and conceptual structure. *American Psychologist* 44: 1469–81. doi: 10.1037/0003-066x.44.12.1469.

Merricks, T. (1998). There are no criteria of identity over time. *Noûs* 32: 106–24.

Nozick, R. (1981). *Philosophical Explanations*. Cambridge, MA: Harvard University Press.

Putnam, H. (1975). The meaning of meaning. In *Mind, Language and Reality*. Cambridge: Cambridge University Press, 215–71.

Pylyshyn, Z. W. (2001). Visual indexes, preconceptual objects, and situated vision. *Cognition* 80: 127–58. doi: 10.1016/s0010-0277(00)00156-6.

Rips, L. J. (1989). Similarity, typicality, and categorization. In S. Vosniadou and A. Ortony, eds., *Similarity and Analogical Reasoning*. New York: Cambridge University Press, 21–59.

Rips, L. J. (2017). Core cognition and its aftermath. *Philosophical Topics* 45: 157–79.

Rips, L. J., Blok, S., and Newman, G. (2006). Tracing the identity of objects. *Psychological Review* 113: 1–30. doi: 10.1037/0033-295x.113.1.1.

Rips, L. J., and Hespos, S. J. (2015). Divisions of the physical world: Concepts of objects and substances. *Psychological Bulletin* 141: 786–811. doi: 10.1037/bul0000011.

Rosch, E., Mervis, C. B., Gray, W. D., Johnson, D. M., and Boyes-Braem, P. (1976). Basic objects in natural categories. *Cognitive Psychology* 8: 382–439. doi: 10.1016/0010-0285(76)90013-x.

Sarnecki, J. E. (2008). Sortals for dummies. *Erkenntnis* 69: 145–64. doi: 10.1007/s10670-007-9094-6.

Shoemaker, S. (1979). Identity, properties, and causality. *Midwest Studies in Philosophy* 4: 321–42.

Soja, N. N., Carey, S., and Spelke, E. S. (1991). Ontological categories guide young children's inductions of word meaning: Object terms and substance terms. *Cognition* 38: 179–211. doi: 10.1016/0010-0277(91)90051-5.

Spelke, E. S. (1990). Principles of object perception. *Cognitive Science* 14: 29–56. doi: 10.1207/s15516709cog1401_3,

Spelke, E. S. (2000). Core knowledge. *American Psychologist* 55: 1233–43. doi: 10.1037/0003-066x.55.11.1233.

Spelke, E. S. (2003). Core knowledge. In N. Kanwisher and J. Duncan, eds., *Functional Neuroimaging of Visual Cognition*. Oxford: Oxford University Press, 29–55.

Spelke, E. S., Kestenbaum, R., Simons, D. J., and Wein, D. (1995). Spatiotemporal continuity, smoothness of motion and object identity in infancy. *British Journal of Developmental Psychology* 13: 113–42. doi: 10.1111/j.2044-835X.1995.tb00669.x.

Spelke, E. S., and Kinzler, K. D. (2007). Core knowledge. *Developmental Science* 10: 89–96. doi: 10.1111/j.1467-7687.2007.00569.x.

Wiggins, D. (1997). Sortal concepts: A reply to Xu. *Mind & Language* 12: 413–21. doi: 10.1111/1468-0017.00055.

Wiggins, D. (2001). *Sameness and Substance Renewed*. Cambridge: Cambridge University Press.

Wilcox, T. (1999). Object individuation: Infants' use of shape, size, pattern, and color. *Cognition* 72: 125–66.

Wilcox, T., and Baillargeon, R. (1998). Object individuation in infancy: The use of featural information in reasoning about occlusion events. *Cognitive Psychology* 37: 97–155. doi: 10.1006/cogp.1998.0690.

Williamson, T. (1990). *Identity and Discrimination*. Oxford: Blackwell.

Wynn, K., Bloom, P., and Chiang, W.-C. (2002). Enumeration of collective entities by 5-month-old infants. *Cognition* 83: B55–B62. doi: 10.1016/s0010-0277(02)00008-2.

Xu, F. (1997). From Lot's wife to a pillar of salt: Evidence that physical object is a sortal concept. *Mind & Language* 12: 365–92. doi: 10.1111/1468-0017.00052.

Xu, F. (2005). Categories, kinds, and object individuation in infancy. In L. Gershkoff-Stowe and D. H. Rakison, eds., *Building Object Categories in Developmental Time*. Mahwah, NJ: Erlbaum, 63–89.

Xu, F. (2007). Sortal concepts, object individuation, and language. *Trends in Cognitive Sciences* 11: 400–406. doi: 10.1016/j.tics.2007.08.002.

Xu, F., and Baker, A. (2005). Object individuation in 10-month-old infants using a simplified manual search method. *Journal of Cognition and Development* 6: 307–23. doi: 10.1207/s15327647jcd0603_1.

Xu, F, and Carey, S. (1996). Infants' metaphysics: The case of numerical identity. *Cognitive Psychology* 30: 111–53. doi: 10.1006/cogp.1996.0005.

Xu, F., Carey, S., and Quint, N. (2004). The emergence of kind-based object individuation in infancy. *Cognitive Psychology* 49: 155–90. doi: 10.1016/j.cogpsych.2004.01.001.

12 What the Study of Psychological Essentialism May Reveal about the Natural World

Susan A. Gelman

I am a psychologist who studies human concepts, with particular interest in early childhood. My research program is aimed at uncovering principles of the mind: the mental representations and developmental processes that underlie our categorization of experience. Thus, for example, I step into raging debates in the field of cognitive development regarding conceptual starting points and directions of conceptual change. In my study of concepts of natural kinds, such as dogs, women, or gold, I have argued against the standard (empiricist) developmental story, that concepts are constructed "bottom-up" from sensory and perceptual primitives. To the contrary, I have argued that from the start, children expect kinds to capture nonobvious similarities, heavily weight causal features, and think about abstractions. Indeed, doing otherwise is (I argue) devilishly difficult for children—or adults.

Although my focus is on the mind, concepts lie at the intersection of world and mind: they are mental representations of a world outside the mind. As such, they inevitably raise the question of how human concepts compare to metaphysical reality. As a psychologist, I have approached this question by starting with assumptions about reality (based on scientific analyses of the biological, chemical, and social world) and then asking how human concepts match or mismatch with what science tells us. In this vein, I have argued for both early conceptual sophistication (preschoolers—like adults—recognize that the world goes beyond surface appearances) and early cognitive biases (preschoolers—like adults—hold to an essentialist view of the natural world that oversimplifies and distorts reality). We can think of the first as supporting the "child as scientist" metaphor (children have a startlingly clear ability to detect important patterns in experience, presumably as an evolved capacity), and the second as supporting a "scientist as child" metaphor (abiding human distortions that are most evident in early childhood but never completely disappear, and continue to influence the language and concepts of adults).

In this chapter, I will consider whether flipping the perspective may yield insights into reality. How might the findings of human conceptual biases be revealing about metaphysics? My argument, in brief, is two-part. First, the biases and distortions that characterize psychological essentialism, alluded to previously, constitute a deflationary account of essentialism as a theory of how the world is structured. Although essentialism has a long and distinguished (albeit controversial) history as a framework for thinking about reality, it makes assumptions that fail to accord with scientific discoveries post-Darwin. And second, given the reach and persistence of essentialist biases in human reasoning, I suggest that even scientific categories may fall prey to these same errors. Basically, scientists are humans just like the rest of us, and prone to some of the same kinds of reasoning biases. Because all of science requires human interpretation, a study of the human theory-making device (i.e., the mind) is critical. Scrutinizing reasoning biases does not directly reveal the true structure of the world, but does suggest where we need to be cautious or skeptical. In the final section of the chapter, I draw some broader conceptual and methodological conclusions.

12.1 PSYCHOLOGICAL ESSENTIALISM

Psychological essentialism is an intuitive belief that certain categories, such as *tiger*, *gold*, or *woman*, share not just observable features, but also a deeper, nonobvious reality, that there exists some inherent, internal, immutable substance or quality (the "essence") that causes the characteristics that category members share, and that words in ordinary language reveal these natural kinds (Gelman 2003; Medin 1989). Thus, tigers share more than a certain size, gait, striped fur, and ferocity, but also internal parts, brain structure, temperament, instincts, as well as an innate, unchanging tiger "essence." That hidden essence might be presumed to be genes, or blood, or something as-yet-unknown.

Psychological essentialism is a realist assumption about categories. Just as an individual tiger exists (it can be seen, measured, weighed; its effects on the world are as real as anything we perceive), so too the category of "tigers" is thought to exist as an entity "out there" for us to pick up and analyze. In contrast to arbitrary categories invented for the sake of human convenience (e.g., the distinction between "jumbo olive" and "colossal olive"), a distinction between "lion" and "tiger" is nonarbitrary, existing outside of human existence.

There is extensive and converging evidence that essentialist beliefs are early emerging, persist in adults, and are found across widely varying cultures. Children from preschool age treat categories as possessing an underlying reality, as having rich inductive potential, and as being innately determined, and they treat category boundaries as sharp and immutable. Tasks assessing these beliefs include children's judgments regarding animal transformations, switched-at-birth scenarios, category boundaries, and growth; as well as their categorizations, inductive inferences, and memory. For example, infants and young children use categories to guide inductive inferences about novel properties. Upon learning that an atypical exemplar is a member of a kind (e.g., that a penguin is a bird), children and adults draw novel inferences from typical instances of the kind to the atypical member—privileging internal features over superficial dissimilarity. Prentice and Miller (2007, 129) summarize a rich body of evidence as following: "a wolf remains a wolf even if it is wearing sheep's clothing . . . , even if a doctor performs an operation that makes it look like a sheep . . . , and even if it eats something that turns it into an object resembling a sheep. . . . Moreover, a wolf will develop wolflike characteristics even if it grows up in a community of sheep." These beliefs are robustly evident long before children have an understanding of biological processes such as genes or reproduction.

I do not dispute that there are discontinuities in the causal and featural structure of the natural world, that evolution has provided us with mental and perceptual tools to detect useful regularities that aid in survival, and that certain categorical distinctions are more motivated than others. Indeed, science rests in part on this foundational expectation that the world has layers of hidden structure that remain for us to discover. However, concepts of natural kinds do not map neatly or unproblematically onto "true" kinds. Below I summarize a variety of evidence that an essentialist construal of natural kinds oversimplifies the complexities of the natural world, and in so doing introduces error, bias, and distortion of a systematic variety (see also Allport 1954; Gelman 2003; James [1890] 1983; Leslie 2013).

12.2 A COMPARATIVE APPROACH: THE LOGIC

My goal here is to characterize the distortions entailed in psychological essentialism, using a comparative approach, that is, by considering a variety of contrasting representations: those held by children vs. adults, those held by novices vs. experts, those underlying language production vs. interpretation, and those involving how information is presented vs. recalled. It may at first seem hopeless to try to get at reality by looking at the ways in which human concepts depart from reality. However, if one can gain purchase on the nature of the distortions, then at minimum this tells us which views of reality we should view with suspicion, and through a reversed process, one can at least begin to make inferences about the reality. By analogy, if I know that I am wearing rose-tinted lenses, then I can at least question whether a seemingly pink object is indeed pink rather than uncritically assuming it to be so. Further, I can potentially gain insight by "subtracting" the color of the lens from the perceived color of the object (e.g., an object that looks orange to the wearer of rose-tinted glasses may be yellow). Some approaches to surveying people about sensitive issues (such as whether they engage in illegal drug use) even *add* bias in order to get closer to the true representation (e.g., having people first flip a coin to decide whether to answer truthfully; Dalal and Hakel 2016).

A comparative approach is useful for detecting distortions and reasoning errors. The field of cognitive developmental psychology can be viewed in part as an attempt to see the mind's logic that distorts reality. This was the foundation of Piaget's theory of epistemology, in which examining children's thinking at progressively sophisticated stages revealed a broad set of confusions and misapprehensions. According to Piaget, children progress through stages in which their thought processes undergo qualitative changes, at any given point fitting within a framework that obeys its own logic but is limited and distorted in measurable and systematic ways. Thus, for example, four-year-olds fail conservation-of-liquid tasks, insisting that pouring water from a squat glass to a skinny glass increases its amount. Strikingly, giving children the right answers doesn't budge their reasoning—a child who provides the wrong answer is persuaded neither by explicit correction, nor even by the sorts of evidence that would sway an adult (pouring the water back, explaining that height and width are compensatory, etc.). Piaget's stage theory has been disconfirmed, in part due to evidence of children's much greater conceptual sophistication than Piaget imagined (Gelman and Baillargeon 1983), but the point remains that children hold stable misconceptions that are remarkably impervious to evidence or correction.

The idea that one's perspective on the world may be not just ignorant but actively and systematically misconstrued—that it fits within a framework theory that has a logic all its own—is an idea that I would suggest applies to adults as well as children. We construct naive "theories" about the world, with ontological commitments, causal principles, and predictive power (Carey 2009; Gopnik and Wellman 2012; Wellman and Gelman 1998). These theories themselves undergo change over development and with new evidence, but they are resistant to change, and indeed remnants of old theories may persist when people are under cognitive load, for example when asked to give split-second responses rather than giving their full attention (Kelemen et al. 2013).

I turn now to the reasoning gaps that correspond to psychological essentialism. I focus here on three: underestimating variability within a kind, overestimating the importance of category boundaries, and assuming a causal essence shared among members of a kind.

12.3 UNDERESTIMATING VARIABILITY WITHIN A KIND

Historians of science have argued that essentialism was one of the major barriers to developing a theory of evolution, in part because essentialism leads to ignoring or undervaluing variation within a species (e.g., Mayr 1982; see Gelman and Rhodes 2010 for review). If members of a species share a hidden, causal essence, then any observed variation across category members is unimportant. Indeed, much evidence demonstrates that variation is among the most difficult concepts in evolutionary theory for children and adults to grasp. Problems involve underestimating variability at multiple levels: variability of observable traits (features, behaviors), variability of unobservable traits (internal parts, DNA), the functional significance of variation (that variation may affect survival and is not merely superficial), and the source of variation (that much variation is heritable) (Anderson et al. 2002). Understanding actual or potential variability is particularly poor in young children, with elementary school children typically reporting that a category member cannot be born without a typical property (e.g., all giraffes have spotted coats, and a giraffe could not be born with a different kind of coat) (Shtulman and Schulz 2008). Although adults are less prone to this error than children, they nonetheless often show the same pattern.

Interestingly, Shtulman and Schulz (2008) find that at all ages (children as well as adults), internal features (e.g., have green blood) are assumed to be invariant more than behavioral features (e.g., make chirping sounds)—consistent with the essentialist view that visible changes are only superficial, and belie a more constant essence. Thus, learning about within-category variation may be insufficient to counter such essentialist beliefs, even resulting in distortions of genetic science (e.g., a belief that all members of a species have identical gene sequences). Adults also underestimate within-category genetic variability and exaggerate between-group genetic differences (e.g., reporting that members of a given race are always more genetically homogeneous than members of different races; Christensen et al. 2010). In truth, there are no distinct genetic markers that reliably indicate race or ethnicity (Bolnick et al. 2007), and the degree of genetic variability within people of a given race is just as high as the degree of genetic variability across races (Templeton 1998).

Exposure to and learning of evolutionary theory often does not correct this essentialist assumption that members of a kind are fundamentally the same (Anderson et al. 2002; Shtulman and Calabi 2013; Spiegel et al. 2012). For example, in a study of high school and college students (95% of whom accepted the premise that species have changed over time and 69% of whom agreed with the statement "Natural selection is the best explanation for how a species adapts to its environment"), few expected changes over time in the proportion of different variants (the correct "variational" theory), with instead the modal response indicating a belief that featural change over generations entails gradual change of the whole species (an incorrect "transformational" theory) (see figure 12.1) (Shtulman 2006). Adults are more prone to these sorts of errors when placed under speeded conditions (Shtulman and Harrington 2016), regardless of whether they had learned the appropriate scientific theory recently or decades earlier, including professional scientists. These findings support the idea that scientific beliefs do not replace intuitive essentialism but rather they coexist (Legare et al. 2012; Legare and Gelman 2008; Shtulman and Lombrozo 2016).

The roots of this problem can be seen in basic reasoning tasks extending back to early childhood. As noted earlier, children have a strong tendency to use category membership (signaled by same-label) as a basis for inductive inferences,

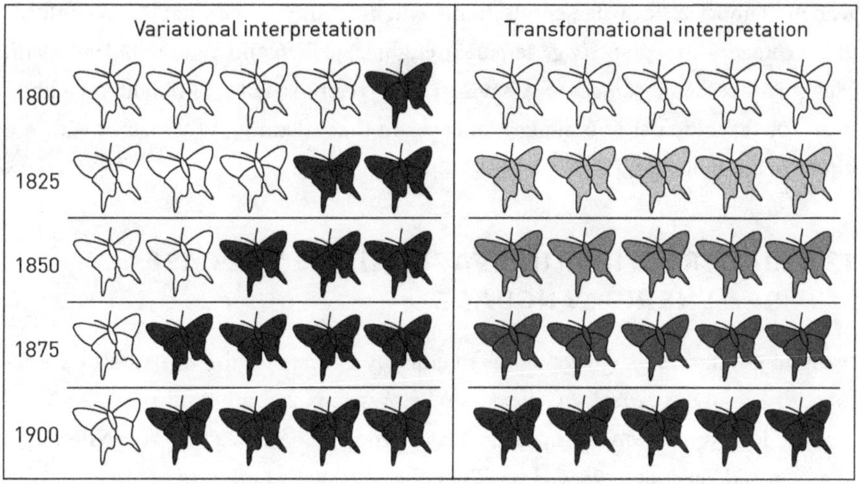

FIGURE 12.1 Variational versus transformational construals of evolutionary change
Source: From Shtulman 2006.

such that category members are assumed to be alike in nonvisible ways, even when they look very different from one another or were reared without a typical environment that might provide role models or teaching opportunities (e.g., a kangaroo raised by goats will still be good at hopping; Gelman and Wellman 1991). Moreover, when asked to reason directly about the distribution of features throughout a novel category, young children often assume that a feature exhibited by a single instance can generalize without exception to the entire kind. In one set of studies (Brandone 2017), children and adults learned about a series of instances within novel categories (e.g., a particular animal called a floom) that had a certain property (e.g., this floom is orange) and were asked how many instances of each category have the relevant property: all, lots, some, or just a few. The youngest children (those below six years of age) showed a tendency to assume within-category homogeneity, often predicting that the property would be present in all members of the category. For many children, the default seems to be the assumption of invariance across the kind. This tendency decreased with age but never entirely disappeared.

Even when children are directly presented with category variation, they have difficulty making use of it to draw inferences about the larger sample. They are as likely to assume that a feature shared by two robins generalizes to birds, as that a feature shared by a robin and an eagle generalizes to birds (Gutheil and Gelman 1997; Rhodes et al. 2008; but see Rhodes et al. 2010). Similarly, when individuating information is deemed by adults to be more informative than category membership, children tend to infer that category information will win out. Thus, when asked to make an inference about gender-stereotypical characteristics, young children overrely on the category for drawing inferences, at the expense of individuating features (Biernat 1991). The tendency of the category to swamp individuating information can also be seen in memory tasks, when children misremember individuals that fail to conform to the category stereotype (e.g., female firefighter) (Liben and Signorella 1993), and even mislabel such instances (e.g., saying "he" to refer to a female firefighter; Gelman et al. 2004). These patterns suggest once again that attending to variation within a kind can be challenging.

12.4 UNDERESTIMATING VARIABILITY: THE CASE OF GENERIC LANGUAGE

Intriguingly, natural language seems to embody a disregard for variation in the semantics and use of *generics*. Generics are expressions that are kind-referring (dogs are four-legged; the lion is a majestic beast; a mosquito carries the West Nile virus) (Carlson and Pelletier 1995; Gelman 2010; Prasada 2000; Leslie 2008). As such, they are distinct from expressions that refer to individuals (those dogs are four-legged; the lion is roaring; a mosquito just bit me). Generics are universally expressed across the world's languages and are frequent in ordinary conversation, including child-directed speech (Gelman et al. 2014). Generics appear to be conceptually intuitive. Thus, from a remarkably early age, children appropriately produce generics ("Does lions crawl?"; "Boys don't ever be ballet dancers") (Gelman et al. 2008), comprehend generics as kind-referring and distinct from specific reference (Graham et al. 2016), and recall whether information was provided using generic or specific language (Gelman and Raman 2003). These patterns have been documented in English, Spanish, Mandarin, and Quechua (Gelman et al. 2016; Mannheim et al. 2011; Perez-Leroux et al. 2004; Tardif et al. 2012).

Children's early capacity to learn generics may at first seem surprising. Generic referents are abstract (one cannot point to a kind, only to instances of a kind), and standard developmental theories assume that children's early concepts are concrete and focused on the "here and now" (Sloutsky et al. 2007). From a morphosyntactic standpoint, generics pose a further puzzle, as there is no single linguistic form or marker to indicate genericity, in any language (Carlson and Pelletier 1995). Instead, each noun phrase that can refer to a kind is also used (in other syntactic contexts) to refer to individuals. Children are sensitive to these semantic and formal aspects of generics by preschool age (Brandone et al. 2012; Cimpian et al. 2011; Gelman and Raman 2003). A further puzzle is that generics are learned before quantifiers (Hollander et al. 2002), even though the semantics of generics are difficult to characterize, whereas the semantics of quantifiers are much more straightforward, using predicate logic. However, these puzzles dissolve if we assume that generics express cognitively default generalizations (Gelman 2010; Leslie 2008; Leslie and Gelman 2012; Sutherland et al. 2015).

One of the most intriguing properties of generics is their treatment of exceptions. Generics imply homogeneity within a kind by glossing over exceptions, permitting a speaker to make a blanket statement without acknowledging the variation that exists. For example, we can say, "Dogs are four-legged" despite three-legged dogs, "Birds lay eggs" although male birds do not lay eggs, and "Mosquitoes carry the West Nile virus" although less than 1% of mosquitoes do so (Leslie 2008). This form of language in a sense treats variation as if it does not exist. Preschool children as well as adults endorse generics of this nature (Brandone et al. 2012). Minimal evidence is required to generate such generalizations. Preschool children and adults alike readily produce generics after seeing just a single exemplar of a novel category, particularly for animals as compared to artifacts even when controlling for item familiarity, complexity, number of features presented, and appearance (Brandone and Gelman 2013). Preschool children also often misremember novel statements about an individual as generic (e.g., after learning "This grasshopper has zevies on its legs," they recalled it as "Grassoppers have zevies on their legs") (Gülgöz and Gelman 2015). These results suggest an expectation that animal kinds have greater homogeneity, coherence, and predictive power. More generally, human representations of kinds seem to privilege conceptually central features (e.g., whether a feature is distinctive, dangerous, or inborn) over an accurate tally of property frequencies (Brandone et al. 2012; Cimpian et al. 2010a, 2010b; Gelman and Bloom 2007; see Tasimi et al. 2017, for how featural properties interact with domain).

There is a further way that generics entail a bias to ignore variation. It may be seen in an asymmetry in their meaning: generics have extremely strong implications but require little evidence to be judged true. On the one hand, one can assert generics on the basis of minimal evidence (e.g., "Mosquitoes carry the West Nile virus"). On the other hand, generics imply a strong link between category and property, and therefore high prevalence. This results in an asymmetry, whereby people tend to exaggerate prevalence, relative to the evidence. For example, if an adult learns that 70% of crullets have spots (where crullets are a novel animal kind), they conclude that it is true to assert, "Crullets have spots." However, if they hear, "Crullets have spots," they on average conclude that 95% of crullets have spots (Cimpian et al. 2010). A similar asymmetry has been found in children ranging from four to eleven years of age (Brandone et al. 2015).

Generics express conceptually central generalizations (Cimpian and Markman 2009) and imply that a category is homogeneous, inductively rich, and stable over time and contexts. Although generics can express core features in any category domain, they are favored for essentialized categories (natural kinds, social kinds) (Brandone and Gelman 2013; Gelman et al. 2008). Individual differences in parental and child use of generics predict parental essentialism (Gelman et al. 2014). When preschoolers learn a novel category by hearing either generic or specific facts, they show greater essentialism of the novel category following generic wording (Gelman et al. 2010; Rhodes et al. 2012).

12.5 INTENSIFYING CATEGORY BOUNDARIES

In addition to underestimating variability within a category, people tend to overestimate differences across categories, and in particular to view category boundaries as sharper, more stable, and more objectively correct than they actually are. An analogy to the perception of speech sounds is informative here. The distinction between the phonemes /ba/ and /pa/ lies on a physical continuum, varying only in the timing between when the lips part and when the vocal cords start to vibrate (also known as "voice onset time," or VOT). In English, a VOT of –10 is a clear "ba," whereas a VOT of +40 is a clear "pa." Interestingly, if one creates VOTs ranging between these values, they are interpreted as clearly "ba" or as clearly "pa," with very little space of disagreement. Moreover, people are unable to hear differences between different values of "ba" or between different values of "pa." This phenomenon is known as "categorical perception," because sounds get converted in human (and some nonhuman, e.g., chinchilla) perception as strictly one category or another (Diehl et al. 2004).

Concepts of natural kinds are not subject to categorical perception in the strict sense as phonemes—people recognize that hybrids exist (e.g., wolfdogs are part dog, part wolf) and are able to distinguish one member of a category from another (e.g., two different dogs). Nonetheless, children and adults alike view membership in a natural kind category as absolute rather than probabilistic (e.g., a penguin is an atypical bird but nonetheless is still 100% a bird; a bat looks very much like a bird but nonetheless is 100% not a bird) (Diesendruck and Gelman 1999; Rhodes and Gelman 2009b), and in-between instances (e.g., lion/tiger hybrid) tend to be viewed as one or the other (e.g., Keil 1989). Interestingly, this tendency to view category membership as absolute extends to reasoning about social kinds. For example, Judaism posits that those with a Jewish mother but non-Jewish father are 100% Jewish, whereas those with a Jewish father but non-Jewish mother are 100% non-Jewish. Similarly, the United States has historically practiced hypodescent, wherein individuals with any black ancestry are considered fully black (but not white; e.g., President Barack Obama). This is not simply a matter of perception, since hypodescent has to be learned—it is not evident in early childhood (Roberts and Gelman 2015), and is strongest among adults who are essentialist about race and have antiblack sentiments (Ho et al. 2015). A preference for sharp boundaries has direct physical consequences in the realm of gender: intersex infants often receive surgery to disambiguate their category membership, and transgender individuals are sometimes legally barred from using the bathroom that corresponds to their gender identity. This assumption of sharp categorical boundaries is not a necessary means of organizing experience, as it is distinctive to reasoning about natural and social kinds; artifacts are not viewed this way (category membership, like typicality, is probabilistic rather than absolute; e.g., a lamp is "sort of" furniture).

The stability of category membership is also inflated in ordinary reasoning. Although the ancient idea of "fixity of the species" has been abandoned even by creationists, many ordinary reasoners limit the idea of species change to nonhuman animals, treating humans as exceptions to this process, in part because humans are unique in possessing a soul (Evans 2001; Evans et al. 2010; Sinatra et al. 2003). Poling and Evans (2004) found that creationist adults were more likely to accept microevolution than macroevolution, and were particularly likely to reject the idea of a common ancestor for dissimilar species-pairs. They suggest that "lay adults from contemporary industrialized societies adopt a view of evolution in which species adapt to novel environments, but remain the same 'kind' despite changes. Therefore, extinction is considered unlikely and the relations between micro- and macroevolution misconstrued" (485).

Another distortion entails viewing category boundaries—even those of social groups—as objectively correct. Social categories vary across cultures; even when focusing on a particular social category (e.g., race), different societies and different historical periods involve different classifications (e.g., race in Brazil; hypodescent in the United States [Baran 2007; Davis 1991]). In contrast to these observations, young children—and many adults—adhere to the belief that there are discoverable, objectively correct category boundaries of race and ethnicity. Marjorie Rhodes developed a "visitor task" in which a group of individuals from a far-away land indicated that they considered two items with contrasting basic-level labels to be the "same kind"—for example, a cat and a dog, or a table and a bookshelf, or a girl and a boy (Rhodes and Gelman 2009a). Young children agreed with the visitors when the two items were artifacts (e.g., table/bookshelf), but disagreed with the visitors when the two items were animals (e.g., cat/dog). Importantly, judgments for social items varied as a function of the children's age and community (see also Diesendruck et al. 2013). Children growing up in a small, homogeneous, politically conservative community in Michigan were more likely to view gender and race categories as objective correct, as compared to children growing up in a larger, more diverse, politically liberal community in Michigan—and these community differences were increasingly strong with age. For example, gender categories were initially viewed as objectively correct by kindergartners in both communities, but such judgments decreased in the liberal community. In contrast, race categories were initially viewed as subjective by kindergartners in both communities, but such judgments increased in the conservative community.

12.6 ASSUMING AN INNER CAUSAL ESSENCE SHARED AMONG MEMBERS OF A KIND

Essentialism is in part a commitment to the reality of natural kinds—that there exist richly structured categories, whose members share a rich host of features, including those that are deep and nonobvious. Additionally, however, essentialism goes one step further to posit that there is some sort of internal, inherent, underlying cause (or "essence") that gives rise to these shared features. Belief in this internal essence can be gleaned *indirectly* by a "switched at birth" task in which people are asked to reason about an individual with one pair of birth parents but a different rearing environment (e.g., a kangaroo raised by goats; a boy raised by females; a child of English-speaking parents who is raised by Portuguese-speaking parents). Such studies show a strong assumption that characteristics of animal kinds are innately determined by birth parents across animal species (Gelman and Wellman 1991; Moya et al. 2015), as well as a tendency for this assumption to be stronger on the part of young children. Thus, for example, preschool children are particularly likely to view gender-stereotypical characteristics or even language spoken as inborn and impervious to environmental influence (Hirschfeld and Gelman 1997; Taylor et al. 2009).

Belief in an essence is often expressed, across cultural contexts, with a blood trope—not only are those to whom one is biologically related said to share blood ("Blood is thicker than water"), but also members of one's religious or ethnic group are said to share blood. For example, in the Kenyan town of Malindi, where there are two ethnic groups (Swahili and Giriama), McIntosh (2009) reported that when asked whether Giriama people could convert to Islam, people often denied that this was possible, providing explicitly essentialist reasons such as "Swahili have Arab blood. Giriama don't. They can't be one of us"; "Swahili identity is about blood; the nature of a person comes from their blood kinship"; "They can practice the culture, but [they] won't become a Swahili."

Similar intuitive beliefs about essences are reflected in folk beliefs about genes. Most formally educated adults in the United States and other industrialized countries learn about genes in school and from the media (Thomas 2000), and are aware of genes as biological entities that are linked to personal characteristics. They are typically exposed to evidence for the heritability of a wide range of human attributes (including disease and temperament, for example), they understand the genetic bases of biological sex (which informs the social construction of gender), and they understand that there are genetic contributions to phenotypic aspects of human variation (which informs the social construction of race). Genes are certainly involved in all of these domains, in the sense that they (along with environmental factors) are always a component of complex human behavior.

Importantly, however, adults also hold a number of misconceptions concerning genes, a set of beliefs referred to as "genetic essentialism" (Dar-Nimrod and Heine 2011). Many adults uncritically apply genetic explanations broadly, beyond the available scientific evidence (Moore 2013). They may overextend genetic explanations for group differences in perceptions, attitudes, and behaviors across a wide range of domains for which the scientific evidence is weak at best, assuming that there is a single "gene for" attributes with complex origins, including criminality, mental illness, gender-linked attributes, race, luck, promiscuity, and success in life (Kendler 2005; Shostak et al. 2009). Adults often mistakenly interpret genetic attributions as implying that a characteristic is fated, deterministic, and beyond environmental manipulation or personal control (Alper and Beckwith 1993; Gould and Heine 2012).

Evidence indicates that genetic attributions often reinforce social inequalities and predict stereotyping and prejudice (Keller 2005; Plaks et al. 2012). Genetic essentialism can foster an overly negative view of the pertinent condition as immutable and untreatable, with implications for health decision-making (Dar-Nimrod et al. 2014; Lebowitz et al. 2013). Similarly, genetic beliefs among parents and teachers may lower children's performance by dampening motivation to invest effort (Uttal 1997), and eighth-graders who hear genetic explanations of race in a classroom context hold more strongly essentialized views of race (Donovan 2014).

In contrast to genetic essentialist views of race, biological analyses show that "race" or "ethnicity" does not go more than skin-deep, thus meaning there is no "essence" shared by group members, other than that which is responsible for the particular (superficial) trait of interest (e.g., skin color) (Templeton 1998). In contrast to the essentialist perspective that each member of a racial group shares an identical inner essence with all other members of that group, there is as much genetic variation within a race as across races (advertisements for AncestryDNA notwithstanding, which promises "Your DNA can reveal your ethnic mix," and includes people who discover that they are "16% Italian"). As Pinker (2009) says, "It's our essentialist mind-set that makes the cheek swab feel as if it is somehow a deeper, truer, more authentic test of the child's ability."

Genetic essentialism also leads to suspicion of genetically modified organisms. Many people seem to believe that a genetically modified plant will take on characteristics of the host species, as if any portion of an organism's DNA is its essence—and that its essence is transferrable (Blancke et al. 2015). For example, in one survey conducted in 2003 and 2004 (Hallman et al. 2004), less than half of participants disagreed with the following statements: "Tomatoes modified with genes from a catfish would probably taste fishy" (27% in 2003; 15% in 2004) and "By eating a GM fruit, a person's genes could also become modified" (17% in 2003; 13% in 2004).

Misconceptions do not focus exclusively on beliefs about genes, which certainly have powerful causal effects (albeit not as conceptualized by the lay public). Other bodily parts (such as hearts, livers, or blood) are often treated as likewise causally powerful and transferrable. For example, over one-third of heart-transplant recipients report a belief that they take on characteristics of the donor (Inspector et al. 2004), and this belief is broadly demonstrated by samples of adults in the United States and India who have not themselves had such transplants (Meyer et al. 2013) as well as young US children (Meyer et al. 2017). Intriguingly, this expectation holds not only for animal species (e.g., that someone who receives a pig heart might become more piglike) but also for human kinds (e.g., that someone who receives the heart of a criminal might become more likely to act like a criminal).

Positing an inner essence includes the idea that the cause of a phenomenon is inherent to the individual entities involved. This is problematic for biological species, as what defines a species can be understood only at the population level (interbreeding populations), not at the individual level (Sober 1994). The essentialist view is what

Sober calls a "constituent definition," which applies at the level of the individual organism. In Sober's words, "Essentialism lost its grip when populations came to be thought of as real" (1994, 227). More broadly, Cimpian and his colleagues have proposed that a precursor to essentialism is a foundational heuristic bias to assume that regularities result from inherent features of the entities involved, rather than in terms of structural features that lie outside the entities involved (Cimpian and Salomon 2014). For example, people might explain why girls wear pink in terms of perceived inherent features of the entities involved (pink is a delicate color, and girls are delicate), rather than in terms of historically contingent facts. Intriguingly, and of particular interest here, they also suggest that this reasoning bias has supported problematic scientific explanations across history, such as theorizing phlogiston (Horne and Cimpian 2016).

12.7 CONCLUSIONS

Natural kinds sit at the intersection of world and mind. They represent the human attempt to organize and make sense of reality, but must be viewed through the distorting lenses of our own cognitive limitations and heuristics. Our concepts of natural kinds have our human fingerprints all over them. The argument in this chapter is that, by identifying properties of the mind that structure our experience of kinds in predictable ways, this may not tell us what reality is, but it promises to shed light on the prism through which we view reality. I have suggested this goes beyond a deflationary account of one model of reality (essentialism), to more broadly provide clues as to where we need to be skeptical when viewing models of the natural world. Young children have a tendency to discount variability within a category, to overemphasize boundaries between categories, and to posit inherent inner essences shared among all members of a kind. These representations of categories find their way as well into adult concepts, and (historically) in the concepts and theories of scientists as well.

A reasonable worry is that using cognitive biases to make claims about reality is circular: we know that conceptual representations are biased *because* of our models of reality (e.g., beliefs that contradict Darwinian evolution are deemed incorrect). However, a comparative approach may provide a mechanism for skirting the circularity. Consider the sorts of contrasting representations that psychologists have examined: those held by children vs. adults, those held by novices vs. experts, those underlying language production vs. interpretation, and those involving how information is presented vs. recalled. To the extent that these contrasts reveal a consistent bias, they may provide a bootstrapping mechanism toward reality. Here I wish to quote from Tversky and Kahneman (1973, 231): "Although the 'true' probability of a unique event is unknowable, the reliance on heuristics such as availability or representativeness, biases subjective probabilities in knowable ways. A psychological analysis of the heuristics that a person uses in judging the probability of an event may tell us whether his judgment is likely to be too high or too low. We believe that such analyses could be used to reduce the prevalence of errors in human judgment under uncertainty."

Although I have argued that psychological essentialism is a biased construal of the world, this does not mean that essentialism is wrong in every respect—and indeed, some have argued that essentialism is basically correct, that we are essentialist because the world supports essentialism, and we have evolved to detect structure in the world. Kornblith (1993, 35), for example, states, "It is precisely because the world has the causal structure required for the existence of natural kinds that inductive knowledge is even possible." Similarly, Bloom (2000, 153) characterizes essentialism as adaptive: "[Essentialism] is adaptive because it is true." Certainly a number of essentialist assumptions seem uncontroversial: appearances can be deceiving; internal entities can have powerful causal consequences; important aspects of animal form and behavior may be inborn. Reconciling the irrationality of essentialism with the basic truths of a framework that posits a deeper reality may start with the observation that heuristics don't have to be 100% right—they just have to be right enough of the time to allow for fruitful predictions. Our reasoning heuristics can lead to errors, yet do the work to get us to survive another generation and even be a boost to learning. Indeed, Gigerenzer and Gaissmaier (2011) propose that in some environmental contexts, decision-making heuristics can be even more accurate than more complex strategies that take into account more information (see also Sanborn et al. 2013, for suggesting that reasoning errors can be reconciled with a view of human cognition as fundamentally rational).

I end by noting that essentialism is a realist assumption about the natural world. Paradoxically, this means that realism about kinds is one of our biggest obstacles to achieving a direct perspective on the natural world. I have suggested that research findings indicate a pervasive human tendency to view kinds as having an objective reality, thus leading people to underestimate variability within a kind, to view category boundaries as objectively correct, and to assume a causal essence shared among members of a kind.

REFERENCES

Allport, G. W. (1954). *The Nature of Prejudice*. Cambridge, MA: Perseus Books.

Alper, J. S., and Beckwith, J. (1993). Genetic fatalism and social policy: The implications of behavior genetics research. *Yale Journal of Biology and Medicine 66* (6): 511–24.

Anderson, D. L., Fisher, K. M., and Norman, G. J. (2002). Development and evaluation of the Conceptual Inventory of Natural Selection. *Journal of Research in Science Teaching 39*: 952–78.

Baran, M. D. (2007). "Girl, you are not Morena. We are Negras!": Questioning the concept of "race" in Southern Bahia, Brazil. *Ethos 35* (3): 383–409.

Biernat, M. (1991). Gender stereotypes and the relationship between masculinity and femininity: A developmental analysis. *Journal of Personality and Social Psychology 61* (3): 351–65.

Blancke, S., Van Breusegem, F., De Jaeger, G., Braeckman, J., and Van Montagu, M. (2015). Fatal attraction: The intuitive appeal of GMO opposition. *Trends in Plant Science 20* (7): 414–18.

Bloom, P. (2000). *How Children Learn the Meanings of Words*. Cambridge, MA: MIT Press.

Bolnick, D. A., Fullwiley, D., Duster, T., Cooper, R. S., Fujimura, J. H., Kahn, J. . . . TallBear, K. (2007). The science and business of genetic ancestry testing. *Science 318*: 399–400.

Brandone, A. C. (2017). Changes in beliefs about category homogeneity and variability across childhood. *Child Development 88* (3): 846–66.

Brandone, A. C., Cimpian, A., Leslie, S.-J., and Gelman, S. A. (2012). Do lions have manes? For children, generics are about kinds rather than quantities. *Child Development 83* (2): 423–33.

Brandone, A. C., and Gelman, S. A. (2013). Generic language use reveals domain differences in children's expectations about animal and artifact categories. *Cognitive Development 28* (1): 63–75.

Brandone, A. C., Gelman, S. A., and Hedglen, J. (2015). Children's developing intuitions about the truth conditions and implications of novel generics versus quantified statements. *Cognitive Science 39* (4): 711–38.

Carey, S. (2009). *The Origin of Concepts*. New York: Oxford University Press.

Carlson, G. N., and Pelletier, F. J., eds. (1995). *The Generic Book*. Chicago: University of Chicago Press.

Christensen, K. D., Jayaratne, T. E., Roberts, J. S., Kardia, S. L. R., and Petty, E. M. (2010). Understandings of basic genetics in the United States: Results from a national survey of black and white men and women. *Public Health Genomics 13* (7–8): 467–76.

Cimpian, A., Brandone, A. C., and Gelman, S. A. (2010). Generic statements require little evidence for acceptance but have powerful implications. *Cognitive Science 34* (8): 1452–82.

Cimpian, A., Gelman, S. A., and Brandone, A. C. (2010). Theory-based considerations influence the interpretation of generic sentences. *Language and Cognitive Processes 25* (2): 261–76.

Cimpian, A., and Markman, E. M. (2009). Information learned from generic language becomes central to children's biological concepts: Evidence from their open-ended explanations. *Cognition 113* (1): 14–25.

Cimpian, A., Meltzer, T. J., and Markman, E. M. (2011). Preschoolers' use of morphosyntactic cues to identify generic sentences: Indefinite singular noun phrases, tense, and aspect. *Child Development 82* (5): 1561–78.

Cimpian, A., and Salomon, E. (2014). The inherence heuristic: An intuitive means of making sense of the world, and a potential precursor to psychological essentialism. *Behavioral and Brain Sciences 37* (5): 461–80.

Dalal, D. K., and Hakel, M. D. (2016). Experimental comparisons of methods for reducing deliberate distortions to self-report measures of sensitive constructs. *Organizational Research Methods 19*: 475–505.

Dar-Nimrod, I., Cheung, B. Y., Ruby, M. B., and Heine, S. J. (2014). Can merely learning about obesity genes affect eating behavior? *Appetite 81*: 269–76.

Dar-Nimrod, I., and Heine, S. J. (2011). Genetic essentialism: On the deceptive determinism of DNA. *Psychological Bulletin 137* (5): 800–818.

Davis, F. J. (1991). *Who Is Black? One Nation's Definition*. University Park: Pennsylvania State University Press.

Diehl, R. L., Lotto, A. J., and Holt, L. L. (2004). Speech perception. *Annual Review of Psychology 55*: 149–79.

Diesendruck, G., and Gelman, S. A. (1999). Domain differences in absolute judgments of category membership: Evidence for an essentialist account of categorization. *Psychonomic Bulletin & Review 6* (2): 338–46.

Diesendruck, G., Goldfein-Elbaz, R., Rhodes, M., Gelman, S., and Neumark, N. (2013). Cross-cultural differences in children's beliefs about the objectivity of social categories. *Child Development 84* (6): 1906–17.

Donovan, B. M. (2014). Playing with fire? The impact of the hidden curriculum in school genetics on essentialist conceptions of race. *Journal of Research in Science Teaching 51* (4): 462–96.

Evans, E. M. (2001). Cognitive and contextual factors in the emergence of diverse belief systems: Creation versus evolution. *Cognitive Psychology 42*: 217–66.

Evans, E. M., Spiegel, A. N., Gram, W., Frazier, B. N., Tare, M., Thompson, S., and Diamond, J. (2010). A conceptual guide to natural history museum visitors' understanding of evolution. *Journal of Research in Science Teaching 47* (3): 326–53.

Gelman, S. A. (2003). *The Essential Child: Origins of Essentialism in Everyday Thought*. New York: Oxford University Press.

Gelman, S. A. (2004). Psychological essentialism in children. *Trends in Cognitive Sciences 8* (9): 404–9.

Gelman, S. A. (2010). Generics as a window onto young children's concepts. In F. J. Pelletier, ed., *Kinds, Things, and Stuff: The Cognitive Side of Generics and Mass Terms*. New York: Oxford University Press.

Gelman, S. A., and Baillargeon, R. (1983). A review of some Piagetian concepts. In P. H. Mussen, gen. ed., J. H. Flavell and E. M. Markman, vol. eds., *Handbook of Child Psychology*, vol. 3, *Cognitive Development*. New York: Wiley, 167–230.

Gelman, S. A., and Bloom, P. (2007). Developmental changes in the understanding of generics. *Cognition 105* (1): 166–83.

Gelman, S. A., and Davidson, N. S. (2013). Conceptual influences on category-based induction. *Cognitive Psychology 66* (3): 327–53.

Gelman, S. A., Goetz, P. J., Sarnecka, B. W., and Flukes, J. (2008). Generic language in parent-child conversations. *Language Learning and Development 4* (1): 1–31.

Gelman, S. A., Heyman, G. D., and Legare, C. H. (2007). Developmental changes in the coherence of essentialist beliefs about psychological characteristics. *Child Development 78* (3): 757–74.

Gelman, S. A., and Markman, E. M. (1986). Categories and induction in young children. *Cognition 23* (3): 183–209.

Gelman, S. A., and Raman, L. (2003). Preschool children use linguistic form class and pragmatic cues to interpret generics. *Child Development 74* (1): 308–25.

Gelman, S. A., and Rhodes, M. (2012). "Two-thousand years of stasis": How psychological essentialism impedes evolutionary understanding. In K. S. Rosengren, S. Brem, E. M. Evans, and G. Sinatra (Eds.): *Evolution Challenges: Integrating Research and Practice in Teaching and Learning about Evolution*. New York: Oxford University Press.

Gelman, S. A., Sánchez Tapia, I., and Leslie S. J. (2016). Memory for generic and quantified sentences in Spanish-speaking children and adults. *Journal of Child Language 43* (6): 1231–44.

Gelman, S. A., Taylor, M. G., and Nguyen, S. (2004). *Mother-Child Conversations about Gender: Understanding the Acquisition of Essentialist Beliefs*. Monographs of the Society for Research in Child Development 69, no. 1. Boston: Blackwell.

Gelman, S. A., Ware, E., and Kleinberg, F. (2010). Effects of generic language on category content and structure. *Cognitive Psychology 61* (3): 273–301.

Gelman, S. A., Ware, E. A., Kleinberg, F., Manczak, E. M., and Stilwell, S. M. (2014). Individual differences in children's and parents' generic language. *Child Development 85* (3): 924–40.

Gelman, S. A., and Wellman, H. M. (1991). Insides and essences: Early understandings of the nonobvious. *Cognition 38* (3): 213–44.

Gigerenzer, G., and Gaissmaier, W. (2011). Heuristic decision making. *Annual Review of Psychology 62*: 451–82.

Gopnik, A., and Sobel, D. M. (2000). Detecting blickets: How young children use information about novel causal powers in categorization and induction. *Child Development 71* (5): 1205–22.

Gopnik, A., and Wellman, H. M. (2012). Reconstructing constructivism: Causal models, Bayesian learning mechanisms, and the theory theory. *Psychological Bulletin 138* (6): 1085.

Gould, W. A., and Heine, S. J. (2012). Implicit essentialism: Genetic concepts are implicitly associated with fate concepts. *PLOS ONE 7* (6): e38176.

Graham, S. A., Gelman, S. A., and Clarke, J. (2016). Generics license 30-month-olds' inferences about the atypical properties of novel kinds. *Developmental Psychology 52* (9): 1353–62.

Graham, S. A., Keates, J., Vukatana, E., and Khu, M. (2013). Distinct labels attenuate 15-month-olds' attention to shape in an inductive inference task. *Frontiers in Psychology 3*. https://www.frontiersin.org/articles/10.3389/fpsyg.2012.00586.

Gülgöz, S., and Gelman, S. A. (2015). Children's recall of generic and specific labels regarding animals and people. *Cognitive Development 33*: 84–98.

Gutheil, G., and Gelman, S. A. (1997). Children's use of sample size and diversity information within basic-level categories. *Journal of Experimental Child Psychology 64* (2): 159–74.

Hallman, W. K., et al. (2004). *Americans and GM Food: Knowledge, Opinion and Interest in 2004*. Food Policy Institute Publication RR- 1104-007. New Brunswick, NJ: Food Policy Institute, Cook College, Rutgers.

Haslam, N., Rothschild, L., and Ernst, D. (2000). Essentialist beliefs about social categories. *British Journal of Social Psychology 39* (1): 113–27.

Hirschfeld, L. A., and Gelman, S. A. (1997). What young children think about the relation between language variation and social difference. *Cognitive Development 12* (2): 213–38.

Ho, A. K., Roberts, S. O., and Gelman, S. A. (2015). Essentialism and racial bias jointly contribute to the categorization of multiracial individuals. *Psychological Science 26*: 1639–45.

Hollander, M. A., Gelman, S. A., and Star, J. (2002). Children's interpretation of generic noun phrases. *Developmental Psychology 38* (6): 883–94.

Horne, Z., and Cimpian, A. (2016). Scientific explanation is influenced by an inherence heuristic. Paper presented at the 57th Annual Psychonomic Society. Boston, MA.

Inspector, I., Kutz, Y., and David, D. (2004). Another person's heart: Magical and rational thinking in the psychological adaptation to heart transplantation. *Israeli Journal of Psychiatry and Related Sciences 41*: 161–73.

James, W. ([1890] 1983). *The Principles of Psychology*. 2 vols. New York: Henry Holt and Company.

Keil, F. C. (1989). *Concepts, Kinds, and Cognitive Development*. Cambridge, MA: MIT Press.

Kelemen, D., Rottman, J., and Seston, R. (2013). Professional physical scientists display tenacious teleological tendencies: Purpose-based reasoning as a cognitive default. *Journal of Experimental Psychology: General 142* (4): 1074–83.

Keller, J. (2005). In genes we trust: The biological component of psychological essentialism and its relationship to mechanisms of motivated social cognition. *Journal of Personality and Social Psychology 88* (4): 686–702.
Kendler, K. S. (2005). "A gene for . . .": The nature of gene action in psychiatric disorders. *American Journal of Psychiatry 162* (7): 1243–52.
Kornblith, H. (1993). *Inductive Inference and Its Natural Ground.* Cambridge, MA: MIT Press.
Lebowitz, M. S., Ahn, W., and Nolen-Hoeksema, S. (2013). Fixable or fate? Perceptions of the biology of depression. *Journal of Consulting and Clinical Psychology 81* (3): 518–27.
Legare, C. H., Evans, E. M., Rosengren, K. S., and Harris, P. L. (2012). The coexistence of natural and supernatural explanations across cultures and development. *Child Development 83* (3): 779–93.
Legare, C. H., and Gelman, S. A. (2008). Bewitchment, biology, or both: The co-existence of natural and supernatural explanatory frameworks across development. *Cognitive Science 32* (4): 607–42.
Leslie, S. J. (2008). Generics: Cognition and acquisition. *Philosophical Review 117* (1): 1–47.
Leslie, S. J. (2013). Essence and natural kinds: When science meets preschooler intuition. *Oxford Studies in Epistemology 4*: 108–65.
Leslie, S. J., and Gelman, S. A. (2012). Quantified statements are recalled as generics: Evidence from preschool children and adults. *Cognitive Psychology 64* (3): 186–214.
Liben, L. S., and Signorella, M. L. (1993). Gender-schematic processing in children: The role of initial interpretations of stimuli. *Developmental Psychology 29* (1): 141.
Mannheim, B., Gelman, S. A., Escalante, C., Huayhua, M., and Puma, R. (2011). A developmental analysis of generic nouns in Southern Peruvian Quechua. *Language Learning and Development 7* (1): 1–23.
Mayr, E. (1982). *The Growth of Biological Thought: Diversity, Evolution, and Inheritance.* Cambridge, MA: Harvard University Press.
McIntosh, J. (2009). *The Edge of Islam: Power, Personhood, and Ethnoreligious Boundaries on the Kenya Coast.* Durham, NC: Duke University Press.
Medin, D. L. (1989). Concepts and conceptual structure. *American Psychologist 44* (12): 1469–81.
Meyer, M., Gelman, S. A., Roberts, S. O., and Leslie, S. J. (2017). My heart made me do it: Children's essentialist beliefs about heart transplants. *Cognitive Science 41* (6): 1694–712.
Meyer, M., Leslie, S. J., Gelman, S. A., and Stilwell, S. (2013). Essentialist beliefs about bodily transplants in the United States and India. *Cognitive Science 37* (4): 668–710.
Moore, D. S. (2013). Current thinking about nature and nurture. In K. Kampourakis, ed., *The Philosophy of Biology: A Companion for Educators.* New York: Springer, 629–52.
Moya, C., Boyd, R., and Henrich, J. (2015). Reasoning about cultural and genetic transmission: Developmental and cross-cultural evidence from Peru, Fiji, and the United States on how people make inferences about trait transmission. *Topics in Cognitive Science 7* (4): 595–610.
Pinker, S. (2009). My genome, my self. *New York Times Magazine, 7,* January 11, 2009.
Plaks, J. E., Malahy, L., Sedlins, M., and Shoda, Y. (2012). Folk beliefs about human genetic variation predict discrete versus continuous racial categorization and evaluative bias. *Social Psychological and Personality Science 3* (1): 31–39.
Poling, D. A., and Evans, E. M. (2004). Are dinosaurs the rule or the exception? Developing concepts of death and extinction. *Cognitive Development 19* (3): 363–83.
Prasada, S. (2000). Acquiring generic knowledge. *Trends in Cognitive Sciences 4* (2): 66–72.

Prentice, D. A., and Miller, D. T. (2007). Psychological essentialism of human categories. *Current Directions in Psychological Science 16* (4): 202–6.

Rhodes, M., Brickman, D., and Gelman, S. A. (2008). Sample diversity and premise typicality in inductive reasoning: Evidence for developmental change. *Cognition 108* (2): 543–56.

Rhodes, M., and Gelman, S. A. (2009a). A developmental examination of the conceptual structure of animal, artifact, and human social categories across two cultural contexts. *Cognitive Psychology 59* (3): 244–74.

Rhodes, M., and Gelman, S. A. (2009b). Five-year-olds' beliefs about the discreteness of category boundaries for animals and artifacts. *Psychonomic Bulletin & Review 16* (5): 920–24.

Rhodes, M., Gelman, S. A., and Brickman, D. (2010). Children's attention to sample composition in learning, teaching and discovery. *Developmental Science 13* (3): 421–29.

Rhodes, M., Leslie, S. J., and Tworek, C. M. (2012). Cultural transmission of social essentialism. *Proceedings of the National Academy of Sciences 109* (34): 13526–31.

Roberts, S. O., and Gelman, S. A. (2015). Do children see in Black and White? Children's and adults' categorizations of multiracial individuals. *Child Development 86* (6): 1830–47.

Sanborn, A. N., Mansinghka, V. K., and Griffiths, T. L. (2013). Reconciling intuitive physics and Newtonian mechanics for colliding objects. *Psychological Review 120* (2): 411–37.

Shostak, S., Freese, J., Link, B. G., and Phelan, J. C. (2009). The politics of the gene: Social status and beliefs about genetics for individual outcomes. *Social Psychology Quarterly 72* (1): 77–93.

Shtulman, A. (2006). Qualitative differences between naïve and scientific theories of evolution. *Cognitive Psychology 52* (2): 170–94.

Shtulman, A., and Calabi, P. (2013). Tuition vs. intuition: Effects of instruction on naive theories of evolution. *Merrill-Palmer Quarterly 59* (2): 141–67.

Shtulman, A., and Harrington, K. (2016). Tensions between science and intuition across the lifespan. *Topics in Cognitive Science 8* (1): 118–37.

Shtulman, A., and Lombrozo, T. (2016). Bundles of contradiction: A coexistence view of conceptual change. In D. Barner and A. S. Baron, eds., *Core Knowledge and Conceptual Change*. Oxford: Oxford University Press, 53–71.

Shtulman, A., and Schulz, L. (2008). The relation between essentialist beliefs and evolutionary reasoning. *Cognitive Science 32* (6): 1049–62.

Sinatra, G. M., Southerland, S. A., McConaughy, F., and Demastes, J. W. (2003). Intentions and beliefs in students' understanding and acceptance of biological evolution. *Journal of Research in Science Teaching 40*: 510–28.

Sloutsky, V. M., Kloos, H., and Fisher, A. V. (2007). When looks are everything: Appearance similarity versus kind information in early induction. *Psychological Science 18* (2): 179–85.

Sober, E. (1994). *From a Biological Point of View: Essays in Evolutionary Philosophy*. New York: Cambridge University Press.

Spiegel, A., Evans, E. M., Frazier, B. F., Hazel, A., Tare, M., Gram, W., and Diamond, J. (2012). Changing museum visitors' concepts of evolution. *Evolution: Education and Outreach 5*: 43–61.

Sutherland, S. L., Cimpian, A., Leslie, S.-J., and Gelman S. A. (2015). Memory errors reveal a bias to spontaneously generalize to categories. *Cognitive Science 39*: 1021–46.

Tardif, T., Gelman, S. A., Fu, X., and Zhu, L. (2012). Acquisition of generic noun phrases in Chinese: Learning about lions without an "-s." *Journal of Child Language 30*: 1–32.

Tasimi, A., Gelman, S. A., Cimpian, A., and Knobe, J. (2017). Differences in the evaluation of generic statements about human and non-human categories. *Cognitive Science 41* (7): 1934–57.

Taylor, M. G., Rhodes, M., and Gelman, S. A. (2009). Boys will be boys, cows will be cows: Children's essentialist reasoning about human gender and animal development. *Child Development 80* (2): 461–81.

Templeton, A. R. (1998). Human races: A genetic and evolutionary perspective. *American Anthropologist 100* (3): 632–50.

Thomas, J. (2000). Learning about genes and evolution through formal and informal education. *Studies in Science Education 35* (1): 59–92.

Tversky, A., and Kahneman, D. (1973). Availability: A heuristic for judging frequency and probability. *Cognitive Psychology 5* (2): 207–32.

Uttal, D. H. (1997). Beliefs about genetic influences on mathematics achievement: A cross-cultural comparison. *Genetica 99*: 165–72.

Wellman, H. M., and Gelman, S. A. (1998). Knowledge acquisition. In D. Kuhn and R. Siegler, eds., *Cognition, Perception, and Language*, vol. 3 of *Handbook of Child Psychology*, 5th ed. New York: Wiley, 523–73.

PART VII
DEBUNKING AND COGNITIVE SCIENCE

13 Debunking Arguments in Metaethics and Metaphysics

Daniel Z. Korman

13.1 INTRODUCTION

A central project of cognitive science is to explain, at a variety of different levels, certain types of mental representations and the cognitive processes responsible for them. What is the evolutionary purpose (if any) of the representations? What sorts of information are the processes operating on to yield the representations? How are the processes and representations implemented in the brain?[1]

In some cases, answers to these questions presuppose that the representations are more or less accurate. The explanation of mental states that represent regions of space as empty or matter-filled is bound to cite the fact that successfully navigating the world involves accurately tracking which regions are and aren't empty, and explanations for why in a particular instance a region is represented as empty will typically cite the fact that it is empty. Likewise for representations of the relative distances of filled regions, and (one might naturally think) for visual representations of three-dimensional midsized objects.

In other cases, it seems quite possible to answer the above questions without presupposing the accuracy of the representations. Cognitive scientists have developed various accounts of the production and adaptive value of theistic belief that in no way presuppose the existence of deities. Color perception can likely be explained in terms of the adaptive value of easily detecting ripe fruit or young foliage, by means of detecting wavelengths in the visible spectrum, and without presupposing that the phenomenal colors represented are actually out there on the surfaces of objects. The processes underwriting our moral beliefs and intuitions can presumably be explained in terms of the adaptive value of detecting certain benefits and dangers to

[1] These correspond roughly to Marr's 1982 three levels of analysis. Needless to say, not all traits will have an evolutionary purpose or adaptive value. Some, for instance, are byproducts ("spandrels") of adaptations; see Gould and Lewontin 1979 for discussion.

ourselves and our kin, and without any reference to a realm of moral facts that the processes are supposed to be tracking.[2]

Suppose that a range of representations can be explained without reference to the putative objects, features, or facts represented. Does the availability of such explanations *debunk* those representations, undermining their justificatory status? This is an epistemological question, one whose answer can be informed by, but not directly read off of, findings in cognitive science. There has been some discussion in metaphysics of such explanations, and there have been attempts to wield them with debunking intent. But such discussions tend to be carried out in isolation from the extensive and nuanced literature on explanatory challenges and debunking arguments in metaethics. My aim here is to forge some connections, drawing lessons from the moral-debunking literature for the metaphysical-debunking literature and vice versa.

As a case study, I'll focus on material-object metaphysics and our beliefs about which midsized objects there are, though lessons learned generalize to explanatory challenges that arise in connection with such other metaphysical domains as color, causation, time, modality, and persistence.[3] In sections 13.3–13.4, I examine and combat the idea that, because our beliefs about midsized objects can be given an "evolutionary vindication," evolutionary debunking arguments won't arise for these beliefs. In section 13.5, I defend an alternative strategy for vindicating them, which involves bootstrapping from the very beliefs targeted by the debunking arguments. In section 13.6, I explain why debunking arguments framed in terms of the insensitivity of targeted beliefs are ineffective against bootstrappers. Finally, in sections 13.7–13.8, I develop and defend an explanatory constraint on bootstrapping, and I identify some existing responses to the debunking arguments that run afoul of the constraint.

13.2 THE DEBUNKING ARGUMENTS

You turn the corner and see some kids setting fire to a cat.[4] You immediately form the belief that they're doing something wrong. You feel inclined to believe this because it looks to you or seems to you intuitively that they are doing something wrong. I'll use the term "moral reactions" to cover this whole range of mental states: the experiences, intuitions, felt inclinations, and beliefs.

[2] See Leech and Visala 2011 on religious belief. See Sumner and Mollon 2000 on the evolution of color vision, and see Schaffer 2016, sec. 2, on the prospects of using the cognitive science of color vision to debunk color beliefs. Much more on moral belief later in the chapter.

[3] See, e.g., Goldman 1992, chaps. 2–3; Stroud 2000; Rea 2002; Chalmers 2006, sec. 6; Paul 2010, 2016; Pautz 2011, sec. 3; Leslie 2013; Gow 2014, sec. 6; Baron et al. 2015; Benovsky 2015; Crisp 2016; Schaffer 2016; and Baron 2017.

[4] The example is from Harman 1977, 4.

Moral-debunking arguments attempt to discredit moral reactions by exposing an explanatory disconnect between the reactions and the associated facts. Here, in its simplest form, is a moral-debunking argument:

(M1) Your moral reactions are not explained by the moral facts nor are the moral facts determined by your moral reactions.

(M2) If so, then you are not justified in retaining your moral beliefs.

(M3) So, you are not justified in retaining your moral beliefs.

Some may want to see the argument restricted in one way or another, so that it targets only realists, or only cognitivists, or only nonnaturalists, or only naturalists.[5] But there is no danger in sticking with this simple version of the argument, so long as we bear in mind that some philosophers—perhaps because they are antirealists, or noncognitivists, or naturalists, or nonnaturalists—may be in a position to wield the argument against their intended targets without opening themselves up to a *tu quoque*. The simple argument also nicely structures the moral-debunking debate, separating nonskeptics into two categories: those who deny M1, affirming the indicated sort of explanatory connection between the facts and the beliefs, and those who deny M2, rejecting the need for any such connection.

There are different routes to M1, and one common strategy involves pointing to an evolutionary explanation of our moral reactions.[6] Those actions that we represent as right or obligatory are ones that tend to keep us and ours alive; representing them as right or obligatory better motivates us to perform them; and these moral reactions now prevail precisely because they enhanced our ancestors' reproductive success. Mutatis mutandis for actions represented as wrong or forbidden.

The idea behind M2 is that if moral facts aren't explaining our moral reactions (or vice versa), then it would take a massive coincidence to end up with beliefs that line up with the facts. Since we have no grounds for believing that such a coincidence occurred, we shouldn't believe that it did, in which case we should not go on thinking that our beliefs line up with the facts.

There is also going to be an evolutionary explanation of the perceptual and cognitive mechanisms that convert the "blooming buzzing confusion" in the retinal image into representations of three-dimensional midsized objects. Accordingly, we ought to consider the prospects of a parallel debunking argument against our "object

[5] See Street 2006; Kitcher 2005, 175–176; Joyce 2006, chap. 6; and Bogardus 2016 respectively. See Bedke 2009, 200–201 and Locke 2014, 224 n. 8 on restricting the argument to those who have actually encountered the relevant explanatory challenges; cf. White 2010, 575, on undermining and blocking debunkers.

[6] Evolutionary moral-debunking arguments have been advanced in one form or another by Ruse 1986, chap. 3; Lillehammer 2003; Kitcher 2005, sec. 4; Joyce 2006; Street 2006; Wilkins and Griffiths 2012, sec. 2; Fraser 2014; and Braddock 2016, 2017.

reactions": beliefs and associated experiences and intuitions concerning which midsized objects there are.

(O1) Your object reactions are not explained by the object facts, nor are the object facts determined by your object reactions.

(O2) If so, then you are not justified in retaining your object beliefs.

(O3) So, you are not justified in retaining your object beliefs.

What was said above in defense of M1 seems to apply equally to O1: our object reactions, like our moral reactions, can be explained by the adaptive value of an ability to represent certain kinds of midsized objects. And like M2, O2 can be motivated by the observation that once you acknowledge the explanatory disconnect, you'd have to admit that it would be a massive coincidence if our beliefs lined up with the facts.

In metaethics, there is near consensus that the O-argument is a complete non-starter. In metaphysics, by contrast, arguments along these lines are deployed constantly against common-sense judgments about how the world divides up into midsized objects.[7] I will argue that, while parties to the object debates have been somewhat overzealous about the prospects of such arguments, they are right to take them seriously. But first let us see why the O-argument is summarily dismissed in the literature on moral debunking.

13.3 THE EVOLUTIONARY VINDICATION

As indicated at the outset, the evolutionary explanation of a range of representations sometimes will and sometimes won't presuppose the accuracy of those representations. Accordingly, in order to get from an evolutionary explanation to the crucial first premise of the debunking argument—that there is no explanatory connection between the reactions and the associated facts—one needs a premise to the effect that the facts do not enter into the evolutionary explanation. To wit:

(M*) The adaptive value of your moral reactions is not explained by the moral facts.

(O*) The adaptive value of your object reactions is not explained by the object facts.

[7] As with the moral arguments, such arguments are sometimes advanced, not with the aim of establishing O3, but rather in support some departure from common sense (e.g., antirealism or mereological universalism) that would enable one to resist the skeptical conclusion. For arguments in the vicinity, see Heller 1990, 41–42; Goldman 1992, 45–46; Merricks 2001, 72–76; Hudson 2001, sec. 3.8; Sider 2001, 156–57; Rea 2002; Hawthorne 2006, vi and 109; Benovsky 2015, sec. 2; Sattig 2015, 25–26; Osborne 2016; Schaffer 2016, sec. 3.2; and Rose and Schaffer 2017. For some discussion of object-debunking arguments within cognitive science, see Hoffman et al. 2015 and Hummel 2015.

And herein, metaethicists are wont to claim, lies the difference between the two arguments: M* is plausible and O* is implausible.[8]

Why is M* supposed to be plausible? Following Street (2006, sec. 6), let's compare two accounts of the adaptive value of our moral reactions: the adaptive link account (ALA) and the tracking account (TA):

(ALA) Representing actions $a_1 \ldots a_n$ as morally right (or wrong) enhances reproductive success because representing $a_1 \ldots a_n$ as such motivates us to perform (avoid) them and performing them enhances (inhibits) reproductive success.

(TA) Representing actions $a_1 \ldots a_n$ as morally right (or wrong) enhances reproductive success because $a_1 \ldots a_n$ really are morally right (wrong).

ALA seems preferable to TA for a number of reasons. First, it is more parsimonious than TA, insofar as the latter invokes a domain of moral truths. Second, ALA is illuminating: it makes good sense of why beings with lives like ours—bearing relatively few offspring and needing to cooperate with neighbors—would have the particular kinds of moral reactions that we do. Third, TA is obscure: it is entirely unclear how, for instance, the putative fact that protecting our children *really is* obligatory is supposed to explain, even in part, why believing it's obligatory helps keep us and ours alive. For these reasons, Street suggests, we should reject TA and accept M*.

Why is O* supposed to be implausible? Because it is hard to see how the adaptive value of our object reactions could be explained without supposing them to be accurate. It's adaptive to represent the world in terms of tigers and cliffs, the idea goes, only because there are tigers and cliffs. As Street (2006, 160–61) puts the point:

> In order to explain why it proved advantageous to form judgements about the presence of fires, predators, and cliffs, one will need to posit in one's best explanation that there *were indeed* fires, predators, and cliffs, which it proved quite useful to be aware of, given that one could be burned by them, eaten by them, or could plummet over them.

It is not as if there is some more parsimonious, tiger-free explanation of the adaptive value of tiger experiences! Thus, the idea goes, we should reject O*. And since the object facts do ultimately get in on the explanatory action, insofar as they explain why our object reactions are adaptive, O1 of the object-debunking argument is false. The full evolutionary story vindicates our object reactions.

[8] See Gibbard 2003, 253–58; Lillehammer 2003, 570–71; Joyce 2006, 182; Street 2006, 130 and 160–61; 2008, 217; Bedke 2009, 202; Enoch 2010, 436; Schafer 2010, 472; White 2010, 587; Wielenberg 2010, 441 n. 4; Dreier 2012, 277; Wilkins and Griffiths 2012, 134–39; Fitzpatrick 2015, 885; and Vavova 2015, 108; for a notable exception, see Clarke-Doane 2015, sec. 4. Alternatively, one might point to differences in *proximate* explanations: midsized objects can cause object reactions, whereas moral objects (properties, facts) are causally inert. See my 2014, sec. 5; 2015, chap. 7.4, on the inadequacy of this sort of response.

Or so it would seem. As we are about to see, there is more to O*—and less to the putative difference between the two debunking arguments—than meets the eye.

13.4 NO EVOLUTIONARY VINDICATION

The envisaged evolutionary vindication of object reactions turns on the idea that we need to assume the accuracy of our object reactions in order to explain their adaptive value. But that's just a mistake. In order to explain the adaptive value of our object reactions, it's enough to suppose that we are reliably tracking what nourishes us, endangers us, or can otherwise feature in adaptive or maladaptive interactions. But reliably tracking such things is entirely compatible with certain ways of misrepresenting which midsized objects there are. In particular, misrepresenting the world of midsized objects will still be adaptive so long as we *reliably* misrepresent it.[9]

To see what I have in mind, let's return briefly to what was supposed to be a paradigm case of representations whose adaptive value presupposes their accuracy: our representations of regions as empty or matter-filled. Not so fast. Even the regions we visually represent as empty are typically filled by gaseous matter. So we misrepresent those regions. Nevertheless, the misrepresentations reliably track the difference that matters to successful navigation, namely, which are and aren't filled by nongaseous matter. They reliably misrepresent such regions as empty, and they are adaptive despite the harmless misrepresentation.[10]

The case for moral debunking is also usefully framed in terms of reliable misrepresentation. The error theorist will say that our moral reactions—which divide actions into the right and the wrong and the obligatory and the forbidden—misrepresent those actions but are nevertheless tracking something important. Those represented as right or obligatory tend to promote reproductive success. Those represented as wrong or forbidden tend to inhibit reproductive success. The moral reactions can be adaptive, despite misrepresenting, precisely because of what they reliably track. Indeed, the error theorist might add, misrepresenting actions by imbuing them with moral features is superior (evolutionarily) to accurately representing them merely as beneficial or detrimental to reproductive success, since the morally charged representations better motivate us to choose the beneficial actions.[11]

Now back to objects. Let's suppose (for the moment) that our own object reactions are generally accurate, and let's contemplate two different ways in which a creature might have erroneous object reactions. First, it might misrepresent the regions of space filled by midsized objects as containing no solid matter at all. Second, it might

[9] I borrow this useful notion from Mendelovici 2013, 2016.

[10] Perhaps you'll think that the sort of "emptiness" we represent the regions as having is compatible with their being gas-filled—just as a refrigerator can be empty, in a sense, despite containing shelves and drawers. Fair enough. My point can instead be made with hypothetical beings whose emptiness representations are more demanding.

[11] Cf. Joyce 2005 on the utility of moral fictions.

misrepresent how a situation divides up into objects, but without misrepresenting how solid matter is distributed. For instance, there could be creatures—let's call them "troglodytes"—who, upon encountering a situation containing an ordinary dog and tree, believe that there is a dog and a trunk and some branches, but do not believe that there is any further object composed of the trunk and the branches. Let's imagine that they don't even experience the region the tree occupies as being filled by a single object, despite experiencing the region as being filled with solid matter. Odder still, they represent the scattered region jointly occupied by the dog and the trunk as containing a single object. It seems to them that they see a *trog*: a partly wooden and partly furry object composed of the dog and trunk. This would be a misrepresentation; there are no trogs.[12]

Importantly, we don't have to imagine the troglodytes to be under any misconceptions about how stuff moves around. They know that the branches are firmly attached to the trunk and don't move (much) independently from one another, and that the dog and trunk aren't attached and won't move together. When there's a particularly vicious dog, they run only from the dog, not from the tree. Or, as they might put it, it's only the furry part of the trog—not its wooden part—that you have to worry about. Just like us, these troglodytes are tracking the presence, changes, and movements of matter that are relevant to their survival and reproductive success. They reliably misrepresent the world of midsized objects.

Imagining such ways of misrepresenting the object facts points to easily overlooked ways in which our own object reactions may be inaccurate. Perhaps we're wrong and the troglodytes are right: there are trogs but no trees. Or perhaps we're both wrong: there are only the atoms arranged dogwise and atoms arranged treewise, but no midsized objects composed of the atoms.[13]

With all this in mind, let us return to the question of whether it is possible to explain the adaptive value of our object reactions without supposing those reactions to be accurate. Of course it is. When we seem to see a tree falling towards us, we step out of the way and successfully avoid a dangerous situation. To explain why seeming to see a tree is adaptive in such dangerous situations, there is no need to suppose that they contain falling trees, as opposed to just some branches and the wooden part of a trog, or just some atoms arranged treewise. What's important for survival and reproductive success is that we respond appropriately to different distributions of matter—pursuing those that tend to nourish us, avoiding those that tend to harm us, and anticipating the behaviors of each. What this requires is that we have representations that reliably track the dangerous and nourishing stuff, and

[12] Or so I shall assume throughout. See my 2015 for a defense of a trogless ontology, and see in particular chapter 7.3 for an argument that postulating trogs and other such extraordinary objects won't help with the debunking arguments.

[13] I am taking for granted that the atoms arranged treewise are not themselves (identical to) a tree. See Merricks 2001, chap. 1 and my 2015, sec. 3.2.2, on the complaint that the very idea that there are atoms arranged dogwise but no dogs is somehow incoherent.

it is neither here nor there whether they accurately represent how the dangerous stuff being tracked divides up into objects. Reliable misrepresentation is up to the explanatory task.

Not only is the assumption of accuracy not *needed* to explain the adaptive value of our object reactions, it's hard to see what an assumption of accuracy *could* add to an explanation of the adaptive value of our object reactions. To see this, compare the following analogues of Street's adaptive link account and tracking account:

(ALA_O) Perceptually and conceptually dividing perceived distributions of matter into midsized objects of kinds $k_1 \ldots k_n$ enhances reproductive success because doing so better enables us to track and anticipate the movements of matter that are relevant to our reproductive success.

(TA_O) Perceptually and conceptually dividing perceived distributions of matter into midsized objects of kinds $k_1 \ldots k_n$ enhances reproductive success because the matter really does divide into objects of kinds $k_1 \ldots k_n$, whose movements we are thereby able to track and anticipate.

It would seem that all of the reasons from section 13.3 for preferring ALA to TA are equally reasons for preferring ALA_O to TA_O. First, TA_O looks to be less parsimonious: it is committed to the existence of midsized objects, whereas ALA_O incurs no such commitment.[14] Second, ALA_O makes good sense of why beings with lives and environments like ours—which give us no reason to track dogs and trunks as one—would wind up with the particular sortal concepts we do, *tree* and not *trog*.[15] Third, TA_O is obscure: it is entirely unclear how the putative fact that matter really does divide up into trees and the like is supposed to help explain why tracking matter under the kind *tree* is selectively advantageous.

One way to see this last point is to imagine discovering that some of our ancestors *were* troglodytes, but proved to be less fit than those of our ancestors who divided up the world into trees and other such familiar kinds.[16] What could explain why they were less fit? One promising explanation is that successfully tracking matter under sortals like *trog* rather than *tree* required more cognitive resources or slowed processing. But in that case, it's not the inaccuracy of the object reactions that explains why they were less fit but rather their cognitive limitations. It's hard even to make sense of the idea that dividing the successfully tracked matter into trogs is less

[14] While there is room for doubt about whether postulating composites in addition to simples renders a theory less parsimonious (cf. Schaffer 2015), requiring the composites to do critical explanatory work does seem to reduce the simplicity of the theory. In any event, ALA_O is no *less* parsimonious than TA_O. Thanks to Elanor Taylor for helpful discussion.

[15] Though see Hirsch 1993, chap. 5, on some difficulties for explaining why strange ways of dividing the world would in principle be maladaptive.

[16] Another way to see it is to imagine that *we* have it wrong and that in fact there are trogs but no trees. In that case, would troglodytes thereby enjoy a selective advantage? It's hard to see why.

adaptive or less efficient than dividing successfully tracked matter into trees because there really are trees but no trogs.

For these reasons, the evolutionary vindication of object reactions is unconvincing. One does not need to assume that object reactions are accurate in order to explain their adaptive value. The adaptive value of our object reactions can be explained—and, it would seem, explained just as well (if not better)—on the assumption that they reliably misrepresent how distributions of matter divide up into midsized objects.

13.5 BOOTSTRAPPING TO VINDICATION

Our object reactions cannot be vindicated on the grounds that their evolutionary explanations make indispensable reference to their accuracy. They don't. There is, however, a more promising way of vindicating our object reactions.

Even when some phenomenon can be explained without reference to a certain range of entities or facts, we may still sometimes have good reason to prefer a less parsimonious explanation that does make reference to them. Suppose, for instance, that I observe the creation of a xion in a particle accelerator. This is typically the result of a single yion colliding with a single zion, though it could in principle result from two yions colliding with a zion and overdetermining the creation of the xion. Which explanation should I prefer? If all I observe is the effect, the newly created xion, it would be unreasonable to prefer the latter explanation. But if I actually observed the two yions simultaneously colliding with the zion, then given my total evidence I should prefer the second explanation. This is so even though the second is a less parsimonious explanation of the appearance of the xion considered in isolation.

Similarly, my object reactions can either be explained simply by reference to the atoms arranged midsized-objectwise or, less parsimoniously, by both the atoms and the midsized objects they compose. Which explanation should I prefer? If my observations include only the effects, the object-representing mental states, then it would be unreasonable to adopt the latter explanation. But this isn't all that I have observed. I have observed the midsized objects themselves. I have seen dogs and trees and even a few tigers, and I can see that they have volumes and masses and everything else it takes to be causally interacting with things. So, given my total evidence—which includes the noninferential, direct perceptual evidence that there are dogs and trees—I should prefer the second explanation, on which my object reactions are caused by midsized objects. This is so even though this is a less parsimonious explanation of my object reactions (the mental states) considered in isolation.

There is, to be sure, a glaring disanalogy between the two cases. In the xion case, I presumably have independent evidence of the reliability of whatever equipment is informing me about the number and behavior of the yions. But when it comes to my object reactions, I have no independent evidence of their reliability, no independent means of confirming that there really are trees out there—not just trogs and branches, or just atoms arranged treewise—answering to my tree experiences. And yet, despite lacking any independent evidence for the reliability of my object reactions, here I am reasoning from those reactions to the conclusion that midsized objects are causing my experiences.

To put a label on it, let's say that one *bootstraps* from the deliverances of some source of information when one reasons from the deliverances of that source to a vindication of that source, while having no independent evidence of its reliability or of the accuracy of those particular deliverances. What makes the object-reaction reasoning different from xion-creation reasoning is the bootstrapping.

There is something unsettling about bootstrapping. It reeks of "question-begging." And yet bootstrapping is rampant in the moral-debunking literature. Moral realists almost invariably respond to debunking arguments by reasoning from their antecedent moral beliefs—the very beliefs that the debunker is calling into question—to the conclusion that evolution is more or less bound to lead us to accurate moral beliefs. For instance, one can find Robert Nozick and David Enoch advancing something like the following response: It's no accident that we evolved to believe that protecting our children is good and obligatory, since such moral beliefs motivate us to perform actions that promote inclusive fitness. Moreover, protecting our children *is* good and obligatory (there's the bootstrapping). So, it's no accident that we have accurate moral beliefs.[17]

Indeed, consider what Street herself has to say on behalf of her antirealist response to the moral-debunking arguments:

> [O]f course it is no coincidence that there is such a striking overlap between the content of evaluative truths and the content that natural selection would have tended to push us toward. Of course it's no coincidence that, say, breaking one's bones *is* bad and that's also exactly what evolutionary theory would have predicted we think. But whereas the realist is forced to offer the scientifically unacceptable tracking explanation of this overlap, the antirealist is able to give a very different account. (2006, 153)

In other words, the antirealist thesis that moral reactions determine the moral facts best explains the striking overlap between adaptive moral beliefs and accurate moral beliefs. What, though, is her reason for thinking that they overlap? Evidently it's that believing bone-breaking to be bad is adaptive *and it's true that it's bad*; believing we have obligations to our children is adaptive *and it's true that we do*; and so forth. What reason is there, though, to believe that it's true that bone-breaking is bad or that we have obligations to our children? Street is silent on the matter, presumably

[17] See Nozick 1981, 342–48, and Enoch 2010, secs. 3–5; 2011, chap. 7. Other bootstrappers include Copp 2008, Schafer 2010, Wielenberg 2010, Skarsaune 2011, and Clarke-Doane 2015, 2016.

because she takes herself to be stating the obvious—the obvious deliverances of her moral reactions. She is bootstrapping from these reactions to an antirealist vindication of her moral reactions.

How can metaethicists be so complacent about bootstrapping? The short answer (which is all that we will have space for here) is that, as incredibly tempting as it may be to dismiss bootstrapping as question-begging, one of the lessons of contemporary epistemology is that we must resist temptation on pain of sweeping skeptical results.[18] For instance, one might complain that the bootstrapping is question-begging because the bootstrapper is relying on beliefs that the debunker does not agree are true. But, if we really are prohibited from relying on beliefs that one's skeptical interlocutors find suspect, then merely *encountering* a skeptic who insists that none of your beliefs are justified would be enough to render you unable to escape their arguments, however bad those arguments may be. So we had better not accept any such evidence neutrality constraint.

Alternatively, one might complain that the bootstrapping is question-begging because the bootstrapper relies on her moral or perceptual faculties without any independent evidence of their reliability. But requiring that one always have independent evidence of the reliability of a source of information before relying on it leads straight to a fairly obvious regress. Each source would have to be validated by some further source, which would have to be validated by another source, ad infinitum. Without infinitely many distinct sources to draw on, the chain of sources must either form a loop, with some source ultimately validating itself, or the chain must come to an end, with an unvalidated source ultimately validating all the others. If relying on self-validating and unvalidated sources is forbidden—as it would be if we require all sources to be independently validated—then we are stuck with the intolerable skeptical conclusion that *no* sources can be relied upon. Thus, on pain of global skepticism, one must allow that there are some basic sources to which we are simply entitled to help ourselves, even without independent evidence of their accuracy. And, once one recognizes the need for this, it is hard to see what would be wrong with believing what follows deductively, inductively, or abductively from the deliverances of those basic sources—including when what follows is some vindication of the sources themselves (e.g., that they are reliable).

I obviously don't expect to have won many over by these brief remarks in defense of bootstrapping. Still, I'm right about this, and I will proceed on the assumption that *if* there's something wrong with bootstrapping vindications of object reactions and moral reactions, it's not merely that they involve bootstrapping.[19] That said,

[18] For the long answer, see Nozick 1981, 197–98; Alston 1986; Bonjour 1998, sec. 5.5; Huemer 2001, sec. 8.2; Van Cleve 2003; Bergmann 2004; Pryor 2004; Feldman 2005, 95; Williamson 2007, sec. 7.3; Kelly 2008; Vogel 2008; Titelbaum 2010; and Alexander 2011 on the skeptical ramifications of universal demands for evidence neutrality, independent evidence, and/or noncircularity.

[19] Those familiar with this literature might wonder whether one can embrace bootstrapping as I have without also licensing Roxanne's track-record reasoning from Vogel 2000. The answer is yes; see Bergmann 2004 and Vogel's more bootstrapping-friendly 2008 for two promising approaches.

whatever entitlement we have to trust these sources, and whatever justification we have for beliefs formed on their basis, is surely defeasible. The remainder of the paper is devoted to examining what might be able to defeat these beliefs. We'll begin with the suggestion that the beliefs are defeated once they are shown to be *insensitive* to the facts in question.

13.6 SENSITIVITY

13.6.1 Merricks's Overdetermination Argument

Trenton Merricks advances an overdetermination argument against midsized inanimate objects.[20] Any phenomenon that we might ordinarily point to as being caused by midsized inanimates is already being caused by the atoms that putatively compose it. Accordingly, the idea goes, we should deny that the midsized inanimates exist, on pain of implausibly and unparsimoniously taking such phenomena to be systematically overdetermined by midsized objects and pluralities of atoms.

Merricks anticipates the bootstrapping response advanced in the previous section, that I am entitled to believe in midsized objects because I can see them and see that they have everything it takes to enter into causal relations. In response, Merricks argues that any perceptual reasons we have for believing in midsized objects are undermined as soon as we acknowledge that such objects would be mere overdeterminers.

How? "[E]ven if there were no statues," he says, "our perceptions of 'each statue' would be the same, caused by atoms arranged statuewise working in concert" (2003, 739). And this realization is meant to act as a defeater for our beliefs:

> [O]nce a perceiver realizes that her perceptual experiences of statues would be the same whether or not statues existed, it seems clear that those experiences no longer justify her belief in statues (in addition to atoms arranged statuewise). It seems clear that this realization "defeats" any justification that might have come from seeing statues. (2003, 739–40)

Extrapolating somewhat, we get the following argument:

(S1) You would have believed that there are statues even if it were false that there are statues, and you realize this.

(S2) If you realize that you would have believed p even if p were false, then you are not justified in believing p.

(S3) So, you are not justified in believing that there are statues.[21]

[20] See Merricks 2001, chap. 3; 2003. See his 2001, chap. 4, on why some composites are spared from elimination.

[21] Merricks never *explicitly* endorses anything as general as S2. It may be that the argument would be more faithfully rendered as an argument from analogy, turning on the premise that there is no

S1 is meant to be supported by the observation that statues (if they exist) are mere overdeterminers. Since the presence and activities of the atoms arranged statuewise is already enough to produce the experience as of a statue, the absence of a statue composed of them would make no difference to whether you end up having such an experience.

S2 says that recognizing that a belief is insensitive—that you would have had it even if it were false—undermines the belief.[22] In support of the claim that recognized insensitivity is a defeater, Merricks presents the following case:

> You gaze upon the Emerald City. Its buildings appear to be green. You are then informed that your glasses have green lenses. Thus you learn that the buildings would appear green to you even if they were some other color. And so you are no longer justified in believing that the buildings are green. (2003, 739)

Recognized insensitivity, the idea goes, explains why you lose your justification upon learning about the colored lenses. And if this is what undermines justification here, it plausibly undermines justification wherever else it rears its head.[23]

In what follows, I'll grant S2 for the sake of the argument, and focus just on S1.[24] But before turning to S1, it will be illuminating to consider an episode from the history of metaethics in which sensitivity also makes a notable appearance. Curiously, in this episode the roles are reversed: sensitivity is wielded *by* the bootstrapper *against* the debunker.

13.6.2 The Harman-Sturgeon Debate

Well before the recent surge of interest in evolutionary moral-debunking arguments, there was the Harman/Sturgeon debate. Gilbert Harman (1977, chap. 1) maintained that moral reactions can be explained simply by appeal to mundane natural facts.

epistemically significant difference between belief in statues and your color beliefs in Emerald City (more on Emerald City shortly). But this premise is questionable for the same sorts of reasons given subsequently against S1. We have good reason to think that our color beliefs in Emerald City are insensitive (given our previous encounters with tinted lenses), whereas we don't have good reason to think that our statue beliefs are insensitive (as I argue later). That would seem to be an epistemically significant difference.

22 The premise really ought to include a restriction to methods, to handle well-known counterexamples (see Nozick 1981, 179–85). I ignore this complication since what I say below applies equally to more complicated formulations of the sensitivity principle.

23 Merricks also offers a second argument in support of what I am calling S2; see my 2015, chap. 10, for discussion of his 2001, 72–76, argument from strange fusions. See Benovsky 2015 and Osborne 2016, 216, for similar attempts to impugn object beliefs on grounds of insensitivity.

24 See my 2015, sec. 7.2.2, for criticism of S2. See Vogel 2012 for a deluge of arguments that insensitivity does not preclude knowledge, arguments that can be repurposed to show that recognized insensitivity does not preclude justification.

This threatens to make moral facts entirely irrelevant to the explanation of our moral reactions, which in turn threatens to show that our moral reactions cannot serve as evidence for moral beliefs formed on their basis.

In reply, Nicholas Sturgeon (1988) suggested a way of checking whether the moral facts are or aren't explanatorily relevant. Explanations tend to support counterfactuals: if A explains B, then had it not been the case that A it would not have been the case that B.[25] Accordingly, to get to the bottom of whether moral facts explain moral reactions, we should ask ourselves whether we would have had the same moral reactions had the moral facts been different—or, in other words, whether our reactions are sensitive to the facts. If the reactions are sensitive, that's reason to think that the moral facts are explaining the reactions; if they're insensitive, that's reason to think that the moral facts aren't explaining the reactions.

Following Sturgeon (1988, 246–48), let's apply the test to Harman's own case. You turn the corner, see the kids setting fire to the cat, and immediately judge them to be doing something wrong. To check whether the fact that they are doing something wrong explains your judgment, we need to assess the following conditional:

(CF) You would have judged that they were engaged in wrongdoing even had they not been engaged in wrongdoing.

If CF is true, then the fact that they're engaged in wrongdoing fails the test for explanatory relevance. If CF is false, then that fact passes the test for explanatory relevance.

We know how to assess conditionals like CF. You imagine a situation in which the kids aren't doing something wrong but that is otherwise as similar as possible to how things actually are. That is, you imagine the "closest world" in which they aren't doing something wrong. Then you check whether in that situation you judge them to be doing something wrong.[26] What, then, is the closest world in which they're not doing something wrong?

Error theorists and realists will predictably reach different conclusions when they ask themselves this question. The error theorist thinks that nothing has any moral properties, in which case setting fire to the cat isn't even actually wrong. Thus, by her lights, the closest no-wrongdoing world is the actual world, a world in which they're setting fire to a cat. Since that world (a.k.a. this world) is one in which you do judge them to be doing something wrong, CF comes out true.

Now for the moral realist. What, by her lights, is the closest world in which the kids aren't engaged in any wrongdoing? Not one in which they are doing the same as what they actually are doing. We'd have to go to pretty far-out worlds to find ones in which the kids are setting fire to a cat and yet (by realist lights) doing no wrong—maybe worlds where cats are unaffected by fire, or where setting fire to the

[25] See Sturgeon 1988, 245 n. 19, on some notable exceptions.

[26] Cf. Stalnaker 1968 and Lewis 1973.

cat will foil some terrorist plot. Rather, the closest worlds in which they're doing nothing wrong are ones in which they're "engaged in some robust but innocent fun" (Sturgeon 1988, 247), like arm-wrestling or racing their bikes. Since we're not absolute curmudgeons—liable to think the kids are up to no good no matter what they're doing—we don't in these close-by worlds judge that they're engaged in wrongdoing. CF thus comes out false by the moral realist's lights, and the moral facts pass the test for being relevant to our moral reactions. So, moral realists should conclude that moral facts do explain our moral reactions.

Is this line of reasoning somehow illicit? True, the moral realist is taking for granted the accuracy of her moral reactions in the face of the explanatory challenge, without offering any independent evidence of their reliability. But, as Sturgeon emphasizes (1988, 237–38), there is nothing dialectically illicit about that. Harman's concern is that even if the moral facts are as the moral realist takes them to be, they are explanatorily irrelevant. Harman is therefore inviting us to start from the assumption that the moral facts are as we think they are and see if he is right about what follows. Nor, I would add, is there anything *epistemically* illicit about the reasoning. As we saw in section 13.5, it had better at least sometimes be legitimate to believe the deliverances of a source of information even when one has no independent evidence of its reliability. Some auxiliary argument would be needed to establish that our moral faculties are not among those sources for which we enjoy this default entitlement.

Here then, as I see it, is the upshot of the Harman/Sturgeon debate. Sturgeon is right. Counterfactuals like CF cannot be assessed without first taking a stand on what the moral facts actually are. And just as you are entitled to rely on your moral reactions in deciding whether to call the cops on the kids, you are entitled to rely on your moral reactions in assessing Harman's concern that moral facts don't explain our moral reactions. Drawing on such reactions, those of us who were realists to begin with ought to think that CF is false and that the facts do explain the reactions.

13.6.3 The Sensitivity of Object Beliefs

Now back to Merricks's argument, which similarly turns on assessing our reactions to CF-like conditionals. Recall S1:

(S1) You would have believed that there are statues even if it were false that there are statues, and you realize this.

Should we accept S1? Certainly a statue eliminativist, who believes only in the atoms arranged statuewise, will think it's true of you. After all, the eliminativist thinks the closest world in which it's false that there are statues is the actual world—in which case, since you (I assume) do actually believe in statues, S1 comes out true.

What, though, should the statue realist think is the closest world in which there are no statues? Not one in which there are atoms arranged statuewise. We'd have to go to pretty far-out worlds to find ones in which there are atoms arranged statuewise but no statues—perhaps all the way out to impossible worlds. Rather, the closest worlds in which there are no statues are perfectly normal statueless worlds in which there are no atoms arranged statuewise either. Since we're not weirdly fanatical about statues—liable to think there are statues when there aren't even atoms arranged statuewise around—we'll judge in such close-by worlds that there are no statues. S1 comes out false by the statue realist's lights.

Just like the moral realist in section 13.6.2, the statue realist is within her dialectical and epistemic rights to rely on her object reactions when reasoning her way to the falsity of S1. Merricks and other eliminativists are of course welcome to accept S1, but noneliminativists shouldn't be convinced by his argument.

Some, I suspect, will complain that Sturgeon and I have missed the point. If you vary the moral facts *while holding fixed* what the kids are doing, they'll say, our moral beliefs would be the same; if you vary the midsized object facts *while holding fixed* how all the atoms are arranged, they'll say, our object beliefs would be the same. Maybe so.[27] But it's hard to see what of interest is supposed to follow from these counterfactuals. To derive any interesting epistemic consequences from them, we'd have to replace S2 with something like the following:

(S2′) If you realize that you would have believed that p even if (i) p were false and (ii) all non-p facts were exactly the same, then you are not justified in believing p.

But this is going to undermine beliefs that moral skeptics and object eliminativists never meant to challenge. You look at a reliable thermometer—or, if you insist, atoms arranged thermometerwise—that reads 87 degrees, and you're justified in believing that it's 87 degrees. But you would have believed this even if (i) it weren't 87 degrees and (ii) the thermometer still read 87 degrees. So S2′ wrongly classifies your belief as unjustified.[28]

13.7 LIMITS ON BOOTSTRAPPING

13.7.1 Explanatory Concessions

We have a default entitlement to believe and reason from the deliverances of our perceptual and moral faculties. The entitlement is of course defeasible. We have just been assessing one candidate defeater—insensitivity—but we did not find any convincing reason to agree that the targeted reactions are insensitive.

[27] Though I have my doubts; see my 2015, 98.

[28] Cf. Sturgeon 1988, 250–53. I leave it as an exercise for the reader to try to find a sensitivity constraint that undermines moral beliefs and object beliefs without overgeneralizing in this way.

I turn now to another putative defeater: the discovery that our moral reactions and object reactions bear no explanatory connection to their subject matter. Plausibly, discovering, or even just conceding (rightly or wrongly), that one is forming one's beliefs about some domain D for reasons having entirely nothing to do with the D-facts is poised to serve as a defeater for any justification one might otherwise have for one's D-beliefs.[29] This suggests an explanatory constraint on justification: one is justified in believing p only if one believes that it is at least in part because of p that one believes p. (We'll see a more adequate formulation of the constraint in section 13.8.) This is why it would be troubling if Merricks were right that our object beliefs are insensitive: if indeed they are, then that would suggest that it's not *because* there are statues that we believe in statues, which in turn undermines our entitlement to such beliefs. This is why it is troubling to find out that you are wearing green lenses in Emerald City: your entitlement to your color beliefs is defeated once (and because) you become convinced that it's because the lenses are green, and not because the buildings are green, that you believe the buildings to be green. Ultimately, it is the explanatory concessions that defeat the beliefs.

Crucially, Sturgeon *does* accept that moral facts play a role in explaining our moral reactions. Indeed, he invokes sensitivity precisely for the purpose of establishing the explanatory relevance of moral facts. So he doesn't run afoul of this defeater. But forget about Sturgeon, and consider Shmurgeon. Shmurgeon concedes that there is no explanatory connection between his moral reactions and the moral facts—maybe because he can't see how moral features could have any kind of causal efficacy, or because he thinks that citing moral facts would be objectionably redundant or unparsimonious. Shmurgeon concedes that his belief that the kids are engaged in wrongdoing is not even partly explained by the action's in fact being wrong. Nevertheless, Shmurgeon stands by his sensitivity beliefs. Like Sturgeon, he bootstraps from his moral reactions to the conclusion that CF is false: the closest no-wrongdoing worlds are ones in which the kids are racing their bikes.

This, I submit, is epistemically impermissible. Once you concede that your moral faculties render the verdicts they do about the moral facts for reasons having nothing at all to do with what the moral facts actually are, you lose any entitlement you had for relying on those faculties. Shmurgeon accepts that it is not because the actions are wrong that he believes or perceives or intuits them to be wrong. He accepts that which moral features things have has nothing to do with why he makes the judgments he does. And once he makes this explanatory concession he thereby loses his entitlement to the moral judgments, and bootstrapping from them is no longer an option.

The lesson here is that sensitivity derives whatever significance it has from its connection to explanation. Sensitivity is a mark of an explanatory connection. The

[29] Cf. Benacerraf 1973; Sayre-McCord 1988, secs. 5–6; Field 1989, 25–30; Gibbard 2003, chap. 13; Locke 2014; and Korman 2015, chap. 7.3.

absence of the former is some evidence for the absence of the latter, and discovering the absence of the explanatory connection is what does the defeating. Accordingly, once one has made the explanatory concession, it is neither here nor there that one can reason one's way from the disconnected beliefs to sensitivity. For whatever justification those beliefs once had has already been undermined by the explanatory concession.

13.7.2 Flouting the Explanatory Constraint

The above observation about explanatory connections and their relation to sensitivity has ramifications for a certain style of response to debunking arguments found both in the objects literature and in the moral literature. I'll start with the former.

In his recent book, Thomas Hofweber (2016, chap. 7.3) offers a response to Merricks's argument that is somewhat different from the one I have advanced. Whereas I maintain that midsized objects and their atomic parts overdetermine our object reactions, Hofweber suggests that we can stand by our object beliefs "even if Merricks is right about the simples doing all the causal work" (195):

> How our beliefs are causally produced in us is not the crucial question for epistemology. Whether they track [i.e., are sensitive to[30]] the facts is relevant for this, and the answer is that they do, or so we have reason to think. (194)

And what reason do we have for thinking that our object beliefs track the object facts? Our object reactions, to which we are defeasibly entitled, and from which we can bootstrap to the sensitivity of the reactions:

> The counterfactual situation relevant to evaluate the counterfactual conditional about my cup not being there [viz. that if the cup didn't exist, then I wouldn't believe it exists] isn't one where the simples are still there, but they somehow don't compose a cup, but rather one where both the simples and the cup are gone. But then nothing would cause me to believe in a cup in front of me. (194)

On a natural reading, then, Hofweber is saying that our object beliefs are sensitive to the object facts even though the facts don't causally explain the beliefs (they do no "causal work").[31]

[30] Hofweber doesn't explicitly define tracking in terms of sensitivity, but he does informally gloss the former in terms of the latter: "My belief that there is a cup in front of me does track the cup in the sense that if the cup didn't exist then I wouldn't have this belief" (2016, 194).

[31] Since I wrote this chapter, Hofweber has informed me that this is a misconstrual. He does say that the object facts do no *causal work* but doesn't thereby mean to deny that they *causally explain* the object reactions (a distinction I'd be interested to see spelled out in greater detail). I've chosen to leave

So understood, Hofweber is a real-life Shmurgeon, bootstrapping from his object reactions to a vindication of those reactions, all the while granting that the sensitivity is not underwritten by an explanatory connection between the reactions and the facts. Thus, he runs afoul of our explanatory constraint, and thereby undermines his entitlement to the object reactions from which he bootstraps. It is neither here nor there that he can reason his way from the discredited object reactions to sensitivity. *Pace* Hofweber, how the beliefs are produced *is* the crucial question, and the answer needs to be: by the associated facts.[32]

The explanatory constraint also has ramifications for the moral-debunking literature. There, one of the dominant strategies—what Dustin Locke (2014) dubs "the minimalist response"—has been to grant M1 and instead deny M2. The strategy is to grant that one's moral reactions are not explained by the moral facts, but show how our moral reactions *could* still turn out to be sensitive, or safe, or in some other way nonaccidentally accurate. Do we have reason to think that this not only could happen, but has happened? Yes, the idea goes, for we can bootstrap from our moral reactions to the conclusion that they are nonaccidentally accurate (recall the Nozick/Enoch reasoning from section 13.5).[33]

What minimalists are presumably thinking is that explanatory connections have epistemic import only insofar as they are decent indicators of nonaccidental accuracy. Justin Clarke-Doane (2015, sec. 6), for instance, insists that if—using our antecedent moral beliefs—we can assure ourselves that our moral beliefs are "bound to be true," it couldn't possibly matter whether one accepts an explanatory connection. But if the explanatory constraint is correct, then this sort of bootstrapping cannot even get off the ground; the explanatory concession undermines our entitlement to the moral reactions before the bootstrapping begins. Indeed, I think the minimalist has things exactly the wrong way round: once we concede that there is no explanatory connection between our moral beliefs and the moral facts, it couldn't possibly matter that we can use the discredited, disconnected beliefs to assure ourselves that the beliefs are bound to be true.[34]

the criticism in the text, with apologies to Hofweber, since I suspect some will still be tempted by the strategy I understood him to be advancing—and I didn't want to try readers' patience by changing it to a critique of Shmofweber.

[32] See my 2014, secs. 3–4; 2015, chap. 7, on other attempts to block the object-debunking arguments while granting that there is no explanatory connection.

[33] Minimalist strategies have been advanced or defended by Nozick 1981, 342–48; Dworkin 1996, 117–26; Huemer 2005, 218–19; Enoch 2010, secs. 3–5; 2011, sec. 7.4; Schafer 2010; White 2010, 588–89; Wielenberg 2010, secs. 4–8; 2014, chap. 4; 2016, sec. 3; Brosnan 2011, 60–63; Parfit 2011, 532–33; Skarsaune 2011, sec. 3; Berker 2014; Clarke-Doane 2015, secs. 4–6; 2016, secs. 2–4; Talbott 2015; Moon 2017; and Schechter 2018, sec. 6. Some on this list are more explicit (or aware) than others about the need for bootstrapping.

[34] See Korman and Locke (forthcoming) for more on minimalist responses to moral-debunking arguments.

13.8 WHAT IS THE EXPLANATORY CONSTRAINT?

What we have seen is that debunking arguments may be resisted by bootstrapping from antecedent beliefs, so long as bootstrappers respect the explanatory constraint on their entitlement to those beliefs. But what exactly is the explanatory constraint? Let's start simple and work our way to a viable formulation.[35] Impatient readers can jump down to EC5 to see where we are headed.

We'll begin with EC1:

(EC1) S is justified in believing p only if: [S believes p] because p.

In other words: only if the fact that p at least partly explains S's believing p. This is obviously too demanding. An unwitting brain in a vat is justified in believing that it has hands even though its belief that it has hands is not explained by its having hands.

This leads us to EC2:

(EC2) S is justified in believing p only if: S believes that [[S believes p] because p].

Unlike EC1, EC2 does not imply that the brain in a vat is unjustified, since the brain at least *thinks* that its beliefs are explained by the associated facts. But EC2 cannot be right as it stands because it cannot accommodate inductive beliefs. Take, for instance, beliefs about the future. The sun's rising tomorrow clearly doesn't explain your belief that it will, and you realize this. Nevertheless, your belief that it will rise tomorrow is surely justified. Other counterexamples to EC2 include cases where the belief and the associated fact have a common cause. I believe that the email I sent you is in your inbox, not because it is in your inbox, but because I clicked *send*.

The fix is to allow beliefs to be justified even when they're not believed to be explained by the associated facts, so long as they're prima facie justified by beliefs that themselves are believed to be explained by the associated facts:

(EC3) S is justified in believing p only if: *either* (i) S believes that [[S believes p] because p], *or* (ii) S's belief that p is based on beliefs $Bp_1 \ldots Bp_n$ that prima facie justify S in believing that p and S believes that [[S believes $p_1 \ldots p_n$] because $p_1 \ldots p_n$].

Your belief that the sun will rise tomorrow is based on your beliefs that it rose today and yesterday and the day before that; those beliefs do prima facie justify your belief that it will rise tomorrow; and you *do* think that its rising today explains your belief that it rose today, and likewise for the rest of the beliefs in your inductive base. So

[35] For some alternative formulations of the constraint, see Locke 2014, 232; McCain 2014, sec. 4.4 and 6.4; Lutz 2018, sec. 2; Schechter 2018, sec. 3; and Korman and Locke forthcoming, sec. 8.

the second disjunct is satisfied. Similarly, my belief that my email is in your inbox is prima facie justified by my belief that I clicked *send*, and I do think that my belief that I clicked *send* is explained by my clicking *send*.

EC3 makes room for justified inductive beliefs. But it threatens to overintellectualize justification (as does EC2). EC3 will entail that a young child—old enough to have beliefs about apples but too young to have beliefs about beliefs—cannot be justified in believing that there is an apple present. This can be handled by weakening the constraint so that what's required isn't belief but rather not withholding belief:

(EC4) S is justified in believing p only if: *either* (i) S doesn't withhold belief that [[S believes p] because p], *or* (ii) S's belief that p is based on beliefs $Bp_1 \ldots Bp_n$ that prima facie justify S in believing that p and S doesn't withhold belief that [[S believes $p_1 \ldots p_n$] because $p_1 \ldots p_n$].

Withholding belief is something more deliberate than failing to believe.[36] It involves entertaining a proposition and then opting not to believe it, either by disbelieving it or by remaining agnostic. A young child who, lacking the requisite concepts, cannot entertain propositions about what she does or doesn't believe won't count as withholding belief in those propositions. Counterexample averted.

EC4 permits justified inductive beliefs without depriving brains in vats and young children of justified perceptual beliefs. But it is still open to counterexamples like the following. Upon naming my dog "Emmylou," it is not her being named "Emmylou" that explains my belief that her name is "Emmylou." Rather, it's my taking her name to be "Emmylou" that explains why she is named "Emmylou." Since I recognize that the explanation doesn't run from the fact to the belief, EC4 entails that I'm not justified in believing her name to be "Emmylou." But obviously I am so justified.

The fix is to allow explanations to run in either direction: world to mind or mind to world. It's fine to acknowledge that my belief that p isn't explained by p so long as I think that my believing p explains why it's true that p. Let's say that the belief that p is *e-connected* to p iff either the belief that p is explained by the fact that p or vice versa. Now we can say:

(EC5) S is justified in believing p only if: *either* (i) S doesn't withhold belief that [S's belief that p is e-connected to p], *or* (ii) S's belief that p is based on beliefs $Bp_1 \ldots Bp_n$ that prima facie justify S in believing p and S doesn't withhold belief that [S's beliefs that $p_1 \ldots p_n$ are e-connected to $p_1 \ldots p_n$].

[36] See Bergmann 2005, 420–21 and Locke 2014, 223, on withholding.

I do believe that my belief that my dog's name is "Emmylou" is e-connected to her being named "Emmylou," since I take the former to explain the latter. So EC5 won't entail that the belief is unjustified.[37] This revision also makes room for antirealist treatments of the debunking arguments, denying O1 or M1 on the grounds that our reactions determine the associated facts.[38]

These refinements won't affect my earlier objections to Hofweber. Hofweber's strategy (as I interpreted it) involves withholding belief as to whether I believe there is a cup before me because there is a cup before me. And he certainly doesn't think that the explanation runs in the other direction, that there is a cup before me because I believe there to be one. So condition (i) of EC5 is not satisfied. What about condition (ii)? The only reasons Hofweber cites for his cup beliefs are his cup experiences. Trivially, then, his cup beliefs are not based on any beliefs that he takes to be e-connected to the cup facts. So condition (ii) is not satisfied either.

One might suggest that there is still a problem with EC5, insofar as it cannot accommodate foundational a priori beliefs about necessary truths. On pain of antirealism or skepticism about mathematics, EC5 looks to require us to accept (or at least not withhold belief) that mathematical facts explain our mathematical beliefs. Some may find this intolerable. But one's modus tollens, as they say, is another's modus ponens. I say that realists cannot stand by the belief that 1 + 1 is 2 once they concede that they believe this for reasons having nothing whatsoever to do with what the sum of 1 and 1 actually is. Benacerraf (1973) and Field (1989) are right that this is an unstable epistemic position. To be sure, I agree that it's unclear how mathematical facts could explain our beliefs. But, unless you're ready to embrace mathematical skepticism or antirealism, you had better be prepared to accept that they somehow manage to do so.[39]

13.9 CONCLUSION

Here is what I hope to have shown. Even what would seem to be some of our most secure beliefs, our perceptual beliefs about midsized objects, are not safe from debunking by cognitive science and evolutionary psychology. But even when cognitive scientists or evolutionary psychologists are able to explain some range of beliefs without reference to the associated range of facts, we needn't accept that these explanations are the full story. For we are often entitled to bootstrap to an expanded

[37] Actually, it's probably more complicated than that. It's my *intention* that her name be "Emmylou" that explains both why her name is "Emmylou" and why I believe that her name is "Emmylou." Accordingly, it's the second clause of EC5 that handles this case: my belief that I intend for her name to be "Emmylou" is explained by my having that intention; I believe that this is so; and my belief that I intended that this be her name, together with the belief that she is mine to name (which I take to be e-connected to the fact that she is mine to name), prima facie justify my belief that this is her name.

[38] Cf. Goldman 1992, chaps. 2–3, for subjectivist "demotions" of objects, color, and modality in response to challenges from cognitive science.

[39] See Bengson 2015 for one proposal.

explanation that does cite the associated facts. That said, one's bootstrapping had better culminate in embracing some such explanatory connection, for conceding that there is no connection would undermine one's entitlement to bootstrap.

I certainly don't take myself to have resolved the dispute between the debunker and the bootstrapper. In truth, I think that we have only just scratched the surface. Among other things, I think that there are further constraints on legitimate bootstrapping than simply allowing some explanatory connection between your beliefs and associated facts. Elsewhere, I have argued that the explanatory connection one takes to hold between beliefs and the facts must be "nondeviant," and that bootstrappers have their work cut out for them showing how a causal connection between midsized objects and object beliefs could be nondeviant.[40] If I am right about this, then bootstrappers cannot dismiss the object-debunking arguments simply on the grounds that midsized objects are poised to cause our object beliefs. And it is worth considering whether the threat of deviance also arises for responses to the moral-debunking arguments according to which moral properties are proximate or supervenient causes of moral beliefs,[41] which is something I hope to explore in future work.

ACKNOWLEDGMENTS

Many thanks to David Alexander, Nir Ben-Moshe, Stephen Biggs, Chad Carmichael, Julianne Chung, Billy Dunaway, Thomas Hofweber, John Hummel, Dustin Locke, Angela Mendelovici, Trenton Merricks, Geoff Sayre-McCord, Elanor Taylor, and Jon Tsou, and to audiences at Iowa State, Ohio State, Prindle Institute, University of Virginia, and Wayne State for helpful discussion.

REFERENCES

Alexander, D. J. (2011). In defense of epistemic circularity. *Acta Analytica* 26: 223–41.

Alston, W. P. (1986). Epistemic circularity. *Philosophy and Phenomenological Research* 47: 1–30.

Baron, S. (2017). Feel the flow. *Synthese* 194: 609–30.

Baron, S., Cusbert, J., Farr, M., Kon, M., and Miller, K. (2015). Temporal experience, temporal passage and the cognitive sciences. *Philosophy Compass* 10: 560–71.

Bedke, M. S. (2009). Intuitive non-naturalism meets cosmic coincidence. *Pacific Philosophical Quarterly* 90: 188–209.

Benacerraf, P. (1973). Mathematical truth. *Journal of Philosophy* 70: 661–79.

Bengson, J. (2015). Grasping the third realm. *Oxford Studies in Epistemology* 5: 1–38.

Benovsky, J. (2015). From experience to metaphysics. *Noûs* 49: 684–97.

Bergmann, M. (2004). Epistemic circularity: Malignant and benign. *Philosophy and Phenomenological Research* 69: 709–27.

[40] See my 2014, secs. 5–6 and 2015, sec. 7.4.

[41] Cf. Shafer-Landau 2012, 27, and Mogensen 2015.

Bergmann, M. (2005). Defeaters and higher-level requirements. *Philosophy Quarterly* 55: 419–36.

Berker, S. (2014). Does evolutionary psychology show that normativity is mind-dependent? In J. D'Arms and D. Jacobson, eds., *Moral Psychology and Human Agency*. Oxford: Oxford University Press, 215–52.

Bogardus, T. (2016). Only all naturalists should worry about only one evolutionary debunking argument. *Ethics* 126: 636–61.

BonJour, L. (1998). *In Defense of Pure Reason*. Cambridge: Cambridge University Press.

Braddock, M. (2016). Evolutionary debunking: Can moral realists explain the reliability of our moral judgments? *Philosophical Psychology* 29: 844–57.

Braddock, M. (2017). Debunking arguments from insensitivity. *International Journal for the Study of Skepticism* 7 (2): 91–113.

Brosnan, K. (2011). Do the evolutionary origins of our moral beliefs undermine moral knowledge? *Biology and Philosophy* 26: 51–64.

Chalmers, D. (2006). Perception and the fall from Eden. In T. S. Gendler and J. Hawthorne, eds., *Perceptual Experience*. Oxford: Oxford University Press, 49–125.

Clarke-Doane, J. (2015). Justification and explanation in mathematics and morality. *Oxford Studies in Metaethics* 10: 80–103.

Clarke-Doane, J. (2016). Debunking and dispensability. In N. Sinclair and U. Leibowitz, eds., *Explanation in Ethics and Mathematics*. Oxford: Oxford University Press, 23–36.

Copp, D. (2008). Darwinian skepticism about moral realism. *Philosophical Issues* 18: 186–206.

Crisp, T. M. (2016). On naturalistic metaphysics. In K. J. Clark, ed., *The Blackwell Companion to Naturalism*. Malden, MA: Wiley Blackwell, 61–74.

Dreier, J. (2012). Quasi-realism and the problem of unexplained coincidence. *Analytic Philosophy* 53: 269–87.

Dworkin, R. (1996). Objectivity and truth: You'd better believe it. *Philosophy and Public Affairs* 25: 87–139.

Enoch, D. (2010). The epistemological challenge to metanormative realism. *Philosophical Studies* 148: 413–38.

Enoch, D. (2011). *Taking Morality Seriously*. Oxford: Oxford University Press.

Feldman, R. (2005). Respecting the evidence. *Philosophical Perspectives* 19: 95–119.

Field, H. (1989). *Realism, Mathematics, and Modality*. Oxford: Blackwell.

Fitzpatrick, W. J. (2015). Debunking evolutionary debunking of ethical realism. *Philosophical Studies* 172: 883–904.

Fraser, B. (2014). Evolutionary debunking arguments and the reliability of moral cognition. *Philosophical Studies* 168: 457–73.

Gibbard, A. (2003). *Thinking How to Live*. Cambridge, MA: Harvard University Press.

Goldman, A. (1992). *Liaisons: Philosophy Meets the Cognitive and Social Sciences*. Cambridge, MA: MIT Press.

Gould, S. J., and Lewontin, R. C. (1979). The spandrels of San Marco and the Panglossian paradigm. *Proceedings of the Royal Society of London* 205: 581–98.

Gow, L. (2014). Colour. *Philosophy Compass* 9: 803–13.

Harman, G. (1977). *The Nature of Morality*. New York: Oxford University Press.

Hawthorne, J. (2006). *Metaphysical Essays*. Oxford: Oxford University Press.

Heller, M. (1990). *The Ontology of Physical Objects: Four-Dimensional Hunks of Matter*. New York: Cambridge University Press.

Hirsch, E. (1993). *Dividing Reality*. Oxford: Oxford University Press.

Hoffman, D. D., Singh, M., and Prakash, C. (2015). The interface theory of perception. *Psychonomic Bulletin Review* 22: 1480–506.

Hofweber, T. (2016). *Ontology and the Ambitions of Metaphysics*. Oxford: Oxford University Press.

Hudson, H. (2001). *A Materialist Metaphysics of the Human Person*. Ithaca, NY: Cornell University Press.

Huemer, M. (2001). *Skepticism and the Veil of Perception*. Lanham, MD: Rowman & Littlefield.

Huemer, M. (2005). *Ethical Intuitionism*. New York: Palgrave.

Hummel, J. E. (2015). Is interface theory a special case of Q-morphs or full-blown solipsism? *Psychonomic Bulletin and Review* (invited commentary). http://www.psychonomic.org/news/297868/Is-Interface-Theory-a-Special-Case-of-Q-morphs-or-Full-blown-Solipsism.htm.

Joyce, R. (2005). Moral fictionalism. In M. E. Kalderon, ed., *Fictionalism in Metaphysics*. Oxford: Oxford University Press, 287–313.

Joyce, R. (2006). *The Evolution of Morality*. Cambridge, MA: MIT Press.

Kelly, T. (2008). Common sense as evidence: Against revisionary ontology and skepticism. *Midwest Studies in Philosophy* 32: 53–78.

Kitcher, P. (2005). Biology and ethics. In D. Copp, ed., *The Oxford Handbook of Ethical Theory*. Oxford: Oxford University Press, 163–85.

Korman, D. Z. (2014). Debunking perceptual beliefs about ordinary objects. *Philosophers' Imprint* 14: 1–21.

Korman, D. Z. (2015). *Objects: Nothing out of the Ordinary*. Oxford: Oxford University Press.

Korman, D. Z., and Locke, D. (forthcoming). Against Minimalist Responses to Moral Debunking Arguments. *Oxford Studies in Metaethics*.

Leech, D., and Visala, A. (2011). Naturalistic explanation for religious belief. *Philosophy Compass* 6: 552–63.

Leslie, S.-J. (2013). Essence and natural kinds: When science meets preschooler intuition. *Oxford Studies in Epistemology* 4: 108–66.

Lewis, D. (1973). *Counterfactuals*. Malden, MA: Blackwell.

Lillehammer, H. (2003). Debunking morality: Evolutionary naturalism and moral error theory. *Biology and Philosophy* 18: 567–81.

Locke, D. (2014). Darwinian normative skepticism. In M. Bergmann and P. Kain, eds., *Challenges to Moral and Religious Belief*. Oxford: Oxford University Press, 220–36.

Lutz, M. (2018). What Makes Evolution a Defeater? *Erkenntnis* 83: 1105–26.

Marr, D. (1982). *Vision*. Cambridge, MA: MIT Press.

McCain, K. (2014). *Evidentialism and Epistemic Justification*. New York: Routledge.

Mendelovici, A. (2013). Reliable misrepresentation and tracking theories of mental representation. *Philosophical Studies* 165: 421–43.

Mendelovici, A. (2016). Why tracking theories should allow for clean cases of reliable misrepresentation. *Disputatio* 8: 57–92.

Merricks, T. (2001). *Objects and Persons*. Oxford: Oxford University Press.

Merricks, T. (2003). Replies. *Philosophy and Phenomenological Research* 67: 727–44.

Mogensen, A. L. (2015). Evolutionary debunking arguments and the proximate/ultimate distinction. *Analysis* 75: 196–203.

Moon, A. (2017). Debunking morality: Lessons from the EAAN literature. *Pacific Philosophical Quarterly*. 98 (S1): 208–26.

Nozick, R. (1981). *Philosophical Explanations*. Cambridge, MA: Harvard University Press.

Osborne, R. C. (2016). Debunking rationalist defenses of common-sense ontology. *Review of Philosophy and Psychology* 7: 197–221.

Parfit, D. (2011). *On What Matters*. Vol. 2. Oxford: Oxford University Press.

Paul, L. A. (2010). Temporal experience. *Journal of Philosophy* 107: 333–59.

Paul, L. A. (2016). Experience, metaphysics, and cognitive science. In J. Sytsma and W. Buckwalter, eds., *A Companion to Experimental Philosophy*. Malden, MA: Wiley Blackwell, 419–33.

Pautz, A. (2011). Can disjunctivists explain our access to the sensible world? *Philosophical Issues* 21: 384–433.

Pryor, J. (2004). What's wrong with Moore's argument. *Philosophical Perspectives* 14: 349–78.

Rea, M. C. (2002). *World without Design*. Oxford: Oxford University Press.

Rose, D., and Schaffer, J. (2017). Folk mereology is teleological. *Noûs* 51(2): 238–70.

Ruse, M. (1986). *Taking Darwin Seriously*. Amherst, NY: Prometheus.

Sattig, T. (2015). *The Double Lives of Objects*. Oxford: Oxford University Press.

Sayre-McCord, G. (1988). Moral theory and explanatory impotence. *Midwest Studies in Philosophy* 12: 433–57. Reprinted in his *Essays on Moral Realism*.

Schafer, K. (2010). Evolution and normative scepticism. *Australasian Journal of Philosophy* 88: 471–88.

Schaffer, J. (2015). What not to multiply without necessity. *Australasian Journal of Philosophy* 93: 644–64.

Schaffer, J. (2016). Cognitive science and metaphysics: Partners in debunking. In B. P. McLaughlin and H. Kornblith, eds., *Goldman and His Critics*. Malden, MA: Wiley Blackwell, 337–65.

Schechter, J. (2018). Explanatory challenges in metaethics. In T. McPherson and D. Plunkett, eds., *Routledge Handbook of Metaethics*. New York: Routledge, 443–59.

Shafer-Landau, R. (2012). Evolutionary debunking, moral realism and moral knowledge. *Journal of Ethics and Social Philosophy* 7: 1–37.

Sider, T. (2001). *Four-Dimensionalism*. Oxford: Clarendon Press.

Skarsaune, K. (2011). Darwin and moral realism. *Philosophical Studies* 152: 229–43.

Stalnaker, R. C. (1968). A theory of conditionals. In N. Rescher, ed., *Studies in Logical Theory*. Oxford: Blackwell, 98–112.

Street, S. (2006). A Darwinian dilemma for realist theories of value. *Philosophical Studies* 127: 109–66.

Street, S. (2008). Reply to Copp: Naturalism, normativity, and the varieties of realism worth worrying about. *Philosophical Issues* 18: 207–28.

Stroud, B. (2000). Evolution and the necessities of thought. In *Meaning, Understanding, and Practice*. Oxford: Oxford University Press, 52–67.

Sturgeon, N. (1988). Moral explanations. In G. Sayre-McCord, ed., *Essays on Moral Realism*. Ithaca, NY: Cornell University Press, 229–55.

Sumner, P., and Mollon, J. D. (2000). Catarrhine photopigments are optimized for detecting targets against a foliage background. *Journal of Experimental Biology* 203: 1963–86.

Talbott, W. J. (2015). How could a "blind" evolutionary process have made human moral beliefs sensitive to strongly universal, objective moral standards? *Biology and Philosophy* 30: 691–708.

Titelbaum, M. G. (2010). Tell me you love me: Bootstrapping, externalism, and no-lose epistemology. *Philosophical Studies* 149: 119–34.

Van Cleve, J. (2003). Is knowledge easy—or impossible? Externalism as the only alternative to skepticism. In S. Luper, ed., *The Skeptics*. Burlington, VT: Ashgate, 45–59.

Vavova, K. (2015). Evolutionary debunking of moral realism. *Philosophy Compass* 10: 104–16.

Vogel, J. (2000). Reliabilism leveled. *Journal of Philosophy* 97: 602–23.

Vogel, J. (2008). Epistemic bootstrapping. *Journal of Philosophy* 105: 518–39.

Vogel, J. (2012). The enduring trouble with tracking. In K. Becker and T. Black, eds., *The Sensitivity Principle in Epistemology*. Cambridge: Cambridge University Press, 122–51.

White, R. (2010). You just believe that because … *Philosophical Perspectives* 24: 573–615.
Wielenberg, E. J. (2010). On the evolutionary debunking of morality. *Ethics* 120: 441–64.
Wielenberg, E. J. (2014). *Robust Ethics*. Oxford: Oxford University Press.
Wielenberg, E. J. (2016). Ethics and evolutionary theory. *Analysis* 76: 502–15.
Wilkins, J. S., and Griffiths, P. E. (2012). Evolutionary debunking arguments in three domains. In G. W. Dawes and J. Maclaurin, eds., *A New Science of Religion*. New York: Routledge, 133–46.
Williamson, T. (2007). *The Philosophy of Philosophy*. Malden, MA: Blackwell.

14 Cognitive Science for the Revisionary Metaphysician

David Rose

Many philosophers insist that the revisionary metaphysician—i.e., the metaphysician who offers a metaphysical theory which conflicts with folk intuitions—bears a special burden to explain why certain folk intuitions are mistaken. But though it is widely agreed that the revisionary metaphysician incurs such an explanatory burden, many philosophers think that the revisionist is unlikely to be able to successfully discharge this explanatory burden (e.g., Paul 2012, 22; Hirsch 2002, 117; and Korman 2009). My plan is to offer some resources to the revisionary metaphysician to meet this explanatory burden. Specifically, my proposal is that evidence from cognitive science can help the revisionary metaphysician discharge the explanatory burden of providing a plausible account of how the folk have gone wrong. In taking this up, I'm not going to engage in a discussion of the abstract prospects for how cognitive science *might* end up helping discharge this explanatory burden. Instead, my strategy will be to show how cognitive science can *actually* help the revisionary metaphysician discharge this explanatory burden by taking up a concrete illustration. In particular, I will discuss metaphysical disputes over composition and persistence and various charges of violating folk intuitions. I'll then provide an overview of a range of empirical evidence suggesting that the folk operate with a benighted teleological view of composition and persistence, and go on to argue that there is a debunking explanation for folk intuitions of composition and persistence. Given an empirically informed understanding of the folk view of composition and persistence, the revisionary metaphysician has the resources to offer a plausible debunking explanation of folk intuitions. In this way, I'll take myself to have illustrated one key way in which evidence from cognitive science can help the revisionary metaphysician discharge the explanatory burden of providing a plausible account of how the folk have gone wrong.

Though my main goal is to illustrate how cognitive science can contribute to metaphysics by helping discharge the explanatory burden faced by the revisionary metaphysician, I have another goal. Recently, Dan Korman and Chad Carmichael (2017) have provided a range of objections against the particular debunking explanation I'll be focusing on. Two objections are particularly troubling. The first is that the teleological view of composition that the folk operate with may well be legitimate. If this is right, then it will undercut the attempt to debunk folk intuitions and leave the explanatory burden faced by the revisionary metaphysician untouched. The second is that, in the studies I'll be discussing, the folk don't report their intuitions about the relevant cases. Instead, they merely provide answers. If this is right, then the relevant folk intuitions don't deserve to be debunked. The studies tell us nothing about what the folk intuit. Since the debunking argument has as its focus folk intuitions and since the studies allegedly don't tell us anything about what the folk actually intuit, the debunking argument fails. If either of these two objections is right, the illustration of how cognitive science can help the revisionary metaphysician discharge the explanatory burden will be an utter failure. So my other goal will be to address these two objections.

The Plan: I will begin, in section 14.1, with some stage setting, discussing how I'll be understanding intuitions and their targets. I'll also set out one main reason why conflict with folk intuitions gives rise to an explanatory burden. I will then briefly provide an overview of metaphysical disputes over composition and persistence and the explanatory burden that arises for the revisionary metaphysician in connection with these disputes. In section 14.2, I'll illustrate how cognitive science can help the revisionary metaphysician discharge the explanatory burden in the case of composition and persistence. Section 14.3 responds to two main objections.

14.1 STAGE SETTING

I want to begin with a bit of stage setting before getting on to the illustration of how cognitive science can help the revisionary metaphysician discharge the explanatory burden (section 14.2). To this end, I want to briefly set out:

- How I'll be understanding intuitions
- How I'll be understanding the target of intuitions
- A rationale for why folk intuitions might be taken seriously and the explanatory burden that arises when a theory conflicts with folk intuitions
- The main metaphysical disputes I will focus on (i.e., composition and persistence)

Having accomplished this, section 14.2 will set out the kind of teleological view I take the folk to operate with (i.e., promiscuous teleomentalism), some of the evidence suggesting that teleological considerations play a key role in folk intuition of composition and persistence, and the debunking argument for folk intuitions of composition and persistence.

14.1.1 Intuitions

What are intuitions? There's an important debate over how intuitions are best characterized with a wide range of proposals on offer (e.g., Bealer 1998; Cappelen 2012; Deutsch 2015; Devitt 2015; Pust 2016; Sosa 1998; Weinberg 2014; Williamson 2007). That debate is a swamp, which I won't wade into here. So, I'm simply going to set out the way I'll be understanding intuitions.

Following Stich 2018 (see also Williamson 2007), intuitions are spontaneous judgments—which immediately arise without any awareness of their origin and without having gone through any conscious process of reflection or reasoning—that some person, object, event, etc. described in a scenario has (or lacks) some interesting or important property or relation. Understanding intuitions in this way reflects how they are understood within dual-processing models of cognition (e.g., Epstein 1994; Sloman 1996; Chaiken and Trope 1999; Kahneman and Frederick 2002; Stanovich and West 2000) and is analogous to how they are understood in linguistics (Stich 2018).

Linguists typically gather data by presenting native speakers with sentences and asking participants, for instance, whether the target sentence is grammatical. Often the target sentence will elicit an intuitive judgment about whether it is grammatical, that is, an intuition about grammaticality that arises spontaneously without the subject being aware of its origin or of having gone through any conscious process of reflection or reasoning. "Colorless green ideas sleep furiously" is a familiar example. We have the intuition that it is grammatical (though meaningless): the judgment that it is grammatical arises spontaneously and immediately without our engaging in any conscious reflection or reasoning. Similarly, philosophers present scenarios or "thought experiments" intended to elicit intuitive judgments from their audience. To take one example, in Trolley cases, a scenario is presented in which a protagonist in the scenario is faced with a decision to flip a switch to divert a trolley onto a sidetrack. If the switch is not flipped, the trolley will kill five people on the main track; if it is flipped, the trolley will kill only one person who is on the sidetrack. The audience is then asked whether it would be morally permissible to flip the switch. Many people offer a judgment which arises spontaneously and immediately without being aware of its origin or of having gone through any conscious process of reflection or reasoning (e.g., Hauser et al. 2008). In short, they intuit whether or not it is morally permissible to flip the switch.

14.1.2 The Target of Intuitions

What is the target of intuitions? Intuitions can have one of two targets. The first target is *mentalist*, where intuitions have as their target in-the-head psychological entities, most notably concepts; the second target is *extramentalist*, where intuitions have as their target outside-the-head nonpsychological entities (Goldman and Pust 1998).

Which view one takes on the target of intuitions depends, in part, on what philosophical project is being undertaken. If one's project is to engage in conceptual analysis, such as an analysis of the concept of causation, knowledge, and the like, then insofar as the theorist takes intuitions to play a role in conceptual analysis, intuitions will be targeted at uncovering the features of the relevant concept, its content or its extension. But if one's project is to understand the outside-the-head phenomena themselves, for instance, the nature of causation, the nature of knowledge, and the like, then intuitions will have as their target outside-the-head nonpsychological entities, the phenomena themselves.

There's some controversy over whether the appropriate target for intuitions is mentalist or extramentalist. Goldman (2007) maintains that the only legitimate target for intuitions is mentalist since it is deeply puzzling how intuitions could provide evidence about the extramentalist phenomena themselves (e.g., knowledge, causation, etc.). So intuitions, insofar as they play a legitimate role in philosophical theorizing, do so largely in the course of conceptual analysis. But as Stich and Tobia (forthcoming) note, "[F]or most of human history, perception posed a comparable mystery, and the appeal to intuition in mathematics still does." So while we may lack an account of how intuitions could be linked to extramentalist phenomena, this shouldn't lead to skepticism that they can be legitimately used in this way. And as Sosa (2007) points out, intuitions are often used in this way. That is, philosophers who rely on intuitions often take the content of the intuition to be about the world, and not merely about what is in one's head, and so on this view the content of the intuition is taken to be true (provided the intuition hasn't been triggered in error or is due to a bias, etc.).

In the metaphysical disputes I'll be discussing, it is not always clear what target for intuitions the theorist has in mind. For present purposes, I'll simply assume that they take the target of intuitions to be extramentalist. There are two main reasons for this. First, given the mereological disputes I'll be discussing, it is more natural to understand the appeal to folk intuitions as being useful insofar as they serve as a guide as to what there is in the world. For instance, mereological nihilism—the view that composition never occurs—holds that there are no composite objects. It's hard to make sense of this view if it is born out of an analysis of our concept of composition. Second, conceptual analysis has fallen out of favor for quite some time,[1] and so it seems reasonable to take contemporary metaphysicians, unclear as they are as to what they take the targets of intuitions to be, to assume that intuitions are targeted at the extramentalist phenomena themselves.

[1] Though see, e.g., Jackson 1998 and Kriegel 2017.

14.1.3 Taking Folk Intuitions Seriously and the Explanatory Burden

Why might folk intuitions be taken seriously? And why think an explanatory burden arises for those who endorse theories that conflict with folk intuitions? A number of different rationales could be offered for why folk intuitions might be taken seriously. But since my main goal is to illustrate how cognitive science can help the revisionary metaphysician discharge the explanatory burden of providing a plausible explanation of how the folk have gone wrong, I'm only going to focus on one rationale. Lewis (1986) tells us that "Common sense is a settled body of theory—unsystematic folk theory—which at any rate we do believe; and I presume that we are reasonable to believe it. (Most of it.)" (134). So if it is reasonable to believe common sense, then a theory that conflicts badly with it would appear to have some explaining to do. Specifically, a theory that conflicts with it should give us reasons for thinking that the relevant aspect of the folk view is wrong and tell us just how it is wrong. Indeed, as Korman notes, "[V]irtually everyone agrees that, even after having presented the arguments for their positions, proponents of revisionary philosophical theories—that is, those that deviate from the pretheoretical conception—are required to provide some sort of account of the conflict between their theories and the pretheoretical beliefs of non-philosophers ("the folk")" (Korman 2009, 242). That is, the revisionary metaphysician, in offering a metaphysical theory which departs from common sense or folk intuitions, accrues an explanatory burden in that she is required to provide a plausible account of how the folk have gone wrong.[2]

14.1.4 Mereology and the Explanatory Burden

To illustrate how cognitive science can help the revisionary metaphysician discharge the explanatory burden, I'm going to focus on two issues connected with mereology. Mereology is roughly concerned with parthood relations, the relations of part to whole and the relations of part to part within a whole. Specifically, I'm going to focus on disputes over the question of when mereological composition occurs—when do some parts make a whole?—and a connected question over how a whole—an object—persists through alterations to its parts.

Metaphysical theories of composition and persistence are often charged with violating folk intuitions. In short: a number of metaphysical theories of composition and persistence are revisionary in that they allegedly conflict with folk intuitions. With respect to composition, one example of a revisionary theory is universalism. The universalist thinks that composition occurs under any circumstance whatsoever. It is completely unrestricted. It always occurs. On this view, there exists, for

[2] This isn't to say metaphysicians are the only philosophers confronted with this kind of explanatory burden. For instance it also arises in ethics (Mackie 1977; see Rose and Nichols 2016 for an overview).

instance, a fusion of my nose and a doorknob, a fusion of a trout and a turkey, and even a fusion of the moon and a piece of cheese. Perhaps unsurprisingly, universalism is typically charged with violating folk intuitions. Presumably the folk do not recognize the kind of fusions that the universalist recognizes. Universalism is thus typically regarded as a revisionary metaphysical theory.[3] And indeed one of the main objections against universalism is that it violates folk intuitions, though the charges are typically disputed (see Rose and Schaffer 2017). Thus the universalist is faced with the challenge of providing some plausible explanation of how it is that the folk have gone wrong.

So too with persistence. Again to take just one example, Burke's (1994) sortal-based account of persistence is typically charged with violating folk intuitions. On Burke's view, when a sculptor takes a piece of copper and fashions it into a statue, the piece of copper is destroyed, and in its place there comes to exist a new piece of copper which is in turn identical to the resultant statue. One main objection against Burke's sortal-based account is that it violates folk intuitions. Nobody, not even children, thinks that an object is destroyed upon merely assuming a certain shape (Lowe 1995). Burke, however, disputes these charges.[4] Insofar as this account does conflict with folk intuitions, it counts as a revisionary metaphysical theory. Given this, the metaphysician who embraces such a view of persistence is faced with the task providing some plausible explanation of how it is that the folk have gone wrong.

In general, if a metaphysical theory is revisionary, then the theorist will often be confronted with the explanatory burden. But if this is the case, how might the revisionary metaphysician discharge the explanatory burden? One way would be to argue that the theory doesn't conflict with the relevant folk intuitions.[5] On this way of proceeding, the alleged conflict between the relevant folk intuitions and the deliverances of the theory would only be apparent, not genuine. Another way would be for the revisionist to embrace the conflict between the relevant folk intuitions and the deliverances of the theory as being genuine and seek out an empirically informed debunking explanation of the relevant folk intuitions. This is my preferred strategy. So, moving on to this, I want to now discuss the particular view of composition and persistence—folk mereology—that the folk operate with and some of the evidence supporting this before getting on to the debunking explanation.

[3] I'm focusing on universalism simply for illustrative purposes. A number of other views have also been charged with violating folk intuitions and so count as revisionary metaphysical theories. For an overview see Rose and Schaffer 2017.

[4] Here again, I've only focused on Burke's view for illustrative purposes. For more see Rose 2015 and Sidelle 2002.

[5] See Rose and Schaffer 2017; Rose, Schaffer, and Tobia, forthcoming; Rose et al., forthcoming; and Rose 2015 for documentation of a number of philosophers claiming that their view doesn't really conflict with folk intuitions.

14.2 PROMISCUOUS TELEOMENTALISM, FOLK MEREOLOGY, AND DEBUNKING

14.2.1. Promiscuous Teleomentalism

My view is that the folk heavily indulge in promiscuous teleomentalist thinking. There are two parts to this view. The first is that the folk are promiscuous teleologists: teleological considerations inform their understanding of artifacts, organisms, and even nonliving natural things like rocks. Second, the specific teleological view the folk adopt is teleomentalism, where teleology is understood psychologically, in terms of intentions, goals, and purposes.[6] In other words, the folk view reality as a whole as being infused with agency and purpose and thus the folk indulge in promiscuous teleological thinking because they are promiscuous telelomentalists.

Why think that the folk are promiscuous teleomentalists? There are two relevant lines of psychological research. One concerns promiscuous teleology; the other concerns teleomentalism underwriting promiscuous teleological thinking. I'll briefly discuss each line of evidence (for further details see Rose 2015 and Rose and Schaffer 2017).

First, regarding promiscuous teleology, it has been found, for instance, that children insist that lions are for "going to the zoo," that clouds are "for raining" (Bloom 2007, 150), that "mountains exist to give animals a place to climb," and that rocks are pointy "so that animals won't sit on them and smash them" (Kelemen 1999, 1444–45). Promiscuous teleological thinking isn't merely confined to children. Instead, it extends into adulthood. For instance, Lombrozo et al. (2007) found that adults with Alzheimer's disease, who tend to suffer from deficits in background causal beliefs, naturally default to accepting promiscuous teleological explanations. Even those without cognitive deficits, such as college-educated adults, accept promiscuous teleological explanations such as "The sun radiates heat because warmth nurtures life," "Fungi grow in forests to help decomposition," and "Lightning occurs to release electricity" with ease (Kelemen and Rossett 2009). In fact, even professional physical scientists naturally accept promiscuous teleological explanations when their cognitive resources are limited, such as when in a speeded task (Kelemen et al. 2013). This research suggests that promiscuous teleological thinking emerges early in childhood and is retained into adulthood. And though it may be masked in certain kinds of cases (e.g., as with professional physical scientists explicitly considering the acceptability of promiscuous teleological explanations), it is retained nonetheless. It represents a deep-seated, natural default perspective on the world.

[6] This contrasts with teleonaturalism, where teleological claims are to be understood in ways that do not refer to the intentions, goals, or purposes of psychological agents' (see Allen and Bekoff 1994, 13, for a discussion of teleonaturalism and teleomentalism). Though teleonaturalism may be a respectable form of teleology, I would note that this is a stipulated/revisionary idea introduced to legitimize certain uses of teleology in biology, and was never intended to correspond to the actual concept the folk operate with.

Second, concerning teleomentalism, the evidence suggests that the folk take an agentive perspective on reality as a whole. Importantly, teleomentalism isn't tied into any particular background religious views. Kelemen and DiYanni (2005) report a strong tendency among children—both from religious and from nonreligious backgrounds—to an "intuitive theism" in which nature is viewed as an artifact of a creator, as well as a significant correlation between viewing something teleologically and regarding it as created. Moreover, in recent research by Rottman et al. (2016), they found that even Chinese people, despite being in a largely atheistic culture, succumb to promiscuous teleological thinking. Though background religious views don't appear to be heavily tied into promiscuous teleological thinking, endorsement of background Gaia beliefs (Kelemen et al. 2013; see also Kelemen 2012)—beliefs that nature is a living, powerful, goal-directed being—has been shown to significantly predict people's tendency to engage in promiscuous teleological thinking. On top of all this, decades of research in science education has found that an agentive perspective on reality is one of the main obstacles students confront—religious and nonreligious alike—in acquiring a proper understanding of natural selection (see Kelemen 2012 for an overview). We have what Pascal Boyer (2001) has called a "hypertrophy of social cognition"—a willingness to attribute purpose, agency, and design, even when it is inappropriate to do so—and are "hypersensitive to signs of purpose, design and agency, so much so that we see purpose where all that really exists is artifice or accident" (Bloom 2007, 150). In short: the evidence suggests that promiscuous teleological thinking is underwritten by teleomentalism.

So research in psychology suggest that the folk are promiscuous teleomentalists. The main question now is whether promiscuous teleomentalism plays a role in folk mereology.

14.2.2 Folk Mereology

I won't rehearse the full details of the work on folk mereology (see Rose 2015 and Rose and Schaffer 2017 for further details). Instead, I will discuss some of the highlights to set out some support for the view that promiscuous teleomentalist thinking underwrites folk intuitions of composition and persistence. First, composition.

Beginning with artifacts, people were given a case featuring an unfamiliar artifact, a "gollywag." They were told about two individuals, Smith and Jones, experimenting with the gollywags. In one case, the no-purpose case, participants were told that Jones superglues some of the gollywags together. In the purpose case, people were again told that Jones superglued the gollywags together. But they were also told that Jones had a sore back, placed the superglued gollywags on his chair, sat on them for the rest of the day, and no longer had a sore back. In both cases a disagreement ensues between Smith and Jones with Smith saying that the gollywags do not compose an object and Jones saying they do. Whether the gollywags had a purpose strongly affected people's intuitions about whether composition had occurred: people were significantly more inclined to think that composition had occurred when the gollywags had a purpose.

It turns out that teleological considerations aren't merely confined to intuitions about composition when considering artifacts. They also play a role in intuitions about composition when considering biological organisms and nonliving natural things like rocks. Concerning biological organisms, people were again told that Smith and Jones are experimenting, but this time with mice. In one case Jones superglues some mice together. In another case he superglues them together, runs them through a maze, and finds that they are very successful at detecting bombs. In the former case, the superglued mice had no purpose; in the latter case, they did. Again, there was an impact of teleology on intuitions about composition, with people being significantly more inclined to think that the superglued mice composed an object when they had a purpose. For the cases involving nonliving natural things, rocks, people were told about an individual who lives on the side of a mountain. One evening an avalanche occurs and rocks are scattered across the individual's front lawn. What was varied was whether the protagonist thought that the scattered plurality of rocks had no purpose, thought they had a purpose, or arranged them for a purpose. The key finding was that when the scattered plurality of rocks had no purpose, people denied that composition had occurred, but when the protagonist thought the arrangement of rocks had a purpose or arranged them for a purpose, people thought that composition had occurred.

Promiscuous teleology also plays a significant role in intuitions about persistence. In one study people were told about an individual, John, who is hiking and spots a glowing rock. The rock houses a special sort of microorganism which feeds off minerals in the rock's interior. But the microorganisms can't access the minerals deep in the rocks interior. The microorganisms begin dying and the rock begins fading. So John tries an experiment. He smashes the rock into three pieces with a hammer. In one case, the microorganisms continue dying and the thing fades to black. In the other case, the microorganisms can access the minerals and the thing resumes glowing. In the former case people thought the rock did not survive the smashing; in the latter case people thought the rock survived the smashing. So even when turning to persistence, we continue to see a pattern of promiscuous teleological intuitions.

Turning now to teleomentalism, one of the main reasons for thinking that promiscuous teleology is underwritten by teleomentalism comes from the background research in psychology. But there is some more direct evidence. In particular the same rock case, where it was smashed into three pieces, was rerun. But this time people's background Gaia beliefs were assessed.[7] Utilizing a causal modeling procedure, it was

[7] Kelemen et al. (2013) utilize the following probes as a measure of belief that reality as a whole is infused with agency (i.e., Gaia beliefs):

(1) I believe Nature is driven to preserve living things.
(2) I believe the Earth is alive.

found that Gaia beliefs caused people's teleologically laden intuitions about persistence. This suggests that the tendency toward promiscuous teleology is underwritten by teleomentalism. Taken together, the evidence from Rose (2015) and Rose and Schaffer (2017) suggests that folk intuitions of composition and persistence are tied into promiscuous teleomentalism. [8]

14.2.3 Debunking Folk Mereological Intuitions and Meeting the Explanatory Burden

The revisionary metaphysician, in offering a metaphysical theory that conflicts with folk intuitions, is often faced with an explanatory burden: she is required to provide a plausible explanation of how the folk could have gone wrong. And one case where the explanatory burden is faced is in disputes over mereology, specifically in discussions of composition and persistence. While one typical strategy for meeting the explanatory burden is to deny that there is any conflict between the deliverances of the theory and folk intuitions, my suggestion was that the revisionary metaphysician might do better to confront the explanatory burden head on and seek out an empirically informed debunking explanation of the relevant folk intuitions. But how might the revisionary metaphysician debunk these aspects of folk mereology and thereby discharge the explanatory burden?

One constraint on a metaphysical theory is that it shouldn't conflict with what our best science says about the world. This isn't to say that every metaphysical question is answerable to our best science. Nor is it to say that metaphysical theorizing is held entirely hostage by our best science (Ladyman et al. 2007). Instead I am only making the more modest claim that metaphysics, insofar as it is appropriate to the issue at hand, should be constrained, in part, by what our best science says about the world. In this way, I view science and metaphysics as enjoying a joint partnership, mutually informing one another (Paul 2012). That said, there is an obvious respect in which promiscuous teleomentalism conflicts with what our best science says about the world. Teleology has, for the most part, been purged from the natural sciences.[9]

(3) I believe that Nature is a powerful being.
(4) I believe the Earth is driven to provide optimal conditions for Life.

In the study under consideration only (1) was used (see Rose 2015, 118). See Dink and Rips 2017 for a critical discussion of whether these measures, especially (3), tap into background beliefs about agentive forces affecting the natural world.

[8] I would also note that some recent work on folk intuitions about causation has also found that Gaia beliefs cause people's promiscuous teleological causal intuitions (Rose 2017). I would also add, though not pursue in any detail here, that it may be that folk mereology is underwritten by causal intuitions. The idea here is that people understand composition and persistence causally and given that causal intuitions are tied into promiscuous teleomentalism, it may be that these promiscuous teleomentalist tendencies run through causal intuitions to affect both composition and persistence intuitions.

[9] Teleological notions play a role in biology and there is a dispute over whether appeals to teleology in biology might be naturalistically respectable (see Allen 2009 and Allen and Bekoff 1994 for

Indeed, the rejection of a teleological perspective on all of the natural world traces at least as far back to the emergence of modern science from medieval Aristotelianism. It's arguably due, at least in part, to the rejection of a teleological perspective on all of the natural world that science has made great strides over the past several hundred years. Not to put too fine a point on it, the rejection of an agentive, teleological perspective on the natural world is also one reason why modern-day intelligent design "science" is widely rejected. If scientists were to accept some version of intelligent design along with the agentive, teleological perspective on all of the natural world that comes along with it, science would slip into the dark ages. In short: it is widely agreed that a teleological perspective whereby all of nature—every rock and cloud—is infused with agency and purpose is part of a "superseded, pre-scientific muddle about how the world works" (Hawthorne and Nolan 2006; see also, e.g., Mayr 1998; Allen and Bekoff 1994).

The metaphysician who views herself as allied to the sciences should reject a teleological perspective on the natural world. Accordingly she should reject folk view of composition and persistence since it is encrusted with the muck and funk of an outmoded teleological perspective. Indeed, as Laurie Paul (2012) notes: "after drawing on experience to develop a theory, in evaluating it we need to look back at the natural science just in case our ordinary experience of the world conflicts with what our best natural science says about the world. If it does conflict, then often the assumptions based on ordinary experience should be rejected" (17). Given the conflict between the teleological commitments of the folk in the cases of composition and persistence and the (presumptive) commitments of the revisionary metaphysician who views herself as allied with the sciences, in the specific cases of composition and persistence, there looks to be a basis to debunk folk intuitions.

Debunking can be fleshed out in different ways depending on one's background epistemological view (see e.g., Schaffer 2016 for an overview). To take just one example, Nichols (2014) explores one option for debunking—process debunking—which is based on identifying epistemically defective processes generating certain intuitions and is naturally viewed as being tied into reliabilism. But whatever one's preferred epistemological view for supplying debunking arguments, debunking should be aided by a metaphysical assessment (Schaffer 2016). And I would add, as I've already noted, that certain metaphysical assessments should be informed and guided by our best science. In the case of folk mereology, our best science helps issue a clear metaphysical assessment: teleologically infused intuitions fail to fit reality. On this basis, they deserve to be debunked.

Operating with a fairly neutral background epistemology, I view the situation as follows: conformity with folk intuitions confers prima facie justification to a theory, perhaps because, as Lewis (1986) points out, it's often reasonable to believe most

overviews). But whether or not there is some revised, naturalistically respectable view of teleology to be had in biology, it's agreed that promiscuous teleomentalism is not one such view.

of the deliverances of common sense. But once we have empirically assessed the relevant folk intuitions, in this case, two aspects of folk mereology—composition and persistence—we see that they are tied to a promiscuous teleomentalist outlook on reality. As such, they fail to fit reality. The assessment that promiscuous teleomentalism—and thereby folk mereological intuitions—fails to fit reality is a metaphysical assessment, one which is aided by what our best science says about the world. Given this, and thinking of folk intuitions as testifying as to what the relevant extramentalist phenomena are like, the intuitional testimony lacks ultima facie justification once we learn from cognitive science that the folk have a promiscuous teleomentalist view of the issue. In other words the evidence from cognitive science, coupled with a metaphysical assessment aided by our best science, helps supply an undermining defeater (Pollock 1987) for folk intuitions of composition and persistence, which substantially strips these intuitions of their evidential credentials with respect to the extramentalist phenomena at hand, thereby cutting them off from achieving ultima facie justification.

Concerning the explanatory burden faced by the revisionary metaphysician in the case of composition and persistence, the revisionary metaphysician has the resources to discharge the explanatory burden. Armed with the results delivered from cognitive science and a background commitment to metaphysical theorizing being constrained, in part, by what our best science has to say about the world, she has an explanation as to how the folk have gone wrong in the case of composition and persistence. She has the resources to supply an undermining defeater for folk intuitions about composition and persistence, thereby debunking the relevant folk intuitions and meeting the explanatory burden. In this way, evidence from cognitive science can help the revisionary metaphysician discharge the explanatory burden of providing a plausible account of how the folk have gone wrong.

14.3 OBJECTIONS

That's the illustration of how cognitive science can help the revisionary metaphysician discharge the explanatory burden. But there are two main objections to consider. If either of these is right, it will undercut the debunking explanation set out above.

14.3.1 Legitimate Teleology

The first main objection is that the teleological view the folk operate with may well be entirely legitimate (Korman and Carmichael 2017). Folk intuitions of composition don't appear to be best explained by claiming that the folk are working with anything like the crude, superstitious view embodied in promiscuous teleomentalism. Rather there is something much more innocuous at work. Folk intuitions of composition, far from being tied into promiscuous teleomentalism, are tied into an assessment of creative intentions. On this view, the intentions of a creator play a significant role in folk intuitions of composition. For instance, if a sculptor was fashioning some copper into a statue with the intention of creating a statue, it would be entirely appropriate to use the creator's intention—to create a statue—as a guide to whether the arrangement of copper composes a statue. And as Korman and Carmichael (2017) note, in all of the cases concerning composition in Rose and Schaffer 2017—cases involving artifacts, biological organisms and even nonliving natural things like rocks—the key difference between the purpose and no-purpose versions of the cases lies in a difference between whether the relevant agent has or lacks the relevant creative intention.

As I understand the objection, at least two considerations must be in play in each of Rose and Schaffer's composition cases. The first is that there should be an agent and that this agent should either have or lack an intention to create some further thing. The second is that in every case where participants are considering whether an agent has an intention to create something, they are treating the thing as an artifact. A few brief remarks on these points. When considering artifacts, consulting an agent's creative intentions is arguably relevant in considering the question of whether the arrangement of some parts compose some further object. But creative intentions would be irrelevant when considering composition if there is either no agent or if the would-be composite is a nonartifact. Since most would presumably agree that the creative intentions of an agent are relevant when considering composition for artifacts, it must be the case that creative intentions play a role in cases involving nonartifacts because people are construing the would-be composite as an artifact.

Concerning the cases of composition, Korman and Carmichael are right that in each case, whether it involved an artifact, organism, or nonliving natural thing, there was an agent with (or without) an intention to create some further thing.[10] Given this, it may well be that participants are construing the would-be composite as an artifact.

I think the creative intentions account fails to provide a satisfactory explanation of folk mereological intuitions. First, it is not even clear that it can provide a general account of folk intuitions of composition. Given that Korman and Carmichael only put it to work in attempting to provide an alternative explanation of the Rose and Schaffer results, it is unclear how the creative intentions account would extend to other cases of composition. How do the folk make composition judgments when no (actual) agent is involved in the case? What about when the would-be composite isn't plausibly construed as an artifact? What about when the agent's intention are unsuccessful? We would need a lot more detail regarding their account to answer these and related questions.

[10] One exception to this is the Avalance Accorded Function case from Rose and Schaffer (2017) since in this case the relevant agent didn't plausibly intend to *create* a rock garden from the scattered arrangement of rocks. He only accorded the plurality of rocks scattered across his lawn the purpose of being a rock garden. Korman and Carmichael (2017) maintain though that the agent at least has an intention that the scattered plurality of rocks be a rock garden (see n. 5).

Perhaps the creative intentions account isn't aimed at providing a general account of folk intuitions of composition. In that case we would need some alternative account of folk intuitions of composition for cases in which the creative intentions account is ill suited. But the view that folk intuitions of composition are tied into promiscuous teleomentalism is aimed at providing a general account of folk intuitions of composition. It thus provides a simpler, more general account of folk intuitions of composition.

Second, the creative intentions account doesn't provide a plausible explanation of folk intuitions about persistence. In those case, there was a human agent involved but the relevant agent lacked creative intentions. For instance, in the rock-smashing cases, John smashes the glowing rock—which is fading, as the microorganisms can't access minerals deep in the rocks interior—into three pieces. In one case (see Rose 2015), the microorganisms all begin dying and the thing fades to black; in the other case, the microorganisms can access the minerals after the smashing and the thing resumes glowing. It's doubtful that in either of these cases John had an intention to create anything. Even assuming that John did possess some creative intention, it seems that in both cases John presumably possesses the intention to create either a glowing rock, a better situation for the microorganisms, or both. Yet people make different persistence judgments in this case. If creative intentions play a role in persistence judgments as well, then given that they are held fixed across these cases, we shouldn't see any difference in persistence judgments. But we do. So there's good reason to think that the creative intentions account fails to fit the pattern of findings from this case.[11]

Perhaps the creative intentions account isn't aimed at providing a more general account of folk mereological intuitions. Korman and Carmichael only focused on folk intuitions about composition in the Rose and Schaffer cases, so perhaps they think that there must be some alternative account of why the folk have teleologically laden intuitions when considering persistence. On my proposal, however, there is a single view underwriting both aspects of folk mereology i.e., composition and persistence. That is, promiscuous teleomentalism underwrites folk mereological intuitions and so provides a unifying explanation of the processes underwriting folk intuitions of composition and persistence. Given this and given that it provides a smooth, simple fit to the data, we're owed some explanation of what might be going on in the persistence cases since the creative intentions account is inadequate. But Korman and Carmichael offer no such proposal.

Korman and Carmichael might point out that nonetheless even in the persistence cases involving rocks there is an agent involved and so plausibly the agent's intentions, whatever they are, may still be guiding folk judgments. Moreover, they

[11] I would flag that I'm assuming creative intentions are held fixed across these cases. But they may not be. Given that a probe assessing creative intentions wasn't used for this study, it's an open empirical question whether people actually view the creative intentions similarly in the cases.

might note, the folk are nonetheless considering the relevant things in terms of artifacts. So agent intentions are playing some yet to be specified role, and moreover in each case, people treat the relevant thing as an artifact. Against this, I would point out two things. First, it is implausible that people treat the glowing rock as an artifact. Nobody created the glowing rock. John found it by the side of a trail. To insist that the folk treat the rock as an artifact would be to saddle the folk with the kind of "uncharitable" view that Korman and Carmichael want to avoid. Second, I doubt that the creative intentions of a human agent are playing a role in the persistence cases because Gaia beliefs cause teleologically laden persistence intuitions (see Rose 2015). So an (actual) agent's creative intentions are not playing a role in underwriting teleologically laden persistence intuitions. Instead, and in line with my hypothesis, promiscuous teleomentalism looks to be underwriting these intuitions. Given that, and its coherence with background work in psychology, promiscuous teleomentalism also plausibly underwrites folk intuitions of composition. In short: promiscuous teleomentalism drives, and plays a general role in, folk mereological intuitions. And the folk don't treat every collection of things as an artifact. Thus maintain that the folk operate with a benighted view of composition and persistence. Folk mereology is unfit for real metaphysics.

14.3.2 Answers, Not Intuitions

The second objection I want to consider is that the folk, in responding to the prompts in the experiments, aren't reporting their intuitions. Instead they are merely reporting their answers.[12] If this is right, then folk mereological *intuitions* wouldn't be debunked (Korman and Carmichael 2017).

This kind of objection tends to be broad in scope.[13] If right, it wouldn't simply threaten the work on folk mereology. Instead it would extend to every claim anyone ever makes about folk intuitions, including any claim philosophers make about folk intuitions. No one would be in a position to speak to the content of folk intuitions. As a general criticism, I think there are good reasons to be suspicious of such a claim. I also think that, in the specific case of folk mereology, there is good reason to doubt that the folk are reporting answers and not intuitions.

Korman and Carmichael (2017) claim that, in contrast to reporting intuitions, giving answers involves reporting considered judgments after talking one's self out of her intuitive reactions. This suggests that reporting an answer, as opposed to an intuition, involves something like the following: upon having an intuition, a subject would, for instance, have to bring some background considerations to bear on the

[12] Of course, reporting an intuition is to report a kind of answer. Korman and Carmichael are using "reporting an answer" in the sense of "not reporting an intuition."

[13] Though Korman and Carmichael only raise it for the Rose and Schaffer studies, it is easily extended to any study (see, e.g., Bengson 2013).

task, leading her to suppress reporting the intuition and instead report an answer which is independent of the intuition. This is a somewhat complicated procedure which would surely require a good deal of cognitive effort to execute. And indeed, there is some reason for thinking that the folk have a difficult time executing such a process. Take the Cognitive Reflectivity Test (CRT) developed by Frederick (2005). Here's one test item from the CRT:

A bat and a ball cost $1.10 in total. The bat costs $1.00 more than the ball. How much does the ball cost?

This question has an incorrect but intuitive answer, $.10. To arrive at the correct answer, $.05, one needs to suppress the intuitive judgment, bring background knowledge of algebra to bear on the case, and report the answer delivered from the algebraic computation. Reporting an answer as opposed to an intuition requires a good deal of cognitive effort. And it turns out that in general, the majority of people fail the CRT (i.e., they report the intuitive but incorrect answer for one or more items). Indeed, even those who score high on the CRT (i.e., give the correct answer to each item) are not immune from giving intuitive responses. For instance, one study found that 38% of those scoring high on the CRT commit the conjunction fallacy, that some thing possessing two properties is more probable than that thing possessing a single property (Ochssler et al. 2009). Given that the majority of people fail the CRT and given that some people who score high on the CRT even naturally default to reporting intuitions in some cases, there is good reason to think that, as a default, in any given study probing folk intuitions, people are reporting intuitions and not answers.

In general, I think the claim that people are reporting answers and not intuitions should be viewed with suspicion. There is good reason for thinking that people do indeed tend to report intuitions and that it takes a good deal of cognitive effort to do otherwise. And in the specific case of folk mereology, I also think that the claim that the folk are reporting answers and not intuitions fails.

One reason it fails is because when one looks at work on folk teleology in psychology, one finds that at a very young age children naturally offer and accept promiscuous teleological explanations, for instance, thinking that "mountains exist to give people a place to climb." It's doubtful that children aren't reporting intuitions in this case. It is implausible that they are going through something like the kind of complicated cognitive procedure suggested above in order to provide answers as opposed to intuitions. And given that adults display teleologically laden responses which reflect these childhood tendencies, there is good reason for thinking that they are indeed reporting intuitions, not answers. Moreover, even though some groups of individuals tend to avoid explicit teleological thinking, such as professional physical scientists, these same individuals naturally default to a teleological thinking when their cognitive resources are limited, such as when in a speeded task. This suggest that teleological thinking is an intuitive default, suppressed only by engaging in a good deal of cognitive effort. Given that folk mereological intuitions reflect this default aspect of intuitive teleological thinking, this suggest that their judgments reflect their intuitions and not merely their answers.

14.4 CONCLUSION

The revisionary metaphysician—in offering a theory that conflicts with folk intuitions—is typically confronted with an explanatory burden in that she is required to provide a plausible explanation of how the folk have gone wrong. My suggestion was that evidence from cognitive science can help the revisionary metaphysician discharge this explanatory burden. I illustrated this by taking up metaphysical disputes over mereological composition and persistence. I then discussed evidence from cognitive science which suggests that the folk are promiscuous teleomentalists and that promiscuous teleomentalist thinking underwrites folk intuitions of composition and persistence. On the basis of this, I went on to argue that there is a debunking explanation for folk intuitions. I then responded to two main objections to the debunking argument I set out, finding both objections to be insufficient to undercut the debunking argument. Thus, I uphold the view that folk mereological intuitions concerning composition and persistence deserve to be debunked. In this way, I take myself to have illustrated one key role cognitive science can play in metaphysics; namely by helping the revisionary metaphysician discharge the explanatory burden of providing a plausible explanation of how the folk have gone wrong.

ACKNOWLEDGMENTS

Thanks to Chad Carmichael, Dan Korman, Shaun Nichols, Jonathan Schaffer, Stephen Stich, and John Turri for helpful comments on earlier versions of this chapter.

REFERENCES

Allen, C. (2009). Teleological notions in biology. In E. N. Zalta, ed., *The Stanford Encyclopedia of Philosophy*. Winter 2009 ed. http://plato.stanford.edu/archives/win2009/entries/teleology-biology/.

Allen, C., and Bekoff, M. (1994). Function, natural design and animal behavior: Philosophical and ethnological considerations. In N. S. Thompson, ed., Perspectives in Ethology, vol. 11, *Behavioral Design*. New York: Plenum Press, 1–47.

Bealer, G. (1998). Intuition and the autonomy of philosophy. In M. DePaul and W. Ramsey, eds., Rethinking Intuition. Lanham, MD: Rowman & Littlefield, 201–39.

Bengson, J. (2013). Experimental attacks on intuitions and answers. *Philosophy and Phenomenological Research* 86: 495–532.

Bloom, P. (2007). Religion is natural. *Developmental Science* 10: 147–51.

Boyer, P. (2001). Religion Explained. New York: Basic Books.

Burke, M. (1994). Preserving the principle of one object to a place: A novel account of the relations among objects, sorts, sortals and persistence conditions. *Philosophy and Phenomenological Research* 54: 591–624.

Cappelen, H. (2012). Philosophy without Intuitions. Oxford: Oxford University Press.

Chaiken, S., and Trope, Y. (1999). Dual-Process Theories in Social Psychology. New York: Guilford Press.

Deutsch, M. (2015). The Myth of the Intuitive. Cambridge, MA: MIT Press.

Devitt, M. (2105). Relying on intuitions: Where Cappelen and Deutsch go wrong. *Inquiry* 58 (7–8): 669–99.

Dink, J., and Rips, L. (2017). Folk teleology and its implications. In D. Rose, ed., Experimental Metaphysics. Bloomsbury, 207–36.

Epstein, S. (1994). Integration of the cognitive and psychodynamic unconscious. *American Psychologist* 49: 709–24.

Frederick, S. (2005). Cognitive reflection and decision making. *Journal of Economic Perspectives* 19: 25–42.

Goldman, A. (2007). Philosophical intuitions: Their target, their source, and their epistemic status. *Grazer Philosophische Studien* 74: 1–26.

Goldman, A. (2010). Philosophical naturalism and intuitional methodology (Romanell Lecture). *Proceedings and Addresses of the American Philosophical Association*, 115–50.

Goldman, A. and Pust, J. (1998). Philosophical theory and intuitional evidence. In M Depaul and W. Ramsey, eds., Rethinking Intuitions: The Psychology of Intuition and Its Role in Philosophical Inquiry. Lanham, MD: Rowman and Littlefield, 179–97.

Hauser, M., Young, L., and Cushman, F. (2008). Reviving Rawls' linguistic analogy: Operative principles and the causal structure of moral actions. In W. Sinnott-Armstrong, ed., Moral Psychology, vol. 2: The Cognitive Science of Morality: Intuition and Diversity. Cambridge, MA: MIT Press, 107–43.

Hawthorne, J., and Nolan, D. (2006). What would teleological causation be? In Metaphysical Essays. Oxford: Clarendon Press, 265–84.

Hirsch, E. (2002), Against revisionary ontology. *Philosophical Topics* 30: 103–27.

Jackson, F. (1998). From Metaphysics to Ethics: A Defense of Conceptual Analysis. Oxford: Oxford University Press.

Kahneman, D., and Frederick, S. (2002). Representativeness revisited: Attribute substitution in intuitive judgment. In T. Gilovich, D. Griffin, and D. Kahneman, eds., Heuristics and Biases: The Psychology of Intuitive Judgment. New York: Cambridge University Press, 49–81.

Kelemen, D. (1999). Why are rocks pointy? Children's preference for teleological explanations of the natural world. *Developmental Psychology* 35: 1440–52.

Kelemen, D. (2012). Teleological minds: How natural intuitions about agency and purpose influence learning about evolution. In K. S. Rosengren, S. K. Brem, E. M. Evans, and G. M. Sinatra, eds., Evolution Challenges: Integrating Research and Practice in Teaching and Learning about Evolution. Oxford: Oxford University Press, 66–92.

Kelemen, D., and DiYanni, C. (2005). Intuitions about origins: Purpose and intelligent design in children's reasoning about nature. *Journal of Cognition and Development* 6: 3–31.

Kelemen, D., and Rosset, E. (2009). The human function compunction: Teleological explanation in adults. *Cognition* 111: 138–43.

Kelemen, D., Rottman, J., and Seston, R. (2013). Professional physical scientists display tenacious teleological tendencies: Purpose-based reasoning as a cognitive default. *Journal of Experimental Psychology*: General 142: 1074–83.

Korman, D. (2009). Eliminativism and the challenge from folk belief. *Noûs* 43: 242–64.

Korman, D., and Carmichael, C. (2017). What do the folk think about composition, and does it matter? In D. Rose, ed., Experimental Metaphysics. New York: Bloomsbury, 197–206.

Kriegel, U. (2017). Metaphysics and conceptual analysis: Experimental philosophy's place under the sun. In D. Rose, ed., Experimental Metaphysics. New York: Bloomsbury, 7–45.

Ladyman, J., Ross, D., Collier, J., Spurrett, D., Spurrett, D., and Collier, J. G. (2007). Every thing must go: Metaphysics naturalized. Oxford University Press on Demand.

Lewis, D. (1986). On the Plurality of Worlds. Oxford: Basil Blackwell.

Lombrozo, T., Kelemen, D., and Zaitchik, D. (2007). Inferring design: Evidence for a preference for teleological explanation in patients with Alzheimer's disease. *Psychological Science* 18: 999–1006.

Lowe, E. J. (1995). Coinciding objects: In defense of the "standard account." *Analysis* 55: 171–78.

Mackie, J. L. (1977). Ethics: Inventing Right and Wrong. New York: Penguin.

Mayr, E. (1998). The multiple meanings of teleological. *History and Philosophy of the Life Sciences* 20: 35–40.

Nichols, S. (2014). Process debunking and ethics. *Ethics* 124: 727–49.

Ochssler, J., Roider, A., and Schmitz, P. (2009). Cognitive abilities and behavioral biases. *Journal of Economic Behavior and Organization* 72: 147–52.

Paul, L. A. (2012). Metaphysics as modeling: The handmaiden's tale. *Philosophical Studies* 160: 1–29.

Pollock, J. (1987). Defeasible reasoning. *Cognitive Science* 11: 481–518.

Pust, J. (2016). Intuition. In E. N. Zalta, ed., The Stanford Encyclopedia of Philosophy. Spring 2016 ed.

Rose, D. (2015). Persistence through function preservation. *Synthese* 192: 97–146.

Rose, D. (2017). Folk intuitions of actual causation: A two-pronged debunking explanation. *Philosophical Studies* 174 (5): 1323–61.

Rose, D., and Nichols, S. (forthcoming). From punishment to objectivity. *Mind and Language.*

Rose, D., and Schaffer, J. (2017). Folk mereology is teleological. *Noûs* 51 (2): 238–70.

Rose, D., Schaffer, J., and Tobia, K. (forthcoming). Folk teleology drives persistence judgments. *Synthese.*

Rose, D., Machery, E., Stich, S. P., Alai, M., Angelucci, A., Berniunas, R., Buchtel, E. E., Chatterjee, A., Cheon, H., Cho, I.-R., Cohnitz, D., Cova, F., Dranselka, V., Lagos, A. E., Ghadakpour, L., Grinberg, M., Hannikainen, I., Hashimoto, T., Horowitz, A., Hristova, E., Jraissati, Y., Kadreva, V., Karasawa, K., Kim, H., Kim, Y., Lee, M., Mauro, C., Mizumoto, M., Moruzzi, S., Olivola, C. Y., Ornelas J., Osimani, B., Romero, C., Rosas Lopez, A., Sangoi, M., Sereni, A., Songhorian, S., Sousa, P., Struchiner, N., Tripodi, V., Usui, N., Vazquez del Mercado, A., Volpe, G., Vosperichian, H. A., Zhang, X., and Zhu, J. (forthcoming). The ship of Theseus puzzle. In T. Lombrozo, J. Knobe, and S. Nichols, eds., *Oxford Studies in Experimental Philosophy*, Volume 3.

Rottman, J., Zhu, L., Wang, W., Schillaci, R. S., Clark, K. J., and Kelemen, D. (2016). Cultural influences on the teleological stance: Evidence from China. *Religion, Brain & Behavior* 7 (1): 17–26.

Schaffer, J. (2016). Cognitive science and metaphysics: Partners in debunking. In H. Kornblith and B. McLaughlin, eds., Goldman and His Critics. Malden, MA: Wiley-Blackwell, 337–65.

Sidelle, A. (2002). Is there a true metaphysics of material objects? *Philosophical Issues* 12: 118–45.

Sloman, S. (1996). The empirical case for two systems of reasoning. *Psychological Bulletin* 119: 3–22.

Sosa, E. (1998). Minimal intuition. In M. DePaul and W. Ramsey, eds., Rethinking Intuition. Lanham, MD: Rowman & Littlefield, 257–70.

Sosa, E. (2007). Experimental philosophy and philosophical intuition. *Philosophical Studies* 132: 99–107.

Stanovich, K., and West, R. (2000). Individual differences in reasoning: Implications for the rationality debate? *Behavioral and Brain Sciences* 22: 645–726.

Stich, S. (2018). Knowledge, intuition and culture. In J. Proust and M. Fortier, eds., Metacognitive Diversity: An Interdisciplinary Approach. Oxford: Oxford University Press, 381–94.

Stich, S., and Tobia, K. (forthcoming). Experimental philosophy's challenge to the "great tradition." *Analytica: Revista de Filosofia.*

Weinberg, J. (2014). Cappelen between a rock and a hard place. *Philosophical Studies* 171: 545–53.

Williamson, T. (2007). The Philosophy of Philosophy. Oxford: Blackwell.

15 Unbunking Arguments

A Case Study in Metaphysics and Cognitive Science

Christopher Frugé

A metainduction on the history of metaphysical debates suggests a healthy skepticism as to their eventual resolution. To some extent, of course, metainductions on most philosophical subjects lead to skepticism, but disputes in metaphysics seem particularly intractable. *When do two objects ever compose another?* The universalist says, "Always." The nihilist says, "Never." *Are the statue and the lump of clay constituting it two things or one?* The multithinger says, "Two!" The one-thinger says, "One!" *Is there an objective passage of time?* The A-theorist says, "Of course there is becoming. Time flows from past to present." The B-theorist says, "Of course there is no becoming. All times coexist in a timeless manner." At most one side can be right, but it is hard to see on what grounds one might decide between them. In this chapter, I outline three ways metaphysical disputes are prone to intractability and then suggest a way forward.

The first way in which resolutions are hard to come by is that similar problems arise for both sides of a metaphysical dispute. The second is that opposing metaphysical views are often equally descriptively powerful. The third is that metaphysical intuitions and beliefs have no source of independent confirmation. The realist metaphysician who wants to treat such disputes as substantive must give a reason to think progress can be made.

The most promising way to do so, I argue, is by giving an *unbunking argument*—the debunking argument's good twin. An unbunking argument for metaphysics would give reason to think that either all or a subset of metaphysical beliefs are caused by one or more processes that are reliable with respect to the relevant metaphysical truths, and it would do so without needing to assume any truths about the domain in question. The most promising sort of metaphysical unbunking argument, I suggest, relies on the *cross-domain strategy*. It argues that a process that is reliable in a given nonmetaphysical domain is also responsible for beliefs about a metaphysical domain and that we have good reason to think this process remains reliable from one domain to the other. By way of demonstration, I examine an unsuccessful attempt to defend our modal beliefs due to Timothy Williamson (2008). I then formulate a more successful unbunking argument that relies on cognitive science.

15.1 INTRACTABILITY OF METAPHYSICAL DISPUTES

The first manner in which metaphysical questions are resistant to answers is that similar problems often arise for both sides of a metaphysical dispute.[1] Take the debate over whether there exist composite objects, those objects that have proper parts. On one side is the realist, who holds that composite objects exist. On the other side is the nihilist, who denies this. One of the nihilist's main arguments is that from causal overdetermination.[2] Let S be a collection of simple noncomposite particles arranged baseball-wise. Suppose S breaks a window. Now suppose that above and beyond the simples arranged baseball-wise there is in fact another object, the baseball, composed out of them. If so, the baseball is a sufficient cause of the window breaking, and so there are two sufficient causes of the window breaking: the baseball and the collection of simples arranged baseball-wise S. If there are indeed composite objects, this sort of causal overdetermination is rampant. Any effect of an event involving a collection of simples would be just as well explained by an event involving a composite object. On the supposition that there isn't rampant overdetermination, then there are no composites—at least assuming that composites can't be epiphenomenal and lack causal powers.

But as Karen Bennett notes, if overdetermination is a problem for the realist, then it is equally a problem for the nihilist (2009, 68–69). Take the collection of simples arranged baseball-wise S. Again, suppose S breaks a window. Now take another set of simples arranged baseball-wise S′, which is just S minus a single simple. S′ exists, hits the window, and is sufficient to break it, and so the window breaking is overdetermined. As before, this overdetermination is rampant. For any effect of an event involving some arrangement of simples, there is another event involving a different arrangement of simples that would also be sufficient to explain the effect. This is just one example of one problem in one dispute, but it is not unique. For many problems that plague one side of a metaphysical debate, there is an analogous problem that plagues the other.

The second issue with metaphysical disputes is that the respective views are often equally descriptively powerful in the sense that they can make all and only the same claims about the world, except with respect to the narrow range of facts about which they disagree. For instance, the compositional realist and the compositional nihilist

[1] The material of this paragraph and the next owes much to Karen Bennett (2009). See her paper for a more detailed discussion and further examples.

[2] Merricks 2001, 65–66. A wrinkle is that Merricks believes that persons exist, but we can ignore that fact for expository purposes.

can agree about everything, aside from putative composition facts. Take any world w. The realist and nihilist can describe this world in the exact same way, except that realist will claim that there are composition facts, whereas the nihilist won't. The realist can say there is a baseball that breaks the window. The nihilist can invoke plural quantification and say there are some simples that break the window. When two metaphysical views are equally descriptively powerful in the preceding sense, then the only direct counterexamples to either view would be one of the disputed putative facts. The only straightforward counterexamples to nihilism would be composite objects, but their existence is precisely what is under discussion. Of course, this sort of phenomenon is common to competing theories, but it is exacerbated in the metaphysical case given that the competing views are often empirically equivalent, and so there is no agreed-upon independent source of evidence that can provide a clear counterexample to either theory.

The third problem is that there is no independent way to corroborate our metaphysical beliefs and intuitions.[3] There is little external support for the verdicts of many metaphysical intuitions and beliefs, because there is no uncontroversial set of metaphysical facts to which we have access. This problem is exacerbated given that we have little initial reason to think we're particularly good intuitors of the joints of nature. As James Ladyman and Don Ross (2007) note:

> Proficiency in inferring the large-scale and small-scale structure of our immediate environment, or any features of parts of the universe distant from our ancestral stomping grounds, was of no relevance to our ancestors' reproductive fitness. Hence, there is no reason to imagine that our habitual intuitions and inferential responses are well designed for science or for metaphysics. (2)

In effect, Ladyman and Ross offer an evolutionary debunking argument of metaphysical beliefs and intuitions: evolution selected mechanisms that are responsible for our metaphysical beliefs and intuitions, but metaphysical truth does not enhance fitness and so there is no reason to think such mechanisms are reliable.

Perhaps the above problems leave the realist metaphysician with only the methodology of appealing to broad considerations of theoretical virtue and consilience with natural science—with little hope of ever coming to a single, indisputable metaphysical theory about a given domain.[4] However, in this chapter I explore whether there might be an argument that local metaphysical beliefs and intuitions are justified. In the next section, I examine the structure of negative debunking arguments before turning in section 15.3 to the structure of positive unbunking arguments. This latter sort of argument seems the best hope for realist metaphysicians who want to preserve some kind of justification for local metaphysical beliefs and intuitions.

[3] For the importance of external corroboration, see Weinberg 2007 and Sinnott-Armstrong 2006.

[4] For an argument that a multiplicity of metaphysical theories is a virtue, see Paul 2012.

15.2 DEBUNKING ARGUMENTS

Debunking arguments aim to show that we have no reason to think that our beliefs about a given domain are true. They rely on three premises. The first premise is *Objectivism*: we have beliefs about a domain that are true or false. This says that there are mind-independent facts about the domain in question, and it leaves open whether it is a fact that realism is correct or a fact that nihilism is correct. The second is the causal premise *Influence*: there is some set of causal processes that is primarily responsible for our beliefs about the domain, where the set may have one or more members. The third premise, *Off-track*, is epistemic: the set of processes do not reliably result in true beliefs. Thus, we have no reason to think our beliefs are true.[5]

We can schematize debunking arguments as follows:

1. Objectivism: beliefs about domain D are either true or false.
2. Influence: S's beliefs about D are due, dominantly, to set of causal processes X.
3. Off-track: set of causal processes X is unreliable with respect to D.
4. Skepticism: Therefore, S's beliefs about D are not justified.[6]

The following evolutionary debunking argument is an instance of this argument schema:

1. Objectivism: moral beliefs are either true or false.
2. Influence: S's moral beliefs are due, dominantly, to evolutionarily selected mechanisms.
3. On-track: evolutionarily selected mechanisms are unreliable when it comes to moral beliefs.
4. Nonskepticism: therefore, our moral beliefs are not justified.

I'll now offer some clarifications.

First clarification. Debunking arguments cannot merely show that one causal contributor to the beliefs about a domain is not reliable, for that is consistent with the *dominant* source or sources being reliable, and hence our beliefs being formed, overall, by a set of reliable processes. Instead, the causal premise must be that the dominant source or sources are not reliable, such that overall our beliefs about a domain are formed by a set of unreliable processes.

Second clarification. Most debunking arguments relativize reliability to a domain.[7] The evolutionary debunker of morality holds that evolution does not select for true

[5] See Goldman 1979 for the original defense of the view that justification and reliability are intimately related.

[6] For detailed discussions of the structure of debunking arguments, see Kahane 2011; Mason 2010; and Vavova 2014.

[7] Thanks to Alvin Goldman for drawing this feature to my attention.

moral beliefs, but presumably they would not want to say that evolution does not select for true beliefs about the trajectories of medium-sized physical objects. Thus, the debunker faces a generality problem of sorts. I will not solve any sort of generality problem here. However, it is worthwhile to note that the problem is less severe for debunkers, given that they are not offering a general theory of justification, but just an argument that we have no reason to think beliefs about a given domain are true. Insofar as we are just interested in beliefs about a given domain, we are licensed to restrict the measure of reliability to just that domain in question.

Third clarification. In this chapter, I connect reliability to justification, but even deniers of the link should find it important that beliefs about a domain are not reliably formed. Even someone who denies that reliability is a necessary condition on justification can agree that reliability is an epistemic virtue.

Fourth clarification. Crucially, debunking arguments are not metaphysical. They are epistemological. They do not directly argue for nihilism, that realist beliefs about the relevant domain are false. Instead, they argue for skepticism, that realist beliefs about the relevant domain aren't justified—whether or not they are true. In establishing Off-track, the debunker cannot assume antirealism without begging the question. This is disanalogous to showing, for instance, that our visual system can be mistaken. Our vision is prone to the Müller-Lyer illusion, where we experience two lines as being of different lengths when they are in fact of the same length. In this case, we can establish that the visual system is mistaken by appealing to independently verified facts, as when we measure the two lines with a ruler. Debunking arguments are particularly important when—as occurs in metaphysics—we have no uncontested domain of facts to which we can appeal in order to tell whether a given set of beliefs are justified or not.

15.3 UNBUNKING ARGUMENTS

In this section, I discuss the general structure of unbunking arguments before turning to a particular variation on that structure that relies on what I call the *cross-domain strategy*. This strategy argues from the reliability of a process or processes in one domain to the reliability of a process with respect to another domain. This strategy is particularly important when it comes to unbunking metaphysics, given that there is no uncontested set of metaphysical facts with which one can directly determine the reliability of a process.

15.3.1 The Structure of Unbunking

The basic structure of a debunking argument can be mirrored by a positive argument that leads to a nonskeptical conclusion. The first two premises remain the same. One assumes Objectivism about a domain, and one assumes that a particular set of causal processes is dominantly responsible for beliefs about that domain. The only premise that changes is Off-track, and it becomes the epistemically positive *On-track*: the set of causal processes leading to belief is reliable with respect to the domain in question. The four clarifications mentioned previously still stand, pending changes making them relevant to this epistemically positive argument.

The following is the schema for unbunking arguments:

1. Objectivism: beliefs about domain D are either true or false.
2. Influence: S's beliefs about D are due, dominantly, to set of causal processes X.
3. On-track: set of causal processes X is reliable with respect to D.
4. Nonskepticism: therefore, our beliefs about D are justified.

To make things concrete, here's a straightforward example of an unbunking argument—which I don't necessarily endorse—that argues from the evolutionary origins of folk physics to its reliability. Folk physical beliefs are either true or false, and our folk physics is the result of evolutionarily selected mechanisms. Evolution selects for mechanisms that are useful to survival. It is useful to survival to accurately track the trajectories of medium-sized objects so that, for instance, we don't get hit by rocks. Therefore, we have good reason to think that our folk physics is largely accurate.

More formally:

1. Objectivism: folk physical beliefs are either true or false.
2. Influence: S's folk physical beliefs are due, dominantly, to evolutionarily selected mechanisms.
3. On-track: evolutionarily selected mechanisms are reliable when it comes to the physics of medium-sized objects.
4. Nonskepticism: therefore, our folk physical beliefs are justified.

Note that nowhere did the argument appeal to an independent set of facts with which one can check the deliverances of folk physics. The argument is not that one checked a thousand folk physical beliefs against cutting-edge theoretical physics and noticed that folk physics mostly got it right. Instead, the argument gave reason to think that folk physics is reliable independently of being able to directly determine its reliability.

If successful, a metaphysical unbunking argument would give us reason to think that our metaphysical beliefs and intuitions are justified. Unbunking arguments are uniquely suited to overcoming the problem that there is no independent source of evidence for metaphysical views outside of intuitions and beliefs. In just the same way that debunking arguments need not rely on truths about the domain in question, unbunking arguments need not either. Even if there is no way to independently confirm beliefs, we can still indirectly show they are generally justified.

15.3.2 Cross-Domain Strategy

How might one go about making a metaphysical unbunking argument? The first thing to do would be to argue for Objectivism. This I'll simply assume, and instead focus on how one might establish Influence and On-track. An immediate dead-end appears to be trying to come up with a wide-domain unbunking argument that shows that all metaphysical intuitions, regardless of stripe, are justified. Given the long history in metaphysics of unresolved disputes and the three problems listed in the first section, there isn't much reason to hope this could be successful. The more promising approach is to make a narrow domain unbunking argument, which argues for the more restricted claim that a certain class of metaphysical beliefs are justified.[8]

But there doesn't seem to be a straightforward unbunking argument available. For instance, appeal to evolutionary or cultural-historical causes of metaphysical beliefs is unable to help because there is no reason to think that either would select for beliefs that carve nature at the joints. True metaphysical beliefs are neither useful for survival, nor do they seem induced by cultural pressures. Therefore, the metaphysical unbunker needs to use a more indirect route.

I suggest unbunkers employ what I call the *cross-domain strategy*. They should argue (1) that a particular set of causal processes X′ is reliable with respect to some nonmetaphysical domain D′, where such reliability is uncontroversial; (2) that the same or a relevantly similar set of processes X is dominantly responsible for our beliefs about a metaphysical domain D; and (3) that D and D′ are relevantly similar so that we should expect X to be reliable with respect to D. Our justification for thinking our beliefs about D are reliable is inherited from our justification in thinking that our beliefs about D′ are reliable.

Here's a schematization of the cross-domain strategy style of unbunking argument, with the simplifying assumption that X and X′ are the same set of processes:

1. Objectivism: beliefs about domain D are either true or false.
2. Influence: S's beliefs about D are due, dominantly, to set of causal processes X.
   Lemma—cross-domain strategy:
   a. Set of processes X produces reliable beliefs about domain D′.
   b. Domain D′ and D are relevantly similar so if X is reliable with respect to D′, then X is reliable with respect to D.
3. On-track: X is reliable with respect to D.
4. Nonskepticism: therefore, our beliefs about D are justified.

Of course, the plausibility of an unbunking argument that uses the cross-domain strategy hinges on whether the two domains and processes are "relevantly" similar.

[8] This distinction maps onto Kahane's (2011) distinction between local and global debunking arguments.

I have nothing both general and useful to say about what makes for such similarity. Let's first just try to get some successful cases of the cross-domain strategy onboard before making an abstract analysis!

15.4 WILLIAMSON'S METAMETAPHYSICS

The contemporary philosopher who has articulated something closest to a metaphysical unbunking argument is Timothy Williamson. In his view, the methodology and forms of thinking used in philosophy are continuous with those used in other areas, including the sciences and ordinary life. There is no uniquely "philosophical" way of thinking. Instead, philosophy is just a specialization of ubiquitous ways of thinking, and so inherits its reliability from them. As Williamson (2008) himself puts it:

> One main theme of this book is that philosophical exceptionalism is false. . . . Although there are real methodological differences between philosophy and the other sciences, as actually practiced, they are less deep than is often supposed. In particular, so called intuitions are simply judgments (or dispositions to judgment); neither their content nor the cognitive basis on which they are made need be distinctively philosophical. In general, the methodology of much past and present philosophy consists in just the unusually systematic and unrelenting application of ways of thinking required over a vast range of non-philosophical inquiry. The philosophical applications inherit a moderate degree of reliability from the more general cognitive patterns they instantiate. (3).

While such a view, in broad outlines, seems correct, Williamson's implementation in the case of metaphysics falls short. He focuses on the metaphysics of modality and argues for its reliability in two stages. First, we are generally reliable in counterfactual thinking. Second, modal thinking about metaphysical possibility and necessity is just a special case of counterfactual thinking. Given these two claims, he argues, our modal thinking is generally reliable. In this section, I summarize the argument and then show how it fails.

As for the first step, Williamson primarily motivates the claim that ordinary counterfactual reasoning is generally reliable by invoking folk physics as an example. Take a rock rolling down a hill by a lake and hitting a bush and stopping. One can correctly judge that "If the bush had not been there, the rock would have ended in the lake" (142). We need not go and actually conduct an experiment, removing the bush and rolling the rock down the hill again. Instead, our folk physics allows us to make counterfactual judgments about different nonactual but possible trajectories of the rock.[9]

[9] Actually, folk physics is hopelessly inadequate when it comes to uncovering fundamental physics—superpositions, for one, are not countenanced at all in folk physics—and is notoriously prone to certain errors even in plotting the trajectories of macroscopic objects. For this last point, see McBeath et al. 2010 and McCloskey et al. 1983.

As for the next step, Williamson shows convincingly that a necessity claim is logically equivalent to a certain sort of counterfactual claim and that a possibility claim is logically equivalent to another sort of counterfactual claim (158). The following are the logical equivalences:

(Necessity) $\Box$A if and only if ($\neg$A > $\perp$)

(Possibility) $\Diamond$A if and only if $\neg$(A > $\perp$)

Necessity says that A is necessary if and only if its denial leads to a contradiction. A is necessary if and only if it is incoherent for A not to hold. Possibility says that A is possible if and only if it is not the case that A leads to a contradiction. A is possible if and only if it is not the case that it is incoherent for A to hold.

On the basis of these two claims, Williamson concludes that modal reasoning is generally reliable. But even granting that folk physics is generally reliable and that necessity and possibility claims are logically equivalent to certain counterfactual claims, his argument fails. To see why, it will be helpful to construe it as an unbunking argument:

1. Objectivism: metaphysical modal beliefs are either true or false.
2. Influence: S's modal beliefs are due, dominantly, to our ordinary capacity at handling counterfactuals.

Lemma—cross-domain strategy:

  a. Our ordinary capacity at handling counterfactuals is reliable with respect to everyday claims.
  b. Assessing everyday claims is not relevantly different from assessing claims of modal metaphysics.

3. On-track: our ordinary capacity at handling counterfactuals is reliable with respect to modality.
4. Nonskepticism: our modal beliefs are justified.

The argument fails because both Influence and On-track fail in this case.

As for Influence, all Williamson has shown is that necessity and possibility claims are logically equivalent to certain kinds of counterfactual claims. This does not show the same cognitive capacity is recruited to assess both modal claims and ordinary counterfactuals, and it does not even show that the cognitive capacity recruited to assess modal claims is relevantly similar to the one used for ordinary counterfactuals. Williamson says, "Whoever has what it takes to understand the counterfactual conditional and the elementary logical auxiliaries $\neg$ and $\perp$ has what it takes to understand possibility and necessity operators" (158). This might be so if one is appealing to explicit reasoning in a proof-theoretic system where the inferential roles of >, $\neg$, $\perp$, $\Box$, and $\Diamond$ are interdefined. But Williamson himself holds that counterfactual reasoning is not inferential and is instead imaginative (145–48). While logically one can move from one form of the statement to the other, the question at hand is whether our psychological capacity at assessing ordinary counterfactuals is used to assess modal claims. There is no reason to think that because two concepts are logically equivalent they are used identically in imagination, for they might be presented under different modes of presentation. The coextensionality claims of "$\Box A$ if and only if $(\neg A > \bot)$" and "$\Diamond A$ if and only if $\neg(A > \bot)$" might be Frege cases, where an agent does not realize that the left side is logically equivalent to the right side.[10] Because they conceptualize the left side differently from the right side, their manner of assessing claims when presented under one mode could very well be different than when presented under the other. Hence, there is no reason to think one and the same cognitive capacity used to assess counterfactuals is used to assess modal claims.

Even if Influence goes through, On-track still fails. It is plausible that our competency with ordinary counterfactuals might be fairly reliable with respect to certain sorts of domains, like plotting the trajectories of ordinary-sized objects, but not in others, like those pertaining to metaphysical claims. Even if we grant Williamson that the same cognitive capacity is recruited to assess ordinary counterfactuals as is recruited to assess metaphysical counterfactuals, there is no reason to think the latter are similar enough to the former for reliability to be inherited. Take Williamson's primary example of folk physics. At best, he has shown that we should be fairly reliable at assessing necessity and possibility claims as they pertain to trajectories of medium-sized physical objects. But there is no reason to think such reliability holds in judgments about, for instance, whether it is possible for physical duplicates of functioning humans to lack phenomenal consciousness. Indeed, the most current cognitive scientific view of folk physics is that it is a simulation engine, much like a graphics engine used to simulate physics in movies and video games (Battaglia et al. 2013). A graphics simulation engine has no way to represent counterfactuals involving colocation—let alone a host of other particularly metaphysical claims.

15.5 UNBUNKING METAPHYSICS USING COGNITIVE SCIENCE

This section is divided into two parts. In the first, I show how cognitive science can help in general to unbunk by giving an example of how work on the simulation theory of mindreading helps to unbunk our capacity to attribute mental states

[10] Frege cases arise between logical equivalents all the time. Take the equivalence of "$A \vee {\sim}A$" and "$(B \rightarrow C) \rightarrow (((C \rightarrow D) \rightarrow B) \rightarrow C)$." Both are tautologies of propositional logic, but most of us have no explicit belief that the second one is a tautology, let alone that the two statements are equivalent. The difference in the modes of presentation of alethic modal claims and counterfactual claims is even more extreme.

to other humans. In the second, I use results from cognitive science to support a cross-domain strategy unbunking argument for metaphysical beliefs about mutual supervenience. Often, cognitive science is used negatively against metaphysics, as when it is used to debunk metaphysical claims (for critical discussion, see Schaffer 2016), lower our credence in a realist view (Goldman 1989, 2015), or just generally make us aware of errors and biases leading to mistaken metaphysical commitments (Paul 2010). While this is a worthy enterprise, I want to explore another approach: using cognitive science to positively support metaphysical methodology.

15.5.1 Cognitive Science and Unbunking

In order for cognitive science to help unbunk it should accomplish three tasks. First, it needs to be able to identify processes dominantly responsible for our beliefs about a domain. Second, it needs to be able to help us assess their reliability. Third, it should allow us to discover when beliefs about one domain are produced by identical or at least similar processes that form beliefs about a second domain. This third task is required for unbunking arguments that rely on the cross-domain strategy, where one uses the reliability of a process with respect to one domain as evidence for the reliability of a process with respect to another. Simulation theory provides an example of cognitive science accomplishing all three tasks. I do not want to take a stand on whether the view is true, but simply want to use it as a clear case of cognitive science's ability to play a role in unbunking arguments.

The first task for cognitive science is to identify a set of one or more processes dominantly responsible for beliefs in a domain. According to simulation theory, one attributes mental states to another by imagining oneself in their situation.[11] One pretends to have the same mental states as the target of attribution and then feeds those states into one's own cognitive processes and attributes the resulting state or states to the target (Goldman 2012). A prime case is that of predicting another person's decision via imaginatively putting oneself in that person's place (Goldman 2006, 19–30). One pretends to have the same initial mental states, like beliefs and desires, and then one runs those beliefs and desires through one's own decision-making process. One attributes the resulting decision to the other person. For instance, while playing chess, I imaginatively put myself in my opponent's shoes and decide what move I would make in that situation. I then predict that my opponent will make that move.

The second task for cognitive science is to help us assess the reliability of belief-forming processes. These assessments can come in the negative and positive variety. As for the negative variety, simulation theorists have identified a certain sort of error in mindreading called "quarantine failure" (Goldman 2006, 29–30, 41–42). Quarantine failure occurs when one does not adequately isolate one's own

[11] In what follows, I rely heavily on the presentations in Goldman 2006 and 2012.

nonpretend mental states while simulating another's mind. If these states are not shared by the target of attribution, they could undermine the accuracy of the simulation. For instance, when trying to predict my opponent's move in chess, if I do not adequately quarantine my beliefs about my own future move, then I might inaccurately simulate my opponent's move as being made—illicitly—in response to my own plans. The phenomenon known as "egocentric bias" provides evidence that quarantine failure is systematic. One instance of this bias is the "curse of knowledge," whereby people have trouble ignoring information they possess that the target of attribution doesn't (Camerer et al. 1989). This bias often makes it difficult for both children and adults to correctly attribute false beliefs to others (Birch and Bloom 2003; Birch and Bloom 2007). Simulation theorists, unfortunately, have done less work of the positive variety. However, cognitive science is certainly capable of doing work that uncovers how often we are reliable simulators and the conditions that promote such accuracy—even if that proves to be a less exciting research program.

The third task for cognitive science is to help discover when beliefs about two distinct domains are produced by similar or even identical processes. Simulation theory is proposed to account for a wide range of mental attributions, such as third-person attributions of sensations, emotions, propositional attitudes, and decisions as well as first-person attributions of future states of the same sort. Simulation theory holds that these different attributions are made using the same general routine. One feeds initial pretend states into one's own cognitive mechanism for the process to be simulated and one attributes the resulting mental state to the target of attribution. Simulation of each of these domains relies on a similar process, and so simulation theory in principle can provide support for a cross-domain strategy. Given that simulation is used to make attributions of sensations and emotions, we should expect roughly the same reliability with respect to the former domain as for the latter. Given that simulation is used to make attributions of both nonpropositional mental states and propositional attitudes, if we are very reliable in making the first sort we should expect to be at least somewhat reliable in making the second. Of course, simulation theorists can just directly test our reliability with respect to these domains and so the cross-domain strategy is not particularly needed here, but it is an important example of how cognitive science can support the strategy in general.

15.5.2 Cognitive Science Unbunking Metaphysics

I turn now to offering an actual unbunking argument for a certain class of metaphysical beliefs. Crucially, this argument relies on results from cognitive science to support an implementation of the cross-domain strategy. The unbunking argument I formulate revolves around identity claims whose justification comes from mutual supervenience claims, which can be understood as special sorts of correlation claims. Work in cognitive science suggests that we are reliable at assessing such correlations.

Correlations often serve as evidence for identity claims. If two things are constantly observed to obtain together, then one potential explanation for this fact is that the two things are in fact one and the same thing. If Superman and Clark Kent always have the exact same haircut, one possible explanation is that Superman *is* Clark Kent. This sort of argument has been used to justify physicalism about the mental. As Brian McLaughlin (2010) formulates his argument for type physicalism, if we are justified in believing

the correlation thesis: for any type of state of phenomenal consciousness C there is a type of physical state P such that it is true and counterfactual supporting that a being is in C if and only if the being is in P (267),

then we are justified in believing

type physicalism for phenomenal consciousness: for every type of state of phenomenal consciousness C, there is a type of physical state P such that C = P (265)

on the basis of inference to the best explanation.

The correlation thesis says that C iff P is *counterfactual supporting*. The best way to understand this so that it supports the identity claim of C = P is that the thesis says C properties and P properties mutually supervene. Set of properties A supervenes on set of properties B if and only if a change in A requires a change in B (see McLaughlin and Bennett 2014). Therefore, A mutually supervenes on B if and only if a change in A requires a change in B and a change in B requires a change in A. The properties of Clark Kent mutually supervene on the properties of Superman. A change in Clark Kent's haircut requires a change in Superman's, and vice versa. The shape properties of a table supervene on its microphysical properties, but do not mutually supervene. A change in shape requires a change in microphysical properties, but not vice versa.

Although the mutual supervenience of the properties of two entities does not entail identity, identity entails mutual supervenience and so mutual supervenience can provide *evidence* for identity. Indeed, metaphysicians often appeal to considerations of mutual supervenience to argue for or against a metaphysical identification. The statue and the clay constituting it are often thought to be distinct because their properties do not mutually supervene. The statue can be destroyed by squishing it, but the clay will remain. The group agent is distinct from the mere aggregate of its human members because there can be a mere aggregate without any relevant unity to the group. The set of properties of the group agent mutually supervenes on the properties of an appropriate interrelation of its members, not the mere aggregate. Because beliefs about mutual supervenience play a role in generating metaphysical commitments, unbunking mutual supervenience beliefs would unbunk an important class of metaphysical beliefs.

Mutual supervenience is a special sort of correlation. It is perfect correlation without possible exception. Therefore, the same processes by which we detect exceptionless correlations are those we use to detect mutual supervenience. If humans are good at assessing whether two items are perfectly correlated across possible scenarios, then they are good at assessing whether two items mutually supervene. The cross-domain strategy can exploit this fact. If we form reliable beliefs about correlations, then we should form reliable beliefs about mutual supervenience. The former can be tested with respect to domains that are uncontroversial and not overtly metaphysical, and so we can gain evidence as to our reliability with respect to the latter.

Putting this all together, here is an unbunking argument for beliefs about mutual supervenience:

1. Objectivism: mutual supervenience beliefs are either true or false.
2. Influence: S's beliefs about mutual supervenience are due, dominantly, to the set of causal processes by which we form beliefs about correlations.

Lemma—cross-domain strategy:

   a. The set of processes by which we form beliefs about correlations is reliable with respect to these beliefs.
   b. Mutual supervenience beliefs are not relevantly different from correlation beliefs, so if the set of processes is reliable with respect to the latter, it is reliable with respect to the former.

3. On-track: the set of processes by which we form beliefs about mutual supervenience is reliable.
4. Nonskepticism: our beliefs about mutual supervenience are justified.

If successful, we have a metaphysical unbunking argument. Given that mutual supervenience claims can be used to justify metaphysical identity claims, we have reason to think local intuitions and beliefs about mutual supervenience are justified, as are identity claims based off them. The plausibility of the argument depends largely on premise 2 and subpremises *a* and *b*.

Though I argued for premise 2 and subpremise *b* above, there has not been cognitive scientific work directly addressing them. However, they are claims clearly amenable to cognitive scientific research. The general methods that cognitive scientists use to get at judgments, for example, of causation and responsibility can be used to gather data about judgments of mutual supervenience. There is nothing that would keep researchers from designing experiments testing how people judge whether two items obtain and change together. Moreover, scientific theorizing and fMRI results can provide evidence for whether the same process used to form beliefs about correlations in general is responsible for beliefs about mutual supervenience in particular—as, for instance, when fMRI results suggest that past, future, and counterfactual thinking involve many common brain processes (Van Hoeck et al. 2012).

Fortunately, there has been a lot of cognitive scientific research pertaining to subpremise *a*. Cognitive scientific study of our ability to make judgments of correlation often goes under the heading of assessments of "covariation" or "contingency." The role of beliefs about covariation has been studied in a wide range of areas in psychology, but what is relevant for my particular unbunking argument is our ability to detect and predict correlations. Studies have tended to provide evidence that we are good detectors in general, though prone to certain biases and errors.

As one example, Well et al. (1988) presented subjects with three sets of sixty pairs of numbers of correlation .9, .6, and .1 respectively. One group had both numbers presented simultaneously, and after all sixty pairs were presented they were asked to judge the "strength of the relationship." Another group was forced to make predictions of the last thirty pairs in each set of the second of the numbers on the basis of the first, and after all sixty pairs were presented they were also asked to judge the "strength of the relationship." Neither group of subjects judged the strength of the relationship in a manner corresponding to standard statistical measures of correlation, but the prediction group was quite good at predicting one variable from another in a manner sensitive to the degree of correlation. In a related experiment, Malmi (1986) encouraged subjects to rely on "intuitive" implicit processes to make covariation judgments and found that people were quite good at estimating degree of correlation. In a series of experiments, subjects were presented with pairs of numbers, pairs of lines of differing lengths, and pairs of words and lines. Malmi encouraged "intuitive" strategies of assessing correlations by presenting the pairs for one second each. He found that people were in general quite good at estimating the sign and degree of correlation.

A survey of research on covariation by Alloy and Tabachnik (1986, 119–23) concluded that humans are prone to errors and biases in detection of covariation but that under many conditions we are quite accurate. Accuracy is promoted if the correlated items

are positively correlated
are neutral with respect to rewards
are not associated with success
are not associated with features characteristic of situations requiring skill
are preceded by something that makes randomness a salient possibility

Often these conditions are met when we encounter mutually supervening items.

Two issues with drawing conclusions about our ability to track mutual supervenience from such studies should be addressed. The first is that the studies testing our ability to detect correlations have only tested subjects' ability to detect nonmodal frequencies. They have neither presented the subjects with modal information nor asked them to make modal judgments. The second issue is that the studies give artificial tasks different from those that subserve our mutual supervenience beliefs. Judging the correlation between two numbers and the lengths of two lines is different from judging whether the team mutually supervenes on the players under some teammate relation.

As for the first issue, the studies give reason to think that if humans are good at picking up nonmodal frequency information, then they would be good at picking up modal frequency information. Moreover, cognitive scientists can devise studies to get at our modal correlation judgments—as they do when they devise studies getting at our modal judgments about causation. As for the second issue, the artificiality of the tasks gives us reason to think humans would be even better at detecting non-artificially constructed correlations. And, as before, testing less artificial judgments as to correlations is something amenable to cognitive scientific research.

It is worthwhile to remember that studies showing positive results have given subjects items that aren't perfectly correlated and have shown them relatively few examples, generally a hundred or fewer. The relevant sorts of correlations for mutual supervenience claims are those that are positive, without possible exception, and, often, ubiquitous. Many of the cases relied on in metaphysical theories are of common objects, such as clay statues. Most people see more than a hundred statues in their lives, and certainly many more materially constituted objects. There is good reason to think that if humans are good at detecting imperfect correlations with only a few examples, then we would be even better at detecting and predicting perfectly correlated items we encounter much more frequently.

Given the cognitive scientific research, it appears that beliefs about metaphysical mutual supervenience are unbunked.

15.6 CONCLUSION

In the first part of the chapter I raised three issues with metaphysical practice that suggested the need for an unbunking argument for metaphysical beliefs. The first is that similar problems often arise for competing sides in a dispute. The second is that metaphysical views can make all the same claims as other metaphysical views, except for the narrow range of propositions about which they disagree. The third is that there are good reasons to think human cognition isn't particularly well suited to uncover the fundamental joints of nature. I suggested that the realist—and optimistic—metaphysician could justify certain classes of metaphysical beliefs by appeal to unbunking arguments, particularly those that rely on the cross-domain strategy. Of course, merely having justified metaphysical beliefs about a domain doesn't guarantee that different metaphysical views about that domain won't face similar problems, nor certainly make it such that different views have differing levels of descriptive power. But it does give us reason to think that we are good enough carvers of nature, at least that realm of nature for which we have an unbunking argument.

Unbunking arguments need not be limited to metaphysics. If a successful one can be found for metaphysics—the philosophical subject seemingly most prone to irresolvable disputes—then we should remain optimistic enough that unbunking arguments can be found for other philosophical subjects.

ACKNOWLEDGMENTS

For their helpful comments and discussion, I thank Carolina Flores, Adam Gibbons, Kyle Landrum, Tyler John, John McCoy, and Brian McLaughlin. I want to especially thank Alvin Goldman, who gave me many detailed comments, as well as Danny Forman, who pointed out my nascent views had structural similarities to a debunking argument.

REFERENCES

Alloy, L., and Tabachnik, N. (1984). Assessment of covariation by humans and animals: The joint influence of prior expectations and current situational information. *Psychological Review 91* (1): 112–49.

Battaglia, P. W., Hamrick, J. B., and Tenenbaum, J. B. (2013). Simulation as an engine of physical scene understanding. *Proceedings of the National Academy of Sciences of the United States of America 110* (45): 18327–32.

Bennett, K. (2009). Composition, colocation, and metaontology. In D. Chalmers, D. Manley, and R. Wasserman, eds., *Metametaphysics: New Essays on the Foundations of Ontology.* New York: Oxford University Press, 38–77.

Birch, S. A. J., and Bloom, P. (2003). Children are cursed: An asymmetric bias in mental-state attribution. *Psychological Science 14* (3): 283–86.

Birch, S. A. J., and Bloom, P. (2007). The curse of knowledge in reasoning about false beliefs. *Psychological Science 18* (5): 382–86.

Camerer, C., Loewenstein, G., and Weber, M. (1989). The curse of knowledge in economic settings: An experimental analysis. *Journal of Political Economy 97* (5): 1232–54.

Goldman, A. (1979). What is justified belief? In George Pappas, ed., *Justification and Knowledge.* Dordrecht: D. Reidel, 1–25.

Goldman, A. (1989). Metaphysics, mind, and mental science. *Philosophical Topics 17* (1): 131–45.

Goldman, A. (2006). *Simulating Minds: The Philosophy, Psychology, and Neuroscience of Mindreading.* New York: Oxford University Press.

Goldman, A. (2012). Theory of mind. In E. Margolis, R. Samuels, and S. Stich, eds., *The Oxford Handbook of Philosophy and Cognitive Science.* New York: Oxford University Press, 402–24.

Goldman, A. (2015). Naturalizing metaphysics with the help of cognitive science. In K. Bennett and D. Zimmerman, eds., *Oxford Studies in Metaphysics*, vol. 9. New York: Oxford University Press, 171–216.

Kahane, G. (2011). Evolutionary debunking arguments. *Noûs 45* (1): 103–25.

Ladyman, J., and Ross, D., with Spurrett, D., and Collier, J. (2007). *Every Thing Must Go: Metaphysics Naturalized.* New York: Oxford University Press.

Malmi, R. (1986). Intuitive covariation estimation. *Memory & Cognition 14* (6): 501–8.

Mason, K. (2010). Debunking arguments and the genealogy of religion and morality. *Philosophy Compass 5* (9): 770–78.

McBeath, M. K., Brimhall, S. E., Miller, T. S., and Holloway, S. R. (2010). The naïve physics curvilinear impetus bias does not occur for locomotion. *Journal of Vision 10* (7): 1021.

McCloskey, M., Washburn, A., and Felch, L. Intuitive physics: The straight-down belief and its origin. *Journal of Experimental Psychology: Learning, Memory, and Cognition 9* (4): 636–49.

McLaughlin, B. (2010). Consciousness, type physicalism, and inference to the best explanation. *Philosophical Issues 20* (1): 266–304.

McLaughlin, B., and Bennett, K. Supervenience. In E. N. Zalta, ed., *The Stanford Encyclopedia of Philosophy*. Spring 2018 ed. https://plato.stanford.edu/archives/spr2018/entries/supervenience/.

Merricks, T. (2001). *Objects and Persons*. Oxford: Clarendon Press.

Paul, L. A. (2010). A new role for experimental work in metaphysics. *Review of Philosophy and Psychology 1* (3): 461–76.

Paul, L. A. (2012). Metaphysics as modeling: The handmaiden's tale. *Philosophical Studies 160* (1): 1–29.

Schaffer, J. (2016). Cognitive science and metaphysics: Partners in debunking. In B. McLaughlin and H. Kornblith, eds., *Goldman and His Critics*. Malden, MA: Wiley-Blackwell, 337–68.

Sinnott-Armstrong, W. (2006). Moral intuitionism meets empirical psychology. In T. Horgan and M. Timmons, eds., *Metaethics after Moore*. New York: Oxford University Press, 339–65.

Van Hoeck, N., Ma, N., Ampe, L., Baetens, K., Vandekerckhove, M., and Van Overwalle, F. (2012). Counterfactual thinking: An fMRI study on changing the past for a better future. *Social Cognitive and Affective Neuroscience 8* (5): 556–64.

Vavova, K. (2014). Evolutionary debunking of moral realism. *Philosophy Compass 10* (2): 1–13.

Weinberg, J. (2007). How to challenge intuitions empirically without risking skepticism. *Midwest Studies in Philosophy 31* (1): 318–43.

Well, A. D., Boyce, S. J., Morris, R. K., Shinjo, M., and Chumbley, J. I. (1988). Prediction and judgment as indicators of sensitivity to covariation of continuous variables. *Memory & Cognition 16* (3): 271–80.

Williamson, T. (2008). *The Philosophy of Philosophy*. Malden, MA: Wiley-Blackwell.

Index